W9-CYY-684

Wiley Series in Materials, Modelling and Computation

Series Editors

Professor C. S. Desai
College of Engineering and Mines
Department of Civil Engineering
and Engineering Mechanics
University of Arizona
Tucson, AZ 85721
USA

Professor E. Krempl
Department of Mechanical Engineering,
Aeronautical Engineering and Mechanics
Rensselear Polytechnic Institute
Troy, NY 12180-3590
USA

Time Effects in Rock Mechanics

N. D. Cristescu
University of Florida, Gainesville, USA

U. Hunsche
Bundesanstalt für Geowissenschaften und Rohstoffe, Hanover, Germany

JOHN WILEY & SONS
Chichester · New York · Weinheim · Brisbane · Singapore · Toronto

Other Wiley Editorial Offices

John Wiley & Sons Inc, 605 Third Avenue,
New York, NY 10158-0012, USA

Wiley-VCH Verlag GmbH, Pappelallee 3,
D-69469 Weinheim, Germany

Jacaranda Wiley Ltd, 33 Park Road, Milton,
Queensland 4064, Australia

John Wiley & Sons (Asia) Pte Ltd, 2 Clementi Loop #02-01,
Jin Xing Distripark, Singapore 129809

John Wiley & Sons (Canada) Ltd, 22 Worcester Road,
Rexdale, Ontario M9W 1L1, Canada

British Library Cataloguing in Publication Data

A Catalog record for this book is available from the British Library

ISBN 0 471 95517 5

Produced from camera-ready copy supplied by the author
Printed and bound in Great Britain by Antony Rowe, Chippenham, Wilts
This book is printed on acid-free paper responsibly manufactured from sustainable forestry, in which at least two trees are planted for each one used for paper production.

Contents

Preface

The use of underground space is increasing, due to the increasing world population, and due to our increasing demand. Storage for liquids, gases, wastes are built in huge dimensions at great depths. Also many large underground spaces like garages, caverns for power stations, and storage facilities are being excavated. Mines are going deeper as are wellbores. In addition, materials more difficult to handle, and to be described by mathematical models, are involved. Due to this, the design and construction is becoming increasingly sophisticated. The short term and long term deformation and stability has to be predicted with increasing accuracy and the time scales involved are becoming much longer. Therefore, a purely empirical approach is not reliable enough. The appropriate theoretical background, especially for the constitutive equations describing the time dependent reactions of the rocks affected by the excavation, is to be involved. This also has a considerable influence on the experimental requirements, since more accurate experimental devices and methods are needed.

Therefore the authors decided to collect their knowledge in the field and to edit it as a single volume in this rapidly progressing field of rock mechanics.

Rock mechanics is related to structural geology, soil mechanics, materials sciences, civil, mining and petroleum engineering, seismology and geophysics. People from these and other fields might benefit from this book and might be stimulated to apply the content and to develop it for some other rocks or possible cases. Engineers, researchers, graduate students, practical and theoretical people will have profit. The mathematical level is on a graduate level. The book includes a long reference list; although it was not intended to be exhaustive.

The content concentrates on elasticity, creep, dilatancy, creep failure, short term failure, viscoplasticity and related effects, its measurement and the description by appropriate constitutive equations.

Chapters 1 to 3 show how the suitable experiments should be carried out, give results for creep, failure, and loosening, and explain the microphysical mechanisms of creep. Chapters 4 to 6 deal with general and specific constitutive equations, with experimental results and with the proper determination of the involved functions. Chapter 7 deals with the formulation of proper initial and boundary conditions for mining and petroleum engineering problems. Chapters 8 to 10 give a large number of important, practical, model calculations for the convergence of underground openings with various shapes, and also exhibit dilatancy and creep failure in their vicinity, how to estimate the locations and timing when creep failure is to be expected.

The first draft of Chapters 1 - 3 was written by U. H., and the remaining by N.D.C., though both authors have cooperated and are responsible for the whole book. U. H. is grateful to B.G.R. for giving him the opportunity to write this book. Also, N.D. C. is grateful to the Geomechanics Laboratory from Bucharest for many experimental data on a variety of rocks mentioned throughout the book. Also for the data on alumina powder N.D. C. is grateful to Engineering Research Center (ERC) for Particle Science and Technology at the University of Florida, the National Science Foundation (NSF) grant #EEC-94-02989, and Industrial Partners of the ERC.

This book is based on the personal experience of the authors and they hope to be able to encourage the practical use, and stimulate further discussion and research concerning the time dependent behavior of rocks and other similar materials. The results incorporated here have been presented by both authors at several international meetings, seminar lectures, short courses, etc. (see References). The present text, as well as another book (Cristescu [1989a]), has been used by N.D.C. as text books for courses on Rock Mechanics taught at the University of Bucharest and University of Florida.

The authors are grateful to Cornelia Cristescu for typing and editing the manuscript, and helping with many numerical examples and graphs given mainly in Chapters 4 - 10.

1 Experimental Foundations

1.1 INTRODUCTION

Knowledge of the mechanical properties of rocks is important for many engineering purposes, e.g., for the design of mines, tunnels, repositories, and deep wells, as well as for application in geophysics, tectonics, and seismology. In general, the elastic properties, creep, dilatancy, damage, fracture development, creep failure, failure, and permeability are of interest for this purpose. Swelling is also a time-dependent property of rocks (e.g., anhydrite:) it is, however, not covered here. For corresponding expert judgement and for model calculations one needs a detailed knowledge of the materials in question. The constitutive laws and the equations describing the dependence of the material properties on various parameters are very important. Also, quite often distribution functions for the material properties have to be determined.

Most of the above mentioned properties are coupled, e.g. creep and dilatancy, dilatancy and creep failure, dilatancy and permeability. Therefore, such dependences are to be evaluated and is expected that most of them are incorporated in a well-behaved constitutive equation. Also, it has to be taken into account that many properties are time- or rate -dependent, e.g., failure strength. Another difficult question is whether the material under question has a flow limit or a yield stress distinct from zero.

For the development of reliable laws, appropriate laboratory and field tests have to be performed. In this chapter experimental techniques for the laboratory tests and for the evaluation of time-dependent properties are shown.

A very important time-dependent property is the time-dependent deformation behavior or creep behavior. And this is one of the main topics of the book. **Creep** is defined here as the irreversible deformation in time without fracturing and is observed mainly in soft rocks like salt, and coal, and more or less in all other rocks. However, this definition is somewhat weakened in the case of time-dependent dilatancy, which is still called "creep" or even "stationary creep". All kinds of hard rock also exhibit creep within long enough time intervals. This can be observed in the geological structures of all mountains in the world. The deformation behaviors of different kinds of rocks have many similarities which will be shown below. Of course, rocks show also differences in the deformation behavior, e.g., granite, sandstone, limestone, salt, clay, and there are great variations within a rock type. The similarities and the differences have to be taken into account in the engineering problems to be treated. For this purpose, laboratory and field tests have to be carried out, during long-term and short-term tests. These have to be interpreted and finaly moulded into constitutive equations, parameters, and moduli. The development of constitutive equations has to be supported by theoretical considerations as described in the following chapters.

Knowledge of the experimental technique and the interpretation of tests is not only important for the experimentalist but also for the theoretician who wants to develop reliable and applicable constitutive equations, since otherwise serious misinterpretations and mistakes are very likely to occur. Therefore, close cooperation between experimentalist and theoretician is an essential condition. It is advisable never to use experimental results without careful inspection of the whole test procedure (included the involved persons).

The development of appropriate constitutive equations can be made on the basis of plasticity theory, rheological models, or microscopic deformation mechanisms. General considerations have been performed by, e.g., Langer [1969,1979,1986,1988]. Also, combinations are possible. The use of purely empirical formulas should be avoided, because the great source of expert knowledge is not included. For the reliable extrapolation of the results for long times or for conditions not covered experimentally, the relevant microscopic mechanisms have absolutely to be included. At present, it seems that plasticity theory and the microscopic theory are increasingly combined.

It is not intended to cover rock mechanics in exhaustive way; this publication is focused on time-dependent effects in rock mechanics. Moreover, it covers mainly the experimental technique used in the laboratory and the interpretation of tests. Some theory of creep deformation is given; theories of dilatation and failure are also considered. The aspects of the evolution of a fracture are not described. In addition, the contents of the publication is strongly influenced by the personal experience of the authors. Therefore, the reader is encouraged to read other publications. The following general books are mentioned: Obert and Duvall [1967], Paterson [1978], Jaeger and Cook [1979], Jumikis [1983], Lliboutry [1987], Cristescu [1989a], Charlez [1991], Guéguen and Palciauskas [1994], Ranalli [1995], and especially for rock salt: Dreyer [1972], Hardy and Langer [1984, 1988], Ghoreychi *et al.* [1996], Habib *et al.* [1997]. As sources for rock data the following books are mentioned: Clark [1966], Vutukuri *et al.* [1974], Lama and Vutukuri [1978], Gevantman [1981], Landolt-Börnstein [1982], Carmichael [1982, 1984], Ahrens [1995]. Moreover, standards, suggested methods or recommendations for experimental procedures and their interpretation are given by several national and international commissions, e.g., ISRM (International Society for Rock Mechanics), ASTM (American Society for Testing and Materials), and DGGT (Deutsche Gesellschaft für Geotechnik = German Geotechnical Society). The reader is reminded, however, that rock mechanics is steadily changing corresponding to the evolution and improvement of experimental methods and the development of theory. There are also recommendations for site selection and the application of experimental results in the design of underground disposal of hazardous wastes (e.g., Langer [1995, 1996]).

In the following the experimental foundations for the evaluation of the time-dependent behavior of rock will be given. It is not possible to show all the testing possibilities, but we will at least show the foundations.

1.2 GENERAL REMARKS

Before designing the tests and the equipment necessary for testing, one has to consider what kinds of experiment have to be carried out and what would be the purpose of the tests. In our case we deal with geomaterials, i.e., rocks and similar materials, and their use in engineering geology. The final aim is the formulation of the constitutive laws and the definition of appropriate parameters for these materials which can be used in model calculations, e.g., finite element models.

Thus the question arises: what is a rock?

Most rocks comprise an aggregate of crystals and sometimes amorphous particles of mineral or organic matter, with sizes in the range of millimeters and centimeters. They are held together with various amounts of cement. If the rocks are weathered, fragmented or not yet compacted they grade into soil mechanics. The grains can consist of different kinds of minerals with different mechanical behaviour. On a larger scale rocks are mostly continuous and homogeneous, but quite often joints, faults, bedding planes, or different strata appear. The result is that rocks do not behave homogeneously as metals usually do. Rocks can also exhibit anisotropic behavior. In addition, pressurized fluid is frequently present. Under certain conditions this micro- or macro-structure results in a rather complicated mechanical behavior, because stress and strain distributions tend to be unequal on a small as well as on a large scale,

Table 1.1: Table of laboratory tests on rocks suitable for the determination of time-dependent properties

TEST	PURPOSE
quasistatic tests: duration from one minute to several hours, control of stress rate or strain rate, uniaxial or triaxial	strength, deformation behavior, volume change, elastic moduli, other (deformation) moduli
creep tests: duration hours to years, constant stress and temperature, uniaxial or triaxial	transient and stationary creep, volume change, creep fracture, tertiary creep
relaxation tests: stop of deformation at a certain strain, uniaxial, triaxial	stress relaxation, deformation and transient creep, recovery
stress drop tests: drop of stress difference after some strain, uniaxial and triaxial	deformation and transient creep, recovery, back-stress
direct tension tests: uniaxial or triaxial	tensile strength, deformation
indirect tension tests: Brazilian test, bending test, fracture toughness test	tensile strength, crack growth resistance
torsion test	strength, deformation
shear test	strength, deformation
hydrostatic compaction test	time-dependent volume decrease, healing
permeability measurement	permeability
microacoustic (AE)	crack formation and location, dilatancy
active sonic test	wave velocity, attenuation, dilatancy, healing
unloading, wave velocities	elastic behavior

easily causing fissuring and loosening. When carrying out tests, we have to take this into consideration. This must be regarded for instance when we choose the sample size. Often a minimum number of 2000 grains is regarded to be sufficient for one specimen. Therefore, a diameter of 10 cm for the specimen and lengths of 20 or 25 cm are quite common for rocks with large grains. One has also to consider what the general aim of the tests is: a certain engineering aim or perhaps a basic research question.

Because not all characteristics of a rock material are time-dependent, let us first list in Table 1.1 what kinds of laboratory tests are possible for the determination of time-dependent properties. Most tests are carried out in testing machines for uniaxial or triaxial tests that are designed for cylindrical specimens (Kármán cells).

Let us start with **uniaxial tests**. These tests are carried out with stress or load control with, say, $\dot{\sigma} = 1$ MPa/min rate of stress increase, or $\dot{\varepsilon} = 10^{-4}\,\text{s}^{-1}$ strain rate. A test rig for this kind of tests is shown in Figure 1.1. In uniaxial tests the sample is loaded by a piston and the oil pressure in a cylinder, which is controlled usually by a servocontrol. Screw-driven test rigs are also rather common. Often the displacement is measured by three or two LVDT-transducers

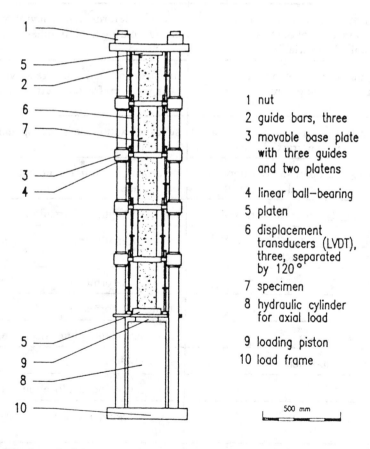

1 nut
2 guide bars, three
3 movable base plate with three guides and two platens
4 linear ball-bearing
5 platen
6 displacement transducers (LVDT), three, separated by 120°
7 specimen
8 hydraulic cylinder for axial load
9 loading piston
10 load frame

500 mm

Figure 1.1 *Rig for uniaxial creep tests on five samples one upon another and loaded with the same force used at BGR (Germany). Specimen size: 250 mm height, 100 mm diameter.*

(transducers working on the basis of a variable transformer for electrical tension) placed at angles of 120° or 180° around the specimen and parallel to its axis; strain gages are not recommended for soft geomaterials (see also Section 1.4). Experience shows that uniaxial quasistatic tests on rocks can yield results with large scatter caused by the inhomogeneity of the material. In order to get rid of some of the scatter it is recommended to reconsolidate the specimen before the test for removing the damage caused by drilling and sample preparation. This procedure is very efficient with rock salt but does not work well with hard rock.

In general, **triaxial tests** — performed at a constant confining pressure p, which re-produces much better the state in which the material is subjected *in situ*— should be preferred for failure tests: the scatter is reduced, the natural state of stress is a triaxial one, and one receives a much more complete picture of the mechanical behaviour of the rock. However, in this case the experimental technique is more sophisticated than for uniaxial tests. In most

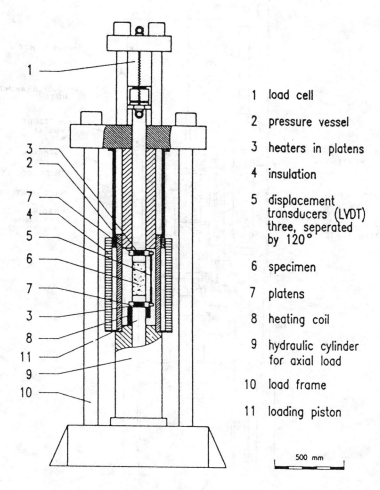

1   load cell

2   pressure vessel

3   heaters in platens

4   insulation

5   displacement transducers (LVDT) three, seperated by 120°

6   specimen

7   platens

8   heating coil

9   hydraulic cylinder for axial load

10 load frame

11 loading piston

500 mm

Figure 1.2 *Triaxial testing equipment for cylindrical samples used at BGR. Specimen size: 250 mm height, 100 mm diameter.*

cases Kármán type rigs are used; very many kinds of this type of testing apparatus have been constructed since the beginning of the century. Figures 1.2 and 1.3 show examples. The rig in Figure 1.2 consists of a pressure vessel for the confining pressure p, a cylinder with a piston for the axial force F or the axial stress σ_1, and three LVDT-transducers for deformation measurement mounted between the platens near to the sample, and has internal and external heaters. In this example the vessel is placed in a frame which supports the axial load. This is not needed for a self-supporting vessel. This rig is equipped with a load cell for the measurement of the axial load. The rig shown in Figure 1.3 has a somewhat different appearance but more or less similar characteristics. Experience shows that, due to the friction at the piston's seal, the determination of the axial load from the oil pressure applied to the piston is less precise than the direct measurement by a load cell. In addition, there are two

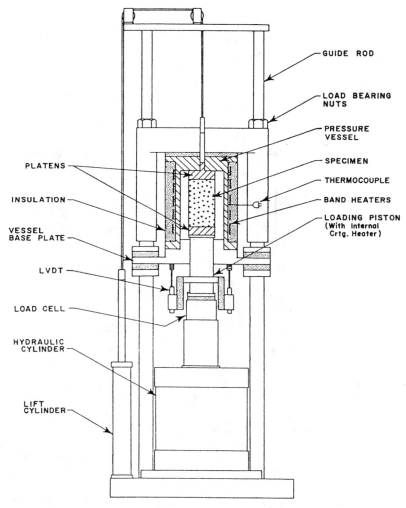

Figure 1.3 *Triaxial testing machine used at Re/Spec and Sandia (USA).*

servocontrol systems for confining pressure and axial load, a temperature control, and a data aquisition system. For long term creep tests one has to take some precautions as shown below.

Some remarks have to be made concerning the accuracy of such systems. Two essential "enemies of precision" exist: temperature fluctuations and friction. In spite of the fact that modern seals have quite low friction they can introduce unexpected high errors in the axial stress difference in triaxial tests. Therefore, for high precision it is recommended to place the load cell as near to the sample as possible, preferably inside the vessel. The same holds for the strain measurement, which should also be placed inside the vessel near to or even on the sample (see Figure 1.2). In this case temperature fluctuations in the room as well as friction do not

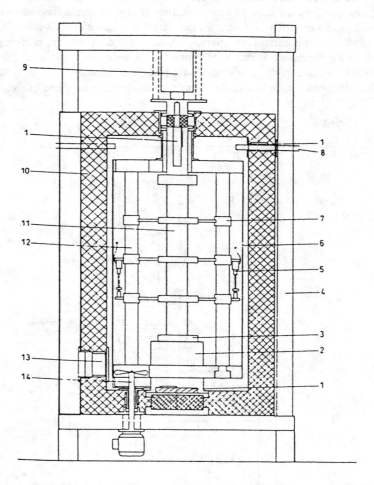

Figure 1.4 *Uniaxial testing machine for four simultaneous creep tests with an external chamber for heating (22 to 80° C) used at BGR. Air humidity (0 to 75% RH) can be controlled as well. 1: heater, 2: hydraulic cylinder, 3: platen, 4: external rod, 5: LVDT, 6: band heater on air guidance, 7: guide with ball bearing, 8: supply for humid or dry air, 9: load cell, 10: chamber with insulation, 11: specimen, 12: guide rods, 13: opening for electric wires, 14: fan.*

distort the measurements. If the load cell and deformation measurement devices are placed outside of the vessel, the construction may be cheaper, but this has the disadvantage that the measurements may be less precise. To avoid the influence of temperature fluctuations and temperature differences on the results, the rig is sometimes placed in a heated chamber. This is shown for a uniaxial testing machine in Figure 1.4 which can be used for temperatures up to 80°C and which allows simultaneous control of air humidity. Anyway, it cannot be stressed often enough that careful calibration is the heart of reliable measurements. That is why very strict rules are given in the national and international standards and recommendations.

In general, the same rigs can be used either for quasistatic tests or for creep tests. However, for long term creep tests one has to consider special precautions (see below). Quasistatic tests are very common. They are used to determine strength and deformation behavior. Quasistatic tests (minutes to hours) are mostly carried out with deformation control where the deformation rate $\dot{\varepsilon}$ is constant within certain test intervals. Also, quite often tests are performed with stress control where a constant stress rate $\dot{\sigma}$ is applied. This needs an active load control.

The usual way of carrying out **quasistatic tests** in a Kármán cell is as follows: an initial small hydrostatic stress is applied to the sample by a confining pressure p and an appropriate

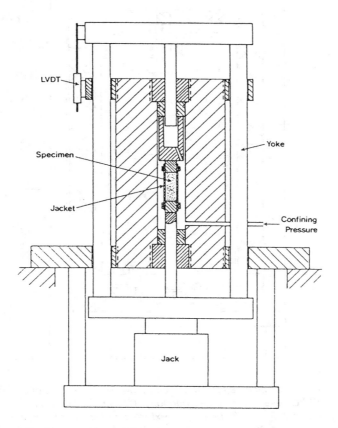

Figure1.5 *Schematic of a triaxial testing apparatus incorporating Grigg's arrangement of compensating piston (Paterson [1978] p.8).*

axial load σ_1. Then the sample is heated to the desired value. Finally, the test itself is started by increasing the hydrostatic pressure to the desired value; then the axial load is raised with stress or strain control, if one wants to perform a compression test. An extension test is performed by decreasing the axial load or by increasing the confining pressure. The specimen is elongated in such a test since the confining pressure is greater than the axial stress. Axial load and confining pressure can also be changed simultaneously in the two kinds of test (see true triaxial tests in Section 1.3). By the way, technically it is not an easy task to switch between load (stress) control and strain control, as is often needed after hydrostatic loading or during a test with several phases. In triaxial tests it has to be taken into account that the confining pressure influences the axial load, which is a function of the cross-section of the sample and of the vessel opening for the piston. If one does not use an internal load cell, this must be taken into account using the following formula for the effective axial stress σ_1:

$$\sigma_1 = \frac{1}{A}[F + p(A - A_g)] \qquad (1.1.1)$$

where $A = A_o(l_o/l)$ is the actual cross section of the specimen and A_g is the cross section of the vessel opening for the piston.

Two assumptions are involved in this formula: (1) no volume change occurs in the specimen, and (2) the specimen keeps its cylindrical shape; as shown from experience this last assumption is the best approximation, because really barrel-shaped samples are very rare. Some rigs have a hydraulic compensation for the effect of p on σ_1 (see Figure 1.5). Some rigs for uniaxial tests (in particular for creep tests) have a mechanical compensation of the

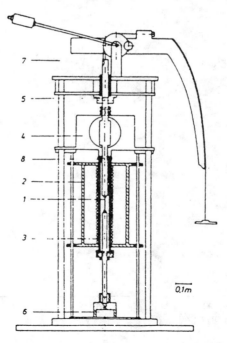

Figure 1.6 *Rig for uniaxial creep tests with a special curved cantilever for keeping σ_1 constant during a test, here for tension (Blum and Pschenitzka [1976]).*

effect of the change of the cross-section of the specimen on σ_1 by a specially curved cantilever as shown in Figure 1.6 (Blum and Pschenitzka [1976], Vogler [1992]). This device adjusts the axial force according to the deformation of the specimen and keeps a constant σ_1 during the test. It can be used for tension or compression.

In Figures 1.7 and 1.8 results for quasistatic tests are given. As an example, Figure 1.7 shows test results for the influence of the confining pressure p on stress-strain curves for rock salt for a certain deformation rate $\dot{\varepsilon}$. Strength and deformation at failure increase considerably with increasing p for this soft kind of rock. Similar results are given by, e.g., Dreyer [1972], Georgi *et al.* [1975], Wawersik and Hannum [1980]. Figure 1.8 shows the result of a test series carried out on marble, a more brittle kind of rock. Again, strength is increasing with increasing p; however, deformation at failure is only slightly dependent because the sample does not have the ability or enough time to deform. Volume change has not been measured in these tests. It must be mentioned that these tests are often combined with the measurement of the elastic "constants" by unloading and reloading or with other special measurements. More test results are given in Chapters 2 and 4.

A very important topic is the specimen preparation. It is essential to prepare the samples with correct dimensions, in particular for triaxial tests. The surfaces must be at right angles to each other and must be plane. A precision of 0.01 mm for the end faces is possible using a lathe. The rubber or metal jacket used for protection of the specimen against the hydraulic oil in a triaxial test must be fixed carefully, especially for extension tests. Usually the stiffness of this material is negligible and does not influence the stress in the specimen. It is also

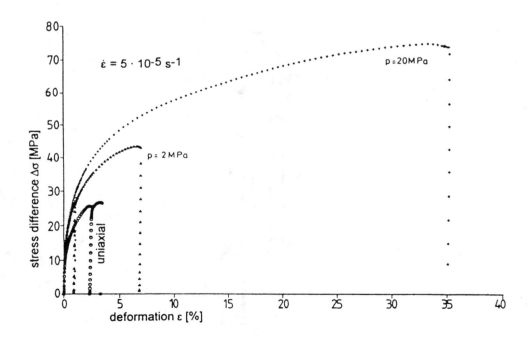

Figure 1.7 *Results of uniaxial and triaxial compression tests on rock salt from Asse Mine (Northern Germany) at the same deformation rate but different confining pressures (BGR).*

important to document the state of the specimens carefully before and after the tests: a geological, petrographic, mineralogical and mechanical description, photos, mineralogical and geochemical analysis, and so on. This makes it possible to understand better the results and to check them. One can also carry out statistical analyses, e.g., of the influence of grain size, petrography, or texture. It also has to be stressed that an accurate documentation of the whole history of the sample must not be forgotten. Numerous good results have been worthless in the past due to this negligence. Sample storage may also be delicate. For instance, changes of moisture content may change the mechanical properties considerably. An effective protection against changes of moisture content is possible only withy a metal box or another metal cover, e.g., a metal foil coated with plastic (as often used for food).

Often the end faces of the samples are not lubricated as long as the pressure cone produced by the friction between the platens and the sample material is small compared with the non-influenced sample. But they have to be lubricated or the results have to be appropriately corrected when the slenderness (length/diameter ratio) of the sample is, say, below 2. A study of the influence of the slenderness of not lubricated specimens on creep has been made by Weidinger *et al.* [1996b]. They give an empirical equation to be used for correction. It is repeated that the sample size has to be large enough.

For a comprehensive description of the behavior of rock it is important also to determine the inelastic volume changes which are not caused by pure elasticity. They are caused by the closing or the opening and growth of microcracks, and occur during hydrostatic loading or during deformation under differential (deviatoric) stress. The volume decrease happens during

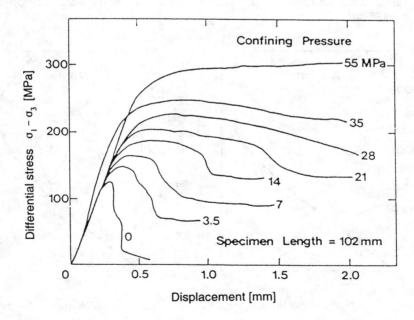

Figure 1.8 *Results of uniaxial and triaxial compression tests at the same displa-cement rate* $v = 10^{-3} \, mm \, s^{-1}$ *($\varepsilon = 1 \times 10^{-5} \, s^{-1}$) but different confining pressures for Tennesse marble (Paterson [1978] p.144).*

hydrostatic loading or at low stress differences, whereas dilatancy happens at higher stress differences. The conditions for one or the other, its time-dependent evolution, and its application are an important feature of this book. Volume increase goes along with the following effects: increase of damage, tertiary creep or creep rupture, permeability increase, acoustic emission. Volume decrease goes along with decrease of damage, permeability decrease. Dilatancy occurs prior to failure. See also Paterson [1978]. Three examples are shown in Figures 1.9 and 1.10. More are given in the following chapters.

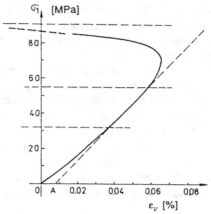

Figure 1.9 *Stress-strain curves for sandstone obtained in uniaxial tests showing dilatancy (Cristescu [1989a]).*

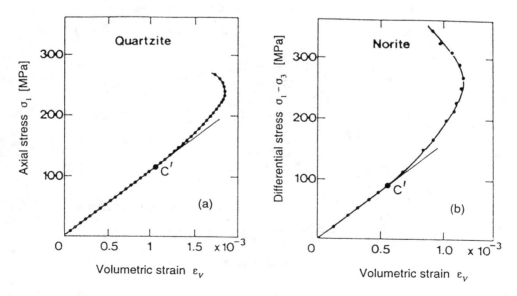

Figure 1.10 *Dilatancy in (a) quartzite in uniaxial compression and (b) in norite in triaxial compression (Bieniawski [1967] , after Paterson [1978], p.117).*

The first measurements of the volume change were made by Baushinger in 1879 (see Bell [1973]). He discovered that sandstone (as well as some metals) are dilatant during plastic deformation in uniaxial compression tests (for the early tests revealing irreversible volume change during plastic deformation see Bell [1973]). Bridgman [1949] found that in uniaxial compression tests the volume of soapstone, marble and diabase is dilatant at high applied stresses. For marble he found that at high stresses dilatancy is produced by "rapid creep". He also suggested that irreversible compressibility is due to closing of pores, while dilatancy (called "retrograde volume change") is due to the openings of pores. Early revealing tests on hard rocks exhibiting compressibility and/or dilatancy are due to Brace *et al.* [1966] and Bieniawski [1967] (for other early papers see Cristescu [1989a, 1996b]).

For the measurement of the volume changes of cylindrical samples one often utilizes the displacement of the oil used for the confining pressure. It is measured by a dilatometer, i.e., a cylinder with a piston which has a rather small diameter, where the piston displacement is measured. However, because the oil is displaced mainly by the movement of the piston for the axial load, the calibration is quite difficult; the effect of the volume change of the sample is much smaller. Temperature effects are also to be considered and suppressed by an appropriate construction. A good solution is the use of Grigg's arrangement of a compensating piston (see Figure 1.5). Thus, the external dilatometer measures in principle the volume change of the

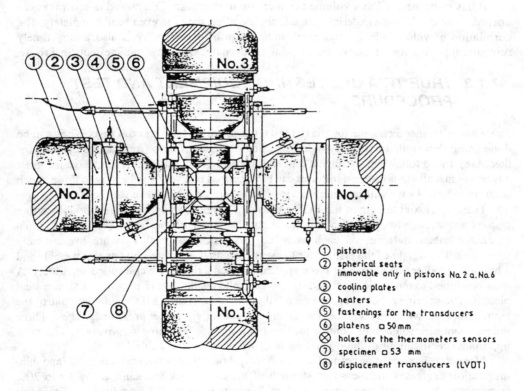

① pistons
② spherical seats
   immovable only in pistons No.2 a.No.6
③ cooling plates
④ heaters
⑤ fastenings for the transducers
⑥ platens □ 50 mm
⊗ holes for the thermometers sensors
⑦ specimen □ 53 mm
⑧ displacement transducers (LVDT)

Figure 1.11 *Arrangement of specimen and pistons in the true triaxial press for cubes at BGR. For specifications of the press see Table 1.2 (Hunsche and Albrecht [1990], Hunsche [1994b]).*

sample itself only. It is also possible to measure directly the change of the diameter or circumference, which can be done by many methods: clips, wires, chains, rings, strain gages, etc. They can be used in uniaxial and triaxial tests. However, they have the disadvantage that they only measure a few diameters or circumferences of the specimen and do not determine the whole volume change itself. Therefore, these measurements have to be corrected or be carried out at several levels of the specimen. Another possibility is the calculation of the volume change from true triaxial tests on cubic samples with the help of the change of the individual edge lengths as shown in Section 1.3. Of course, the measured results have to be corrected for the elastic volume change. Therefore, it is important to measure also the elastic bulk modulus K, e.g., by unloading cycles.

Volume changes go along with a considerable amount of microacoustic emission (AE), which should also be measured. Usually the events are only counted. But with the help of a complete recording and evaluation of the time history one receives much more information and can even locate and analyze the seismic sources. A great number of papers exists on this topic (e.g., Hardy [1995]). This method can also be used satisfactorily in underground, where one also can locate the source of the microacoustic emissions, with the help of several transducers, placed in the space to be evaluated (Spies et al. [1997]). Deformation of rocks can also produce electromagnetic waves, mainly at failure. Their evaluation can yield valuable information on the microphysical mechanisms (e.g., Yamada et al. [1989]).

It has to be stressed that volume can increase or decrease. Cracks and microcracks are opened or closed. In some materials like salt or clay or ice, they can even heal completely. The correlation of volume change (damage) and change of permeability is also a very timely scientific topic because it allows the coupling of mechanical and hydraulic effects in a rock.

1.3 TRUE TRIAXIAL TESTING EQUIPMENT AND TEST PROCEDURE

There exists another technique for triaxial tests: the so-called **true triaxial tests**. They can be done using thin-walled hollow cylinders or using cubic specimens (see Paterson [1978] for literature). In the following, equipment for cubes is described. Tests of this kind have the great advantage that all kinds of stress histories can be simulated in a very similar way to those which occur in nature. As we will see later, this is of considerable importance for rocks.

The true triaxial test rig of the BGR consists of a rigid frame in which six double-acting pistons are arranged opposite each other about the center of the frame (Figure 1.11). To guarantee balanced forces on each axis, the two pistons of each axis are hydraulically connected. The weight of the pistons for the vertical axis is equalized by counter weights and a hydraulic counter balance. The force applied on each axis can be controlled separately. A load is applied to the cubic specimens (up to 20 cm to a side) by six pistons via square steel platens. The specimens are prepared on a lathe. Six independent PID regulators control the temperature to within 1 K. The forces are regulated via three pressure gauges. Three independent electronic units control load and deformation via three servovalves. The most important specifications of the rig and the tests are as follows in Table 1.2.

Many tests have been performed with different lubricants. It was found that good (and thin) lubrication decreases the measured strength of the cubic rock salt samples by 10 to 20% compared with non-lubricated ones; therefore, lubrication is essential for reliable results in this kind of tests since friction between platens and sample seemingly increases strength, which

Table 1.2 Specifications of the true triaxial tests on cubic samples at BGR

–Maximum force (per axis):		2000 kN
- Sample size (edge length):		53 mm
	maximum:	ca. 200 mm
- Platen size in standard tests:		50 mm
- Temperature	maximum:	400 °C
- Friction of piston:		< 1 % of load
- Independent load or deformation control along the 3 axes is possible.		
- Loading rates in standard tests:		
hydrostatic phase:		$\dot{\sigma} =$ 7.6 MPa/min
deviatoric phase		$\dot{\tau} =$ 21.4 MPa/min
- Deformation rate at failure, standard tests:		$\dot{\varepsilon}_{eff} \approx$ 0.007 s^{-1}
- Digital data acquisition, standard interval:		1 s
- Lubrication with a thin film of paraffin wax (for tests at room temperature) or graphite (elevated temperatures)		

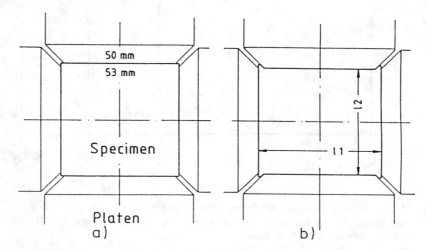

Figure 1.12 *Cross-section of the arrangement of specimen and platens in the true triaxial press of BGR (to scale); a) at beginning; b) at failure (for rock salt).*

can hardly be corrected. Paraffin wax and graphite have very small friction coefficients of $\mu \approx 0.01$. A large number of tests have been carried out to determine the influence of sample size in the true triaxial tests using rock salt. The main results are: (1) the sample size has no influence on the measured strength; (2) the ratio (K) of the edge length of the sample to that of the platens has an influence on the measured strength. A value of $K=1.06$ proved to be small enough and was chosen for the tests. Figure 1.12 shows a cross-section of the arrangement of specimen and platens. A precise calibration makes it possible to determine the absolute volume change of the sample during a test with an accuracy of 0.3% of the total volume of the sample.

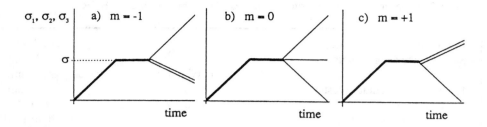

Figure 1.13 *Schematic of the test procedure in true triaxial tests on cubes for three load geometries m with mean stress σ.*

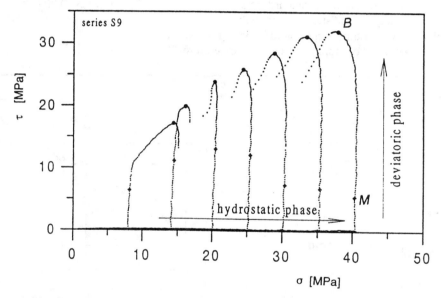

Figure1.14 *Seven true triaxial compression tests of test series S9 on rock salt from the Gorleben salt dome (Northern Germany) performed at room temperature at BGR. Spacing of points: 1 s. M: minimum volume; B: failure.*

It is advisable to carry out the true triaxial tests in a deviatoric manner. The following procedure is therefore recommended; it can be also used in Kármán-type tests. In the tests, the load is first applied hydrostatically (hydrostatic phase). When the desired mean stress level σ has been reached, the three principal stresses are changed linearly in time so that σ is held constant and the octahedral shear stress τ is also increased linearly (deviatoric phase). The procedure is shown in Figure 1.13 schematically for three load geometries m; see also Figure 2.10. This is done until τ reaches the failure strength $τ_B$ of the sample and fracturing occurs as shown in Figure 1.14 for one set of compression tests. $τ_B$ is defined as the local maximum of τ in a single triaxial test. It should be noticed that in triaxial tests the shear stress does not reduce to zero at failure; the fractured sample has a considerable residual strength $τ_R$ which can be determined by unloading and reloading as shown in Figure 1.15 (Hunsche [1992, 1993a], Hunsche *et al.* [1994]). This kind of tests can be combined with the measurement of acoustic emission and active wave velocity measurements (see below). More information about the technique of the true triaxial tests on cubic samples is given in Hunsche and Albrecht [1990] and in Hunsche [1992, 1994a]. Experimental results received from this device have also been published by Hunsche [1984a,b, 1989, 1990, 1991, 1992, 1993a, 1994a, 1995, 1996], Hunsche *et al.* [1994], Xiong and Hunsche [1988].

Thus, some important tools for the performing uniaxial and triaxial tests on rocks have been revealed. Analysis and results are shown in Chapter 2. It has to be emphasized that many kinds of rigs have been constructed and that other testing procedures are possible. Anyway, it is important to consider carefully what kinds of tests are needed and with which test procedure the results under question can be derived.

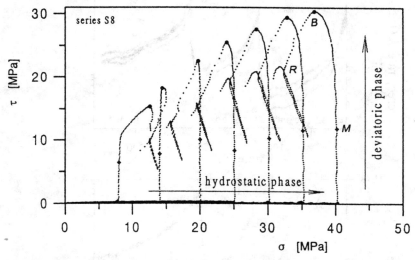

Fig.1.15 *Seven true triaxial compression tests of test series S8 on rock salt from the Gorleben salt dome (Northern Germany) performed at room temperature at BGR. Spacing of points: 1 s. M: minimum volume; B: failure; R: residual strength.*

1.4   UNIAXIAL AND TRIAXIAL CREEP TESTS

Knowledge of **creep** behaviour is an indispensable condition for reliable time-dependent model calculations, especially in soft rocks like rock salt and coal. It is that part of geo-mechanical behavior which influences the most the time-dependent strain and stress evolution in a system. Therefore, it is very important to determine the creep behavior with sufficient accuracy.

Creep tests are carried out in uniaxial or triaxial rigs at constant values of stress σ and temperature *T*. These can be changed stepwise during the tests in order to create phases with constant conditions. Examples for tests on rock salt are given in Figures 1.16. It demonstrates the effect of stress on creep. Figure 2.24 shows the effect of temperature (see also Section 2.2,

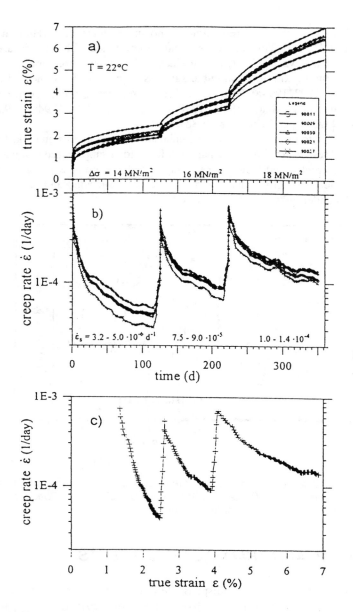

Figure 1.16 *Five simulations uniaxial creep tests at room temperature on rock salt specimens of the same type with three succeeding stresses (BGR). Rock salt (z3OS/LS) from the Asse mine (Northern Germany). For the rig see Figure 1.1. a) strain vs. time; b) creep rates vs. time, ε_s: steady state creep rates; c) creep rate vs. strain for one of the tests.*

and Chapters 3 and 4). For uniaxial or triaxial creep tests the same apparatus are generally used as for quasistatic tests described above. However, for creep tests some additional technical conditions must be taken into account. Because creep tests on geomaterials usually have a very long duration of several months or even years, all parts of the system must be stable during this long period of time. Special attention must be given to the following items. LVDT-transducers or optical systems have proved to be long term stable for deformation measurement whereas strain gages directly glued on the sample may be unstable due to creep effects in the glue. Pressure gauges and load cells must be of high quality; their drift and

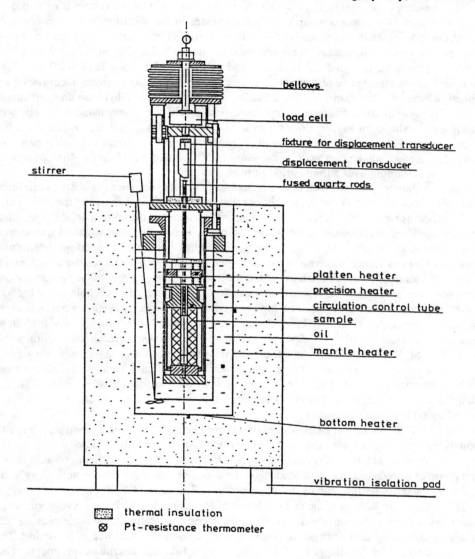

bellows

load cell

fixture for displacement transducer

displacement transducer

fused quartz rods

stirrer

platten heater

precision heater

circulation control tube

sample

oil

mantle heater

bottom heater

vibration isolation pad

thermal insulation

⊗ Pt-resistance thermometer

Figure 1.17 *Cross-section of a high-precision device for uniaxial creep tests for temperatures up to 350°C and a maximum load of 95 kN at BGR (Hunsche [1988]). Strain rates down to $\dot{\varepsilon} = 10^{-12} s^{-1}$ can be measured with this device. Deformation is measured in the center along the axis of the specimen with a precession of 0.1 µm. Temperature is controlled within less than 0.01 K. Total height: 200 cm. Specimen size: 250 mm height, 100 mm diameter.*

temperature effects must be minimized. Some thermometers can exhibit drifts as well. Therefore, resistance thermometers (e.g., PT 100) are recommended; thermocouples are perhaps not free of drift. Also, electronic signal conditioners can exhibit drift and have to be of high quality. The influence of temperature on all mechanical and electronical parts of the system must be minimized since fluctuations of room temperature are nearly unavoidable during long term experiments. Periodical calibration of all parts of the system is essential.

It is of particular importance to control the stresses in the specimen with high accuracy during the tests, since the influence of the equivalent stress (stress difference) $\Delta\sigma = \sigma_1 - p$ on strain rate is very non-linear. A stress exponent around $n = 5$ and even higher values (i.e., $\dot{\varepsilon}_s \approx (\Delta\sigma)^n$) have been found for rock salt and other materials (see Tables 2.2 and 2.4). Therefore stress changes have a great influence on creep rate, and so an accuracy of the applied stresses of better than 1% is required. Because it is not advisable to use an everrunning servovalve system for long term creep tests, one should use a mechanical system with cantilevers as shown in Figure 1.6, or a hydraulic system with a load (or pressure) control which comes only into action when the load (or pressure) leaves a certain window (control on demand). Another weak point may be the stability of a computerized data collection system when a test is running over a long period of time.

It is important to use fixed platens and inflexible guides for the platens in order to minimize an inclination of the endfaces of the specimen relative to the sample axis during the tests which would otherwise probably occur due to non-homogeneous deformation in the specimen. Because small inclinations are nevertheless inevitable, two or three deformation transducers must be placed around the sample. Their averaged readings will further be used for determining strain. Another possibility is the use of a dilatometer system which measures the displacement between the centers of the two endfaces of the specimen outside or even inside the sample. An example is shown in Figure 1.17, where deformation is measured in a borehole along the axis of the specimen. Figure 1.17 shows a high precision creep testing device for uniaxial creep tests at temperatures up to 350°C which has been designed for measuring very low strain rates, down to about $\dot{\varepsilon} = 10^{-12}\,s^{-1}$. Special measures had to be taken to reach this accuracy (see Geisler [1985], Hunsche [1988]). Again, precise temperature control and minimization of friction (the two essential "enemies of precision") were most difficult. Temperature fluctuations can be kept smaller than 0.01 K over long periods of time and friction is less than 80 N. The usual limit of strain rate resolution in common rigs for creep tests is about $\dot{\varepsilon} = 10^{-10}\,s^{-1}$, depending on test duration.

Figure 1.1 shows a rig at BGR which allows one to carry out uniaxial creep tests simultaneously on five samples with the same load, which is controlled by a precise load control (digital control on demand). Because the stress difference $\Delta\sigma$ has to be kept constant during each phase of a creep test, the axial load has to be adjusted from time to time according to the change of the cross-section due to strain, according to Equation (1.1.1).

The influence of friction of the end faces and of loosening due to fracture evolution on the creep rates also has to be considered in creep tests. A compromise has to be made because long samples avoid the influence of end friction and short samples the influence of loosening. This has been analyzed by Weidinger et al. [1996b] with the help of numerous experiments. They also give an elaborate empirical function for correction. A number of analyses have been performed to study the influence of the sample size itself on failure and creep. Special tests for creep have been performed in the Asse mine (Germany), where long term creep tests on two large pillars (1.5 m edge length, 4 m height) with controlled stress conditions have been performed. The results have been compared with laboratory tests on small samples, and it was considered that the results are in good agreement. The experiments and the results are described by Hunsche *et al.* [1985], Hunsche and Plischke [1985], Plischke and Hunsche [1989].

In Section 2.2.2 the technique of relaxation tests and stress drop tests is introduced, and results for tension tests are given in Section 2.1. This chapter has presented the most important techniques for the experimental evaluation of time-dependent properties of rocks. The following two chapters are more focused on the results.

2 Results and Background

2.1 QUASISTATIC TESTS

It has been discussed in the previous chapter how quasistatic failure and deformation tests should be carried out. In this chapter the results which are obtained with such experimental devices will be given, and some theoretical considerations. However, it is not possible to give a complete overview since a huge amount of literature exists in this field. The final aim of the experiments is to gain a general knowledge and to determine constitutive equations and parameters for the mechanical behavior of the rock type of interest, to be used for practical purpose, e.g., expert judgement and in model calculation. Between various concepts involved in these equations are the functions involved in describing the failure, dilatancy, compressibility, yielding, plastic potential, etc. Another aim is to attain information about the time dependence of deformation, which finally must lead to the formulation of a general creep law as part of a comprehensive constitutive equation. Quasistatic deformation tests, creep tests or relaxation tests only illuminate the same inherent time-dependent properties from different viewpoints. It has therefore to be considered that most of the time-dependent mechanical properties are related.

In most cases, quasistatic experiments are carried out with deformation control, where $\dot{\varepsilon}$

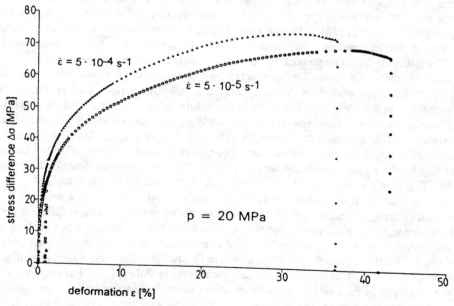

Figure 2.1 Triaxial compression tests on rock salt from Asse mine (Northern Germany) at the same confining pressure p but at different deformation rates $\dot{\varepsilon}$ (BGR).

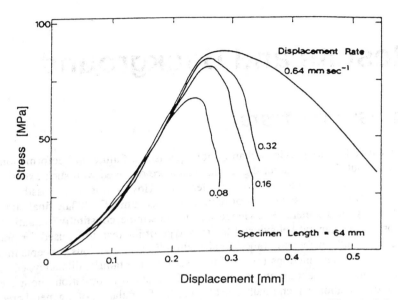

Figure 2.2 *Uniaxial compression tests on Arkose Sandstone at different deformation rates $\dot{\varepsilon}$ from 1.25×10^{-3} to 1×10^{-2} s $^{-1}$ (Paterson [1978], p.145).*

is held constant. At a chosen fixed confining pressure p, but for various $\dot{\varepsilon}$, one obtains $\Delta\sigma - \varepsilon$ curves as shown in Figure 2.1. For rock salt, failure strength (ultimate stress) as determined by the peak stress, is only slightly dependent on $\dot{\varepsilon}$. This is somewhat different for more brittle types of rock where strength increases considerably with increasing $\dot{\varepsilon}$ as shown in Figure 2.2 (Paterson [1978], p.14; see also Chapter 4). However, due to the ductile behavior of rock salt the deformation at failure is much more dependent on $\dot{\varepsilon}$ than for brittle rock.

The following has to be pointed out: if a test is carried out at a sufficiently low value of $\dot{\varepsilon}$ then the experiment becomes finally a creep test with constant stress state and without short term failure (but perhaps with tertiary creep after a longer period of time). This is shown in Figure 2.3a in a more complete picture of the failure behaviour of rock salt derived from a great number of compression tests on one rock salt type, with various values for constant $\dot{\varepsilon}$ and at various confining pressures p (Hunsche [1994a]). In this diagram strength is given as a function of p and $\dot{\varepsilon}$ for room temperature. The failure domain shown is bounded on the left hand side by the steady-state creep curve for 22°C. This has also been described by Wallner [1983]. This is often denominated as the brittle-ductile transition. This is discussed in Section 2.3 in more detail. It is obvious from Figure 2.3a that this type of rock salt has a weak maximum in strength at about $\dot{\varepsilon} = 10^{-4}$ s^{-1} for all confining pressures. This is different for hard rocks, where strength steadily increases with $\dot{\varepsilon}$. This result has also been derived by Klepaczko et al. [1991], who have performed uniaxial compression tests on rather impure rock salt between $\dot{\varepsilon} = 3 \times 10^{-5}$ s^{-1} and 4×10^{3} s^{-1} as drawn in Figure 2.4. They have found a minimum in strength at about 2 s^{-1}. This minimum has also been found by Alheid [1997] as shown in Figure 2.5 for two types of rock salt. The minimum of the strength at higher strain rates is probably caused by dynamic processes in combination with the inhomogeneities of the material. It has to be mentioned that it is just in the loading range of seismic events. This example is shown

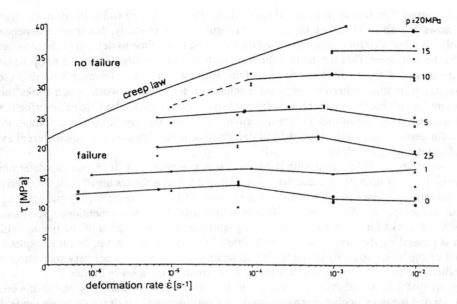

Figure 2.3a *Strength of rock salt from the ASSE mine (Northern Germany) determined from triaxial compression tests on cylindrical specimens at a great number of confining pressures p and deformation rates ε̇ at room temperature (BGR). The line denoted with "creep law" represents the ε̇–τ conditions for steady-state creep.*

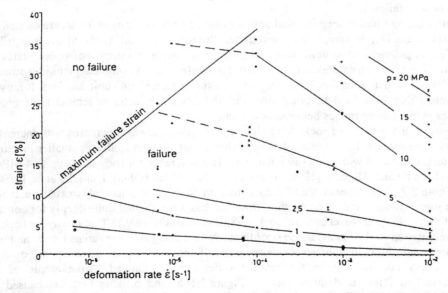

Figure 2.3b *Strain at failure for the same tests as shown in Figure 2.3a, which were performed on virgin samples. There was practically no failure above the line denoted "maximum failure deformation".*

to demonstrate that failure does not always follow the anticipated rules. In addition, Figure 2.3b shows the strain at failure for the tests in Figure 2.3a. Obviously, this strain is dependent not only on p but also on $\dot{\varepsilon}$ since the specimen has just more time to deform in the slow tests than in the fast ones. This means that the deformation up to failure itself does not possess a deep meaning and that there is no criterion of this kind for soft rocks. However, we will present in Chapter 6 another criterion based on mechanical deformation work which looks more consistent for soft rocks. However, a comprehensive creep law must be able to describe these deformations. The "maximum failure deformation" gives only the limit for deformation at failure for certain confining pressures and deformation rates. These strains are different even for different types of rock salt.

If one carries out failure tests with constant stress rate change $\dot{\sigma}$ or $\dot{\tau}$, one can determine the dependence of strength on these variables. Figure 2.6 shows the results of such tests derived from compression tests on cubic rock salt specimens. With increasing $\dot{\sigma}$, the strength τ_B is only slowly increasing. Again, this is different in hard rock. It has to be mentioned that in such tests on soft rocks failure is finally associated with the maximum speed of the piston, which again is caused by the maximum pumping rate of the hydraulic system, i.e., the results can depend on the limit capability of the testing apparatus. Therefore, it is not very surprising that the influence of $\dot{\sigma}$ or $\dot{\tau}$ on strength is found to be small in the tests in Figure 2.5.

From quasistatic deformation tests, one can also draw conclusions about the creep behavior and the microscopic mechanisms, since a considerable part of the deformation is due to (transient) creep in soft rocks. A number of attempts for this purpose have been made; these have been described by Mecking and Estrin [1987] and by Hertzberg [1983] (see also Hunsche [1994a]). Finally, a well formulated and complete creep law must be able to describe creep in practically the whole deformation domain including quasistatic tests (with the exception of the range near to failure).

After all, one has to keep in mind that most of the common tests on rocks are carried out in compression, i.e., the sample becomes shorter. But nature uses all kinds of stress distributions (stress geometries) as defined by the Lode parameter m. Examples of stress states are compression ($m = -1$), extension ($m = +1$), or torsion ($m = 0$). A complete picture of strength behavior can be determined only by using true triaxial tests, e.g., on cubic samples. It must be mentioned that model calculations show that the Lode parameter m around underground cavities in most areas ranges between $m = 0$ and $m = +1$.

Tests on many kinds of rocks have shown that strength always increases with increasing hydrostatic pressure. In addition it has been shown that strength is always smaller in extensional tests compared with compressional tests. The difference for rock salt is up to 4 MPa or 30% (Hunsche and Albrecht [1990], Hunsche [1992, 1993a, 1994a], Hunsche et al. [1994]); see Figure 2.7. The dependence of failure strength on all three principal stresses, i.e. mainly on the intermediate stress, or on the third invariant, has been studied quite deeply for concrete (e.g., Chen [1982], Schreyer and Babcock [1985]; see also Figure 9.1). That is why for rock salt a representation of the results in the coordinate system using stress invariants τ, σ, and m is revealing, and shows this difference more clearly. If one uses $\sigma_1, \sigma_2, \sigma_3$ or $\sigma_1 - \sigma_3, \sigma_2, \sigma_3$, or other stress invariants, then the relative difference is changed. An example of this transformation effect is demonstrated in Figure 2.16a, and b. This fact has caused unnecessary discussion on the question, whether the dilatancy boundary —defining the long term safety boundary— is about one half or one third of the short term failure. Principally all systems are equivalent, and it does not seem to be a reason for preferring one or another. But considerable mathematical problems can arise if one does not use a set of invariants.

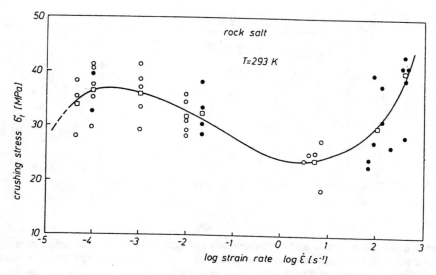

Figure 2.4 *Rate spectrum of crushing stress σ_f for MDPA rock salt.* o: *experimental data of Metz University;* •: *experimental data of Ecole Polytechnique;* □ : *mean values (Klepaczko et al. [1991]).*

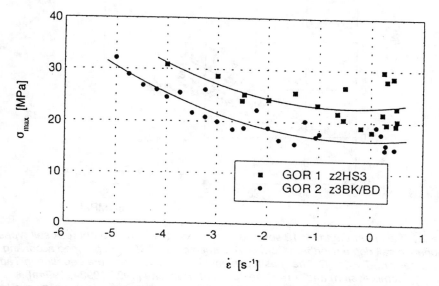

Figure 2.5 *Failure strength of two types of rock salt derived from uniaxial tests performed with strain rates between $\dot{\varepsilon} = 1 \times 10^{-5}$ and 4×10^{0} s^{-1} (Alheid [1997]).*

Important is also the tensile strength of a rock, because tension is always present in the vicinity of openings. Generally, the short term uniaxial tensile failure strength of rock ranges between 5 and 10% of the uniaxial compressional strength. For rock salt one receives values

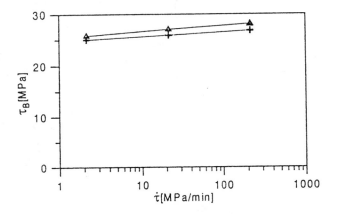

Figure 2.6 *Strength of rock salt from Gorleben salt dome (Northern Germany) determined in triaxial compression tests on cubic samples for three rates of stress change* $\dot{\tau}$ *at room temperature (BGR). Mean stress at failure was* $\sigma = 25$ *MPa.*

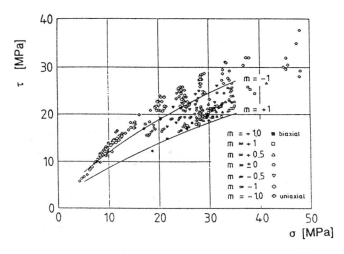

Figure 2.7 *Failure strength for 12 series of true triaxial tests on different rock salt types from the Gorleben salt dome (Northern Germany), carried out at BGR, and grouped according to the load geometry m; T = 30° C. The lines represent the conservative failure equation in Table 2.1 for m = −1 (compression) and m = +1 (extension) (Hunsche [1992, 1993a, 1994a]).*

between 1.5 and 2.4 MPa (uniaxial strength about 25 MPa); see Hunsche [1993b], Gessler [1983]. Brazilian tests yield about the same results. To study the tensile strength under triaxial conditions Hunsche [1993a,b, 1994a] has performed a number of tests om dogbone-shaped samples ($m = +1$). The results for one type of rock salt are shown in Figure 2.8. The strength (σ_z, tension stress along the sample axis at failure) first ncreases with increasing confining pressure p, then decreases and becomes zero at $p = 45$ Mpa ($\sigma = 30$ Mpa). This coincides with

Table 2.1   Conservative failure equation for rock salt (Hunsche [1992, 1993a, 1994a])

$$\tau_B = f(\sigma) \cdot g(m) \cdot h(T) \qquad \underline{\text{Failure strength}}$$

$$f(\sigma) = b \cdot \left(\frac{\sigma}{\sigma_\star} \right)^p$$

$$g(m) = \frac{2k}{[(1+k)+(1-k)J_m]}$$

$$h(T) = \begin{cases} 1 & 20^\circ C \leq T \leq 100^\circ C \\ 1 - c(T - 100^\circ) & 100^\circ C \leq T \leq 260^\circ C \end{cases}$$

$$b = 2.7 \text{ MPa} \qquad\qquad p = 0.65$$
$$k = 0.74 \qquad\qquad\quad \sigma_\star = 1 \text{ MPa}$$
$$c = 0.002 \text{ K}^{-1}$$

$$\tau_R = \tau_B(m = +1) \qquad \underline{\text{Residual strength}}$$

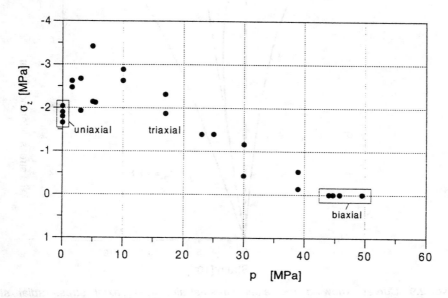

Figure 2.8 *Tensile strength* σ_z *as a function of confining pressure p. The uniaxial and the triaxial test were performed on dogbone-shaped samples, the biaxial tests on cubes. Salt from the Asse mine (Germany).*

with the failure strength in biaxial tests. At further increased confining pressure, failure occurs without tensile stress but with compressive stresses from all sides (see also Paterson [1978]). There are certain relations between strength, tensile strength and fracture toughness or fracture resistance (see, e.g., Whittaker *et al.*[1992]).

However, the final aim cannot be to determine strength of rock in terms of stresses only. A comprehensive picture should also include the prefailure behavior and perhaps also postfailure, and this means not only deformation but also volume change and its related phenomena: permeability, creep failure, damage, microacoustic emissions, and others. Because this is especially substantial in repositories, the determination of dilatation is becoming an increasingly important field in rock mechanics. Dilatation has been observed in many kinds of rock and soil (e.g., Paterson [1978], Jaeger and Cook [1979], Cristescu [1989a, 1994a], Hunsche [1992, 1994a, 1996], Thorel and Ghoreychi [1996]; more examples are given in several chapters of this book).

Depending on the stress state, the volume of a specimen increases (dilatancy) or decreases (compressibility, contractancy) due to the overall microcracking opening or closing. Figure 2.9 shows an example for granite (Brace *et al.* [1966], where more examples can be found; see also Paterson [1978]). In Figure 2.10 the relative volume change ε_v is shown for the various test phases for one of the true triaxial tests on cubic samples, described in Section 1.3 (Figures 1.14 and 1.15), together with the acoustic emission rate (AE), which is obviously a good indicator

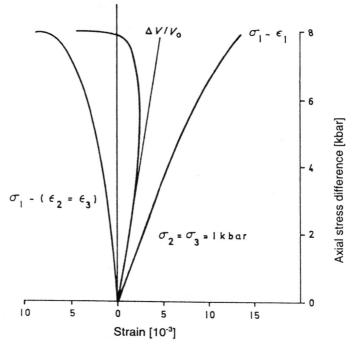

Figure 2.9 *Curves showing the axial stress-strain,* $\sigma_1 - \varepsilon_1$, *axial stress-radial strain,* $\sigma_1 - (\varepsilon_2 = \varepsilon_3)$, *and axial stress-volumetric strain,* $\Delta V/V_0$, *behavior of Westerly granite in a triaxial test at a confining pressure of 1 kbar (= 100 MPa). The deviation of the* $\Delta V/V_0$ *curve from the straight line represents dilatancy (after Brace et al. [1966]).*

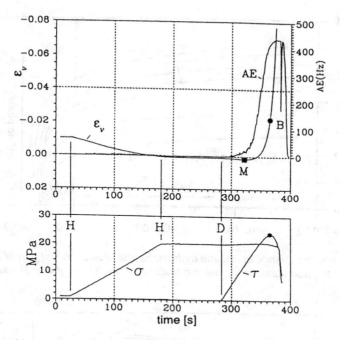

Figure 2.10   *Volume change (ε_v), acoustic emission rate (AE), mean stress (σ), and octahedral shear stress (τ) vs. time in one of the true triaxial compression tests on cubic rock salt samples. The initial compaction is caused by the hydrostatic loading to $\sigma = 20\,MPa$. M: minimum volume; B: failure; H: start of hydrostatic loading phase; D: start of deviatoric loading phase (see also Figures 1.13, 1.14, 1.15, 2.11).*

for microcrack opening and volume increase. Figure 2.11 gives a number of examples for volume change in rock salt (Hunsche [1996]). More examples for measurements of volume changes are given in Chapter 4 for various rocks. They show that for all tests there exists a minimum of the volume. The stress states where this happens are located on the so-called dilatancy boundary. Below this boundary the volume decreases (compression); above this boundary the volume increases (dilatancy). For a precise determination of this boundary the elastic volume change has to be removed, because it is only related to the inelastic volume change (see Section 4.3).

The boundary between these two domains is defined as the **dilatancy boundary** or **compressibility/dilatancy surface** and is shown in Figure 2.12 for rock salt as determined from the above described true triaxial tests. The shape of the compressibility/dilatancy boundary is not yet clearly determined for high mean stresses σ. Moreover, it is more like a band in that range, since the minimum volume is less pronounced at increasing values for σ (see Figures 2.12 and 4.27). The dashed line means: measurable dilatancy occurs only above this horizontal line, which has been determined from tests not shown in this diagram. Between the dashed line and the full line for the dilatancy boundary (or the horizontal axis) practically no dilatancy occurs. At stress states above the compressibility/dilatancy boundary, rock salt dilates, becomes more and more permeable, and may increasingly loosen and fail after a period of time due to creep failure. Creep failure evolves faster in the vicinity of the failure boundary,

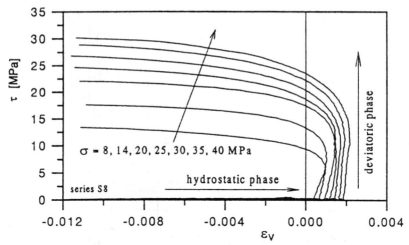

Figure 2.11a   *Volume change during the hydrostatic phase and the first part of the deviatoric phase for most of the true triaxial compression tests of series S8. σ: mean stress at minimum volume.*

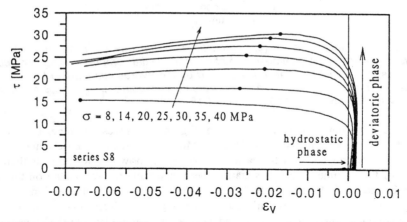

Figure 2.11b   *As Figure 2.11a. Here, volume changes are drawn up to failure (●).*

and slowly just above the dilatancy boundary. This is discussed together with the damage evolution in Chapter 6. Failure occurs when a certain amount of damage has been exceeded (see Chapter 6 and Figure 2.21). It appears from the experiments, that the volume increase is about proportional to the axial deformation. Below the compressibility/dilatancy boundary, rock salt is compressed until all voids are closed. Figure 2.13 gives a nice example for a true triaxial test with a number of creep phases below and above the dilatancy boundary. One can easily observe that the volume decreases during the two phases below the boundary whereas it increases during the two phases above the boundary. The rate of volume increase depends on the distance from the dilatancy boundary (and also from the failure boundary). Figure 2.13 compares with the calculated curve in Figure 5.26 (which has, however, different stresses), and with the tests for marlclay and limestone in Figures 4.9a and 4.9b. This boundary has been studied for many rocks (see Chapter 5) but has probably been examined most carefully for rock salt. It was found that the boundary between the compressible domain and the dilatant domain is practically dependent on the stress state only, more precisely on σ and τ. For rock salt it is not dependent on the deformation rate because creep tests give the same dilatancy boundary, as has been shown by Hunsche [1996], Hunsche and Schulze [1996]. Also it was found that there is practically no influence of the load geometry (or the third stress invariant) and of the type of rock salt on the dilatancy boundary, as shown by Hunsche [1996]. This is somewhat surprising since load geometry and salt type do influence strength considerably. The effect of load geometry has been analyzed experimentally also by Thorel and Ghoreychi [1996], who found a small influence. This means that fortunately the dilatancy boundary is a very stable surface which is quite easy to handle and to implement in a computer program. Formulas are given in Chapter 4. The dilatancy boundary (for compression) has been determined by a number of researchers by performing different kinds of experiments. A collection of results is given in Figure 2.14. Most of them as well as others, and newer results obtained by us, are close to that of Cristescu and Hunsche [1993a, 1996]. The definition of Spiers *et al.* [1989] is different, yielding a shift of the C/D boundary upwards. The dilatancy boundary for rock salt has been found (down to $\sigma = 3.3$ MPa) by tests carried out by Kern and Popp [1997], who measured simultaneously volume change, seismic velocities v_p and v_s, and permeability.

It has to be stressed at this point that the dilatancy boundary is an important long term **safety boundary**, because —with our present knowledge— creep failure and increasing permeability are inevitable under stress conditions above this line, whereas this is not the case below. It has therefore been used for the safety assessment of repositories in rock salt. It is important to determine the short term and the long term behavior of rocks not only in the compressible region, but also in the dilatant region. And this is the most important topic of this book.

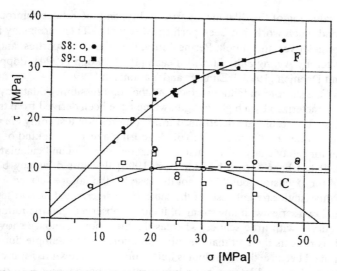

Figure 2.12   *Failure surface F and dilatancy boundary C for two types of rock salt with high strength. Practically no dilatancy occurs between the full line and the dashed line for the dilatancy boundary (Cristescu and Hunsche [1992, 1993a, b, 1996], Hunsche [1994a, 1996]).*

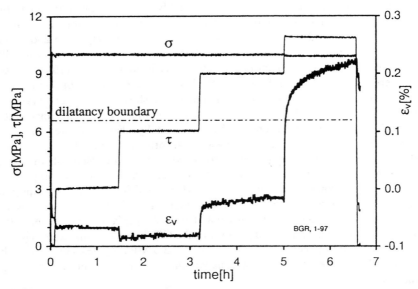

Figure 2.13 *Result of a true triaxial creep test on a cubic rock salt sample with constant mean stress σ and four steps of the octahedral shear stress τ (BGR).It shows the influence of the dilatancy boundary on the evolution of dilatancy. During the initial hydrostatic loading the volume compression was about 0.23%. The first two steps (3 and 6 MPa for τ) were below the dilatancy boundary (small volume decrease) and the last two steps (9 and 11 MPa for τ) were above the dilatancy boundary (volume increase). Compare with the dilatancy boundary in Figure 2.12.*

Therefore, the evaluation of the dilatancy, permeability, and the related microcracking is an important topic and much work has been performed recently. There are many references about this topic spread over this book. Permeability, the elastic properties, and the wave velocities are evaluated by Ayling *et al.* [1995], Peach [1991], Popp [1994], Popp and Kern [1994], Stormont and Daemen [1992], Zimmer and Yaramanci [1993].

Figure 2.12 shows the short-term failure surface and the compressibility/dilatancy boundary for a rock salt type characterized by a high strength, which have been derived from true triaxial tests on cubic samples. The results are determined from compression tests only ($m = -1$). The dilatancy boundary is defined by the minimum volume during a test. It is a kind of safety line and is very important for safety analyses and also for the consideration and calculations in the following chapters. It has been shown by Hunsche [1996] that this dilatancy boundary is practically also valid for extension, but this is not the case for the failure surface. Figure 2.15 shows the results of quasistatic strength tests of the same kind, determined on two series on salt with high strength and two series with low strength. It can be observed that the failure surfaces are rather smooth curves with little scatter, whereas the dilatancy boundary shows a higher degree of scatter. This reflects the fact that the minimum volume of a sample during a test is difficult to be determined because the minimum is quite smooth as shown in Figure 2.11. This gives rise to the picture that the dilatancy boundary is rather a band than a well defined line. Further evaluation shows that this band widens towards higher values of σ (see Figure 4.28).

Figure 2.16a shows the comparison of the mean failure curve and the dilatancy curve with

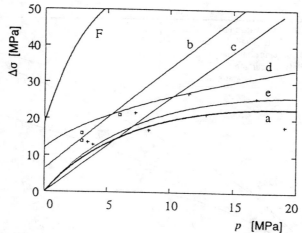

Figure 2.14   *Dilatancy boundaries (i.e., boundary between the dilatant and compress-ible domain) for rock salt determined by several authors (after Hunsche and Schulze [1996]): $\Delta\sigma$: stress difference $\sigma_1 - p$; σ_1: axial stress, p: confining pressure; a: Cristescu and Hunsche [1993a, 1996]; b: Spiers et al. [1989]; c: Van Sambeek et al. [1993]; d: Thorel and Ghoreychi [1996]; e: nonlinear fit of the data by Van Sambeek et al. [1993] by Hunsche; F: failure surface for rock salt determined by Cristescu and Hunsche [1996]; □: Measured values for the beginning of humidity induced creep (Hunsche and Schulze [1996]); +: Measured values for the dilatancy boundary (Hunsche [1992]).*

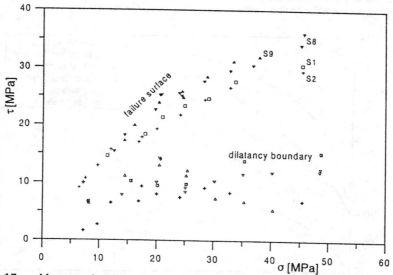

Figure 2.15   *Measured values for the failure surfaces and the dilatancy boundary for compression tests of four types of rock salt from the Gorleben salt dome (Hunsche [1996]).*

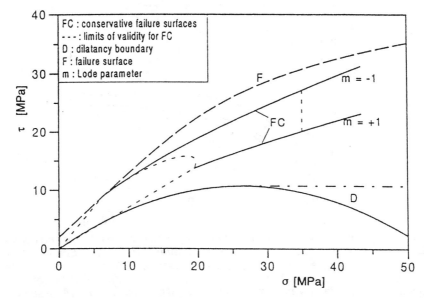

Figure 2.16a *Plots of various equations for rock salt from the Gorleben salt (Northern Germany) for 30 °C. F: failure equation (Cristescu and Hunsche [1996]) . FC: conservative failure equation for compression (m = -1) and extension (m = +1) in Table 2.1. D: dilatancy boundary (Cristescu and Hunsche [1996]) ----:limit of validity for FC below which tension has to applied for failure and/or limit of tests which were used for the fit.*

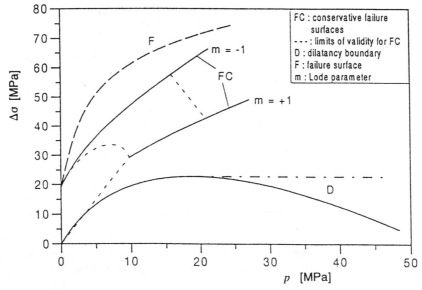

Figure 2.16b *The same curves for various equations for rock as in Figure 2.16a but simply transformed from one coordinate system (σ,τ,m) to another coordinate system (Δσ,p,m) for the stresses. Observe the large distortion of relative distances between the curves.*

a conservative failure law which has been developed by Hunsche and Albrecht [1990], Hunsche [1991, 1992, 1993a, 1994a]. This conservative failure law was fitted to the lower boundary of a great number of test series on rock salt as shown in Figure 2.7 for $m = -1$ and $m = +1$. Its equation and the parameters are given in Table 2.1; the equation is drawn in Figures 2.16a,b and 2.17. The equation involves the dependency of strength on σ, m and T and also gives the residual strength after failure. It has been used effectively in a number of safety analyses for repositories in salt.

   The individual evaluation of the quasistatic and deviatoric true triaxial tests can be done as follows: Figure 2.10 (and Figure 1.13) shows the course of the stresses of a compression test including the micracoustic emission rate. The different phases of the test can easily be observed. Figure 2.18 shows the corresponding deformations in the three directions and the volumetric change as a function of τ. In addition one can calculate from the above test results the irreversible deformation energies related to shape change W_D^I and to volume change W_V^I as explained in Chapter 4. In Figure 2.19 the different kinds of irreversible stress work are shown for one test (see Cristescu [1986, 1994a], Hunsche [1992, 1994a]). W_D^I is caused by the stress deviator and is related to shape change by creep deformation; W_V^I is related to the initiation and propagation of microcracks, i.e., is the energy released by microcrack opening (including surface energy) expressed by the decreasing values of W_V^I. The energy is supplied by the shear stress τ, causing shape change. When W_V^I is increasing, most of the microcracks close, and the energy is stored and dissipated in the rock. W^I is the sum of both.

   W_d^I, W_v^I and W^I are important in the formulation of a comprehensive constitutive equation as shown in Chapter 4. The values of W_v^I are successfully used in this model as a damage variable $d(t) = -W_v^I(t)$ which can predict creep failure (see Chapter 6). It is also interesting to see how large the volume change is and how the slow developing damage leading ultimately to failure can be defined $[d_f = -W_v^I$ (at failure)]. Figures 2.20 and 2.21 give values for volume change and for damage measured for rock salt. It has been stated in many tests that the volume changes at failure are quite large, begin between 2 and 7%. It has to be mentioned that the

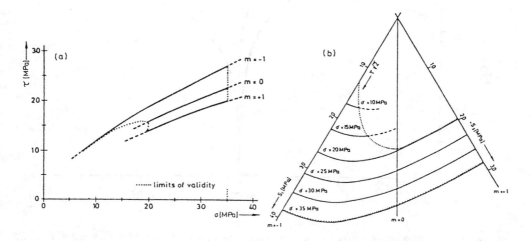

Figure 2.17   *Plots of the conservative failure equation for rock salt (Table 2.1) for $T \leq 100°C$: a) in the τ - σ plane; b) in the octahedral plane (Hunsche [1992, 1993a, 1994a]).*

volume change is not really isotropic because the microcracks are more or less perpendicular with respect to the smallest principal stress. This causes a pronounced dilatation mainly into this direction. Because it is rather complicated to consider this fact in a material law and because the effect does not seem be important, this is not regarded in most cases.

Further the volume decrease at hydrostatic pressure has to be analyzed. This is also a time - dependent process —volumetric creep— and is important for the description of the healing or the compaction of a disturbed (dilated) area of a rock, combined with the decrease of permeability. Figure 2.22a,b show a purely hydrostatic compaction test ($\tau = 0$) with several hydrostatic loading steps of σ and a final unloading step. Under this stress condition the volume of the sample shrinks gradually during each loading step of more than one hour duration. Note also the slow volume increase in Figure 2.22a after unloading and the rather non-linear unloading curve in Figure 2.22b. Interesting, but not yet well understood, healing processes are involved in the volume decrease. Healing of salt has been studied by Kenter *et al.* [1990], Peach [1991], Stormont and Daemon [1992] by the measurement of permeability. An example of the time-dependent decrease of permeability is shown in Figure 2.23. The general dependence of permeability for other rocks has been evaluated by Walsh [1981]. Volumetric creep is much more emphasized and complicated in crushed salt which is used as backfill material in mines (see Liedtke and Bleich [1985], Zeuch [1990], Spiers and Brzesowsky [1993], Hansen *et al.*, [1996], Korthaus [1996], Stührenberg and Zhang [1996], Zhang *et al.*, [1996]). Crushed coal has been studied by Cristescu and Duda [1989].

It has already been mentioned that acoustic emission measurements are a valuable tool to support deformation tests and their analysis especially if volumetric changes are also involved. Measurements of the wave velocities and the attenuation are also a good tool for the analysis

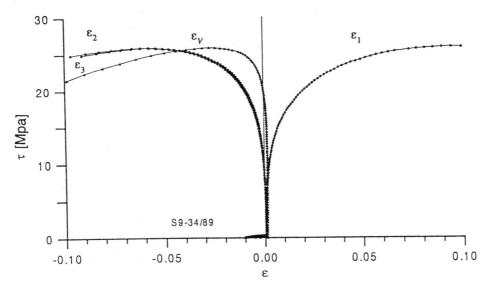

Figure 2.18   *Result of a quasistatic true triaxial test on rock salt. Mean stress σ was 25 MPa. The plot shows the deformations in the three principal directions and the volume change. The maximum value for τ is the measured strength.*

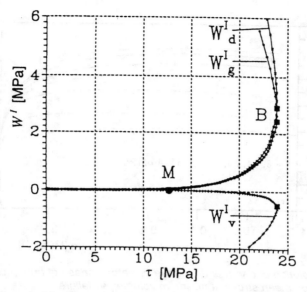

Figure 2.19   *Irreversible stress work of a cubic rock salt specimen during the deviatoric phase of a test (Hunsche [1992, 1993a, 1994a]). W_v^I: irreversible stress work related to volume change; W_d^I: irreversible stress work related to shape change; $W^I = W_v^I + W_d^I$.*

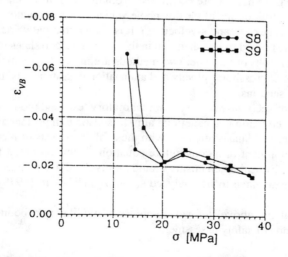

Figure 2.20   *Volume increase at failure determined from the same true triaxial failure tests on rock salt as shown in Figure 2.11 and 2.12 (Hunsche [1992, 1993a, 1994a]).*

of the gradual loosening of the rock. The possibility to detect the damaged zones around underground openings with different seismic methods was successfully tested in granite and salt, e.g., by Alheid and Knecht [1996]. Downhole velocity measurements (measurement of the

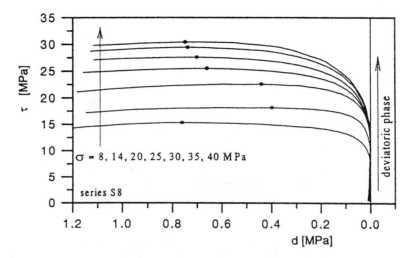

Figure 2.21   *Evolution of damage during the deviatoric phase of true triaxial compression tests on rock salt. σ: mean stress at minimum volume; •: failure.*

velocities between the mouth and different depths of the borehole) and interval velocity measurements (measurement of the velocities within intervals of 10 cm length within the borehole) have been carried out in radial boreholes by Aldheid and Knecht [1996]. In addition refraction seismic measurements were performed from the wall of the underground cavities. Figure 2.24 shows the result of one of the downhole velocity tests in granite, and the calculated interval velocities (average of p- and s-velocities). It is shown that the velocity reduction (and the damaged zone) is restricted to less than 1 m in this case. This correlates with corresponding permeability test. The results of the other two methods are in good agreement with this result. Another tool is the use of crosshole velocity and attenuation measurements for the construction of tomographic cross sections.

The combination of such *in-situ* tests with laboratory tests and model calculations using a suitable constitutive equation is a beneficial aim. It is also an important aim to combine the results of different kinds of quasistatic tests with those of creep tests of medium or long term duration for the development of a constitutive equation on this basis. At the same time our knowledge about the physical mechanisms should be involved as much as possible. For this aim fracture mechanisms have to be analyzed (e.g., Ashley *et al.* [1979], Whittaker *et al.* [1992]).

It is repeated that the above defined compressibility/dilatation boundary is of special importance for all kinds of safety assessment.

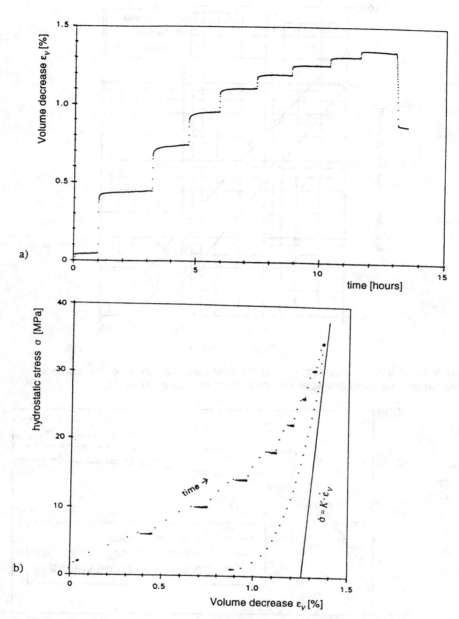

Figure 2.22   *Result of a purely hydrostatic compaction test ($\tau = 0$) on a cubic rock salt specimen from the Gorleben salt dome (Northern Germany) with several hydrostatic loading steps of σ (BGR). Between the steps the stress was held constant for about 1.5 hours. a) volume change ε_v vs. time: b) mean stress σ vs. ε_v; $\sigma = K\varepsilon_v$: Elastic volume compression.*

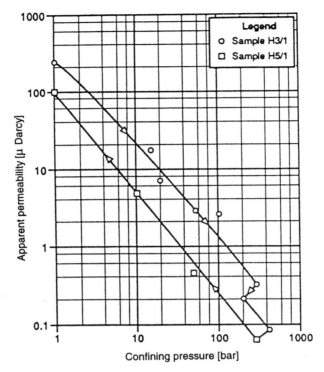

Figure 2.23 *Apparent permeability as function of consolidation time. Parameters: Difference between fluid pressure and hydrostatic pressure (30 MPa).*

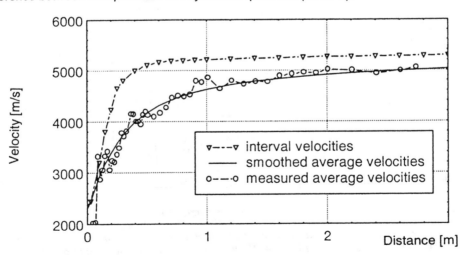

Figure 2.24 *Seismic velocities measured in a borehole in granite (Grimsel, Switzerland) showing the reduction of the velocity near to the wall of the drift due to the damaged zone. ○ - average velocities (mean of v_p and v_s) measured from the surface to the depth on the x-axis; ▽ - interval velocities calculated from the smoothed measured curve (Alheid and Knecht [1996]).*

2.2   CREEP OF ROCKS

2.2.1   Long term creep

The creep behavior of rock is a very important phenomenon for many engineering geological problems, and this is the subject of numerous studies. As already mentioned, creep tests are a very common tool for the investigation of the time-dependent properties of materials including metals, rocks, ceramics, polymers, etc. If one carries out creep tests on rock salt one obtains curves of the kind already shown in Figure 1.16. In addition, the creep rates are shown. In Figure 1.16b it can be observed that the deformation starts with a high deformation rate which decreases continuously until it reaches the stationary creep rate (also called secondary

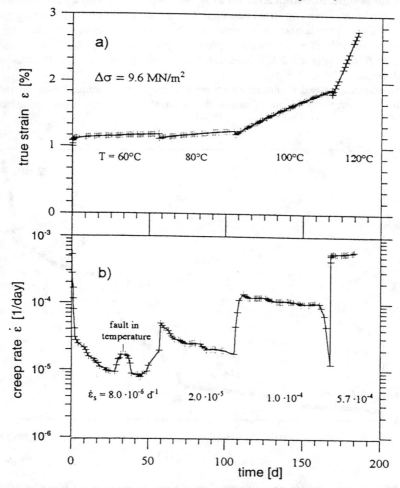

Figure 2.25   *Uniaxial creep test on rock salt (z3OSU) from the Gorleben salt dome (Northern Germany) with four succeeding temperatures (BGR). a) strain vs. time; b) creep rate vs. time, $\dot{\varepsilon}_s$ - steady- state creep rates.*

creep or steady-state creep). After another stress increase, one finds another transient creep phase which ends with another steady-state creep. This is the case for experiments under different temperatures and confining pressures. Figure 2.25 shows an example of stepwise variation of the temperature. For rock salt an increase of 10 K causes an increase of strain rate by about a factor of two. In Chapter 3 examples are given with varying confining pressure. It can be observed that temperature increases the strain rate considerably due to thermally activated micromechanical mechanisms (see Chapter 3). It seems that confining pressure has no apparent influence on creep; however, in the case of dilatancy it may have an indirect effect, which becomes apparent in humidity-induced creep (see Section 3.2). Figure 2.26 shows a triaxial test where the stress-difference was increased stepwise with succeeding normal transient creep phases with decreasing creep rates and finally a steady-state creep. After the lowering of stress-difference from 20 down to 17 MPa we observe in the following an inverse transient creep, with increasing strain rate up to the new steady-state creep. The steady-state strain rate for a given stress and temperature state is generally unique. In a third phase of the test we observe sometimes a tertiary creep phase, with an increasing strain rate which finally ends in creep failure (Figure 2.27). This can occur only if the stress state belongs to the dilatancy domain.

The temperature change does not cause a transient creep. This statement must be understood in the following way: only if one changes the stress slightly according to the temperature-

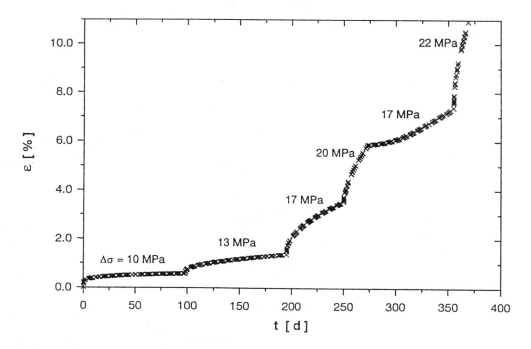

Figure 2.26 *Creep curve (deformation vs. time) from a triaxial laboratory test at BGR on a rock sample performed at 45°C and with a confining pressure of 15 MPa. Stress differences are indicated. Note the inverse transient creep during the phase with reduced stress.*

dependent change of the shear modulus G such that the ratio σ/G stays constant is this statement valid. The physical reason is that the dislocation motion is sensitive to the ratio σ/G and not the stress alone (Vogler [1992]).

It is, however, not sufficient to rely only on experimental results for a satisfactory description of the mechanical characteristics. Their extrapolation over long durations as well as to stress conditions which are not attainable in laboratory studies are possible only if the physical principles of the microscopic deformation mechanisms are known and can be integrated into the material laws, as described by Frost and Ashby [1982]. In this respect, results from the material sciences provide an important foundation (e.g., Cottrell [1953], Seeger [1956], Ilschner [1973], Haasen [1974], Nix and Ilschner [1979], Frost and Ashby [1982], Weertman and Weertman [1983a, b], Oikawa and Langdon [1985], Poirier [1985], Caillard and Martin [1987], Biberger and Blum [1989], Vogler and Blum [1990], Blum [1991], Blum et al. [1991], Hofmann and Blum [1993], Sedlacek [1995]).

Therefore one has to ask: How can the described phenomena be explained?

The irreversible and nearly crack-free change of shape in time of a solid body is known as "creep", or sometimes also as "flow". The creep of rock salt, which is often used as a model material for other rocks or for metals, because it has brittle and ductile deformation behavior (semi-brittle), as well as other crystalline materials, occurs primarily by transcrystalline dislocation movement, i.e., glide, climb, and cross-slip in which (because of compatibility) at least five independent slip systems must be activated concurrently (Haasen [1974]). Otherwise the opening of microcracks is caused and maybe followed by recrystallisation, which also produces deformation. The movement at grain boundaries alone, in other words the gliding between rigid crystal grains, always results in the formation of cracks within the grain mass.

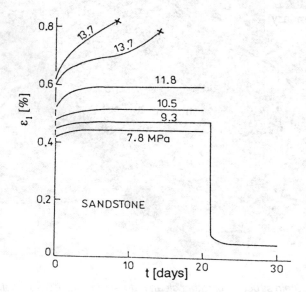

Figure 2.27 *Uniaxial creep curves for sandstone for various constant loading stresses shown (Cristescu [1989e]).*

Nevertheless, grain boundary slide is observed microscopically (e.g., in rock salt) and constitutes a rather small part of the whole deformation (e.g., Ilschner [1973], Wawersik [1988]). The "flow" of unconsolidated deposits (e.g., clays and sands) is governed by a completely different deformation mechanism.

For example, Figure 2.26 illustrates the deformation-time diagram for a triaxial creep experiment performed at a number of stress levels, on a cylindrical sample of rock salt in a Kármán cell. The shape of the curve results from two competing fundamental processes: strain hardening and recovery. If hardening and recovery are not in equilibrium, then the rate of creep varies with time and the rock is in the state of **transient creep**. This is always the case subsequent to a change in load conditions, which is in most of the cases predominantly a change in differential stress, and can also occur as a result of variations in the humidity, confining pressure or temperature. Hardening predominates after an increase in stress, and the initially enhanced rate of creep decreases with time —this is **normal transient creep**. During strain hardening the dislocation density increases by certain multiplication mechanisms. Therefore, the interference between dislocation increases. The resulting dislocation network or forest causes that the number of free dislocations and their average velocity decreases, which causes

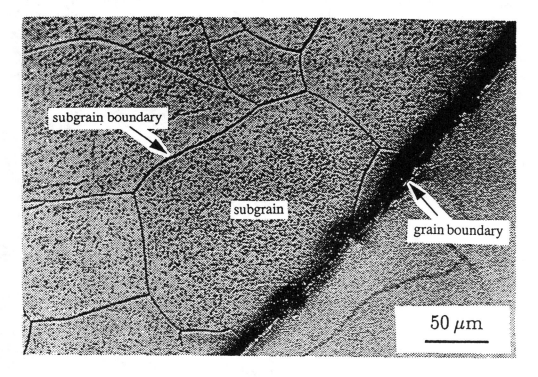

Figure 2.28   *Subgrain structure of rather clean natural rock salt ("Speisesalz") from the Asse mine (Northern Germany) before deformation in the laboratory. The subgrains were caused by the natural deformation during the growth of the salt structure. The points, where the dislocations intersect the specimen surface were made visible by etching (etch pit technique) (after Vogler [1992]).*

a rapidly decreasing deformation rate (see Equation 3.1.3). This process is augmented by the interaction with other lattice defects like impurities or grain boundaries. Since the state of strain hardening is an energetically unstable order, the system has the tendency of decreasing the stored elastic energy. This is achieved by a number of thermally activated processes, which are more efficient at high temperature: climb or cross-slip of edge or screw dislocations into other crystallographic planes for passing obstacles or other reasons, mutual annihilation of dislocations with opposite sign, rearrangement of dislocations of the same sign in subgrain walls (polygonization, see Figure 2.28). Sometimes, recovery is accompanied by recrystallization (see Poirier [1985], Ranalli [1995]). Since not only strain hardening but also recovery is associated with deformation-controlled process and not a purely time-dependent one.

If the loading conditions are kept constant, the creep rate approaches asymptotically a constant positive value —this is the **steady-state creep** whereby the $\varepsilon - t$ creep curve becomes a straight line. Hardening is now in equilibrium with recovery. The amount of preliminary transient deformation necessary for attaining the condition of steady-state creep alone is dependent on stress and temperature as well as on the overall "loading" history. Munson and Dawson [1984], Munson *et al.* [1989, 1996], and Aubertin [1996] have made an attempt to describe it by using an internal state variable. However, their results and those of others are not yet good enough. An overview of the modelling of rock salt has been given by Munson and Wawersik [1993] and Hunsche and Schulze [1994]. The identical steady-state creep rate is attained in tests with constant deformation rate, if it does not experience short term failure as explained in Section 2.1. Steady-state creep rate depends on stress state and temperature alone for a certain material, if the structure is not changed, i.e., by dilatancy, impurities, or recrystallisation.

The reaction of a sample to a reduction of stress depends on the degree of the change in stress as well as the level of the previously attained strain hardening (an internal state variable describing the history). The rate of creep decreases further if the momentary hardening is smaller than the equilibrium level that corresponds to the new, reduced stress level (normal transient creep). However, if the hardening is already greater than that corresponding to the new, reduced differential stress, then recovery predominates as shown in one test phase in Figure 2.26: **inverse transient creep**. The formerly developed denser and more entangled dislocation structure within the hardened material, which restrains deformation, is now thinning and reorganizing: the rate of deformation, that at first spontaneously decreased, now slowly increases until the new rate of steady-state creep is attained. This acceleration of creep after stress reduction is not the result of tertiary creep (see below) but of recovery. The same observations and mechanisms are valid for metals, ceramics, and other rocks. In detail, the individual processes are complex, and have been described, for example, by Eggeler and Blum [1981], Blum [1991], and Vogler [1992]. New results for the steady-state and transient creep of rock salt are given by Weidinger *et al.* [1996a], and Hampel *et al.* [1996] on the basis of the composite model, initially developed for metals and ceramics, and are nearly completely based on physically based microscopic observations —which is an important step forward. It should be possible to apply it for the description of creep in all kinds of crystalline materials including rocks (see Section 3.1).

Tertiary creep is a possible phase of a creep test with increasing strain rate. It is caused by increasing damage of the material and ends at creep rupture, and is only possible in the dilatant domain, where creep and dilatancy are working simultaneously. For example, see Figure 2.27 and Chapter 4. Tertiary creep comes after steady-state creep or directly after transient creep.

The situation around an underground cavity is further complicated because, in addition to the creep rate, the stress state also varies with time and consequently, generally, the system never attains a fully real steady-state condition. If the stress changes are slow enough, in long time intervals one can neglect the transient effect and it is a good approximation to use steady-state creep only.

The special case of the particularly large amount of transient creep, which occurs in previously non-hardened "virgin" samples, is known as primary transient creep. In this case it is difficult to determine the previous level of the weak "natural" strain hardening, although this is an essential requirement for model calculations. However, it was possible to attain reliable values by microscopic observation of the substructure of the undeformed material (see Figure 2.29). Cristescu [1989a] discussed the possibilities for determining the initial conditions in natural rocks with respect to his model.

During steady-state creep, dislocations form a quite regular network of subgrains within

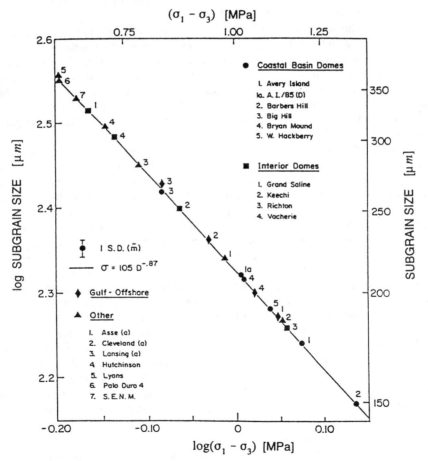

Fig.2.29 *Subgrain sizes of natural rock salt from the indicated salt structures. The solid line shows the relation between stress and subgrain size (with $k = 214 \mu m \times MPa$). It can be used to determine the natural deviatoric stress (after Carter et al. [1982]).*

the grains in all crystalline materials. There is, however, some discussion whether the subgrains are always regular at low temperatures. The subgrains have measured diameters from about $10 \, \mu m$ to $1000 \, \mu m$ and their boundaries consist of the linear arrangement of dislocations with the same sign (see Ilschner [1973], Haasen [1974], Poirier [1985], Vogler [1992]). An example is shown in Figure 2.29. The average diameter L_{sk} of the subgrains is determined from the etch-pit diagram or sometimes by decoration with radioactive irradiation. For rock salt, as well as other crystalline materials, this diameter is in steady state inversely proportional to the (differential) stress σ:

$$L_{sk} = k \sigma^{-1} . \qquad (2.2.1)$$

Numerous references exist in the literature for the value for k for halite (see Hunsche and Schulze [1994]). The values determined for rock salt from different locations are rather consistent, and, generally, a value of about $k = 200 \, \mu m \times$ MPa is therefore a good approximation. Conversely, the previous stress in the underground rock can be deduced from measurements of the subgrain size, as shown in Figure 2.29 for various salt deposits (Carter *et al.* [1982]). The determination of polarostresses in the Earth's crust is performed by the diameter of grains or subgrains rotated or moved during dynamic recrystallization (neoblasts) that obey rather similar relations (see Ranalli [1995]). More examples of measured subgrain sizes are given in Figure 3.8. This chapter also gives the theoretical aspects of creep in more

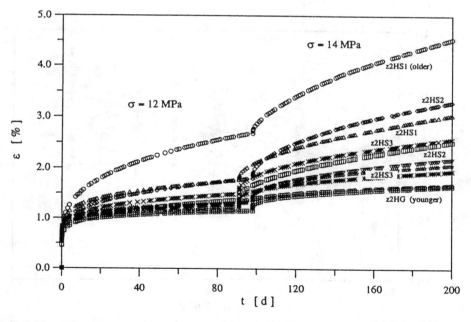

Figure 2.30 *Creep curves for ten rock salt specimens from successive stratigraphic layers in the Gorleben salt dome (Northern Germany) exhibiting systematic changing steady state creep rates (BGR). The older, i.e., earlier sediment, salt types creep faster than the younger types due to the systematic changing differences of the impurities in the salt matrix. Test performed at room temperature (Hunsche et al. [1996]).*

detail.

As a result of the physical principles already discussed, the creep rate for rock salt under a particular load condition (σ, T) —just as for metals— always approaches a definitive and appropriate steady-state creep rate that is usually strongly dependent on the type of salt. A different behavior from that depicted by the shape of the curves in Figure 2.27 occurs if dilatancy and therefore creep fracture (tertiary creep) results from enhanced stress as already shown in Figure 2.27. Furthermore, the amount of transient deformation for previously undeformed samples is usually larger than formerly assumed. A transient deformation of over 10% (depending on $\Delta\sigma$ and T) is commonly observed in rock salt before the condition of steady-state creep is approximately attained (Blum and Fleischmann [1988], Vogler [1992], Hampel *et al.* [1996]; see also Section 3.1).

Over the years many equations have been developed for the description of steady-state creep and transient creep. Most of then have an empirical character and were determined on the basis of a fit to the experiments. This causes, however, great uncertainties in the extrapolation to long time intervals and to conditions not covered in the laboratory. Therefore, a reliable law for creep to be used for long term predictions must be based on a physically based modeling of the microscopically acting deformation mechanisms. This kind of constitutive equation has to be further developed in the future.

Very many examples for the description of creep behavior of geomaterials are found in the literature as given in the more general papers mentioned in Section 1.1 or elsewhere in the

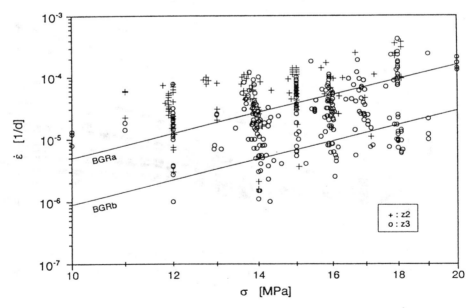

Figure 2.31   *Compilation of a great number of steady state creep rates $\dot{\varepsilon}_s$ from tests at room temperature (mainly uniaxial) performed at BGR on different types of rock salt from the Gorleben salt dome (Northern Germany). z2 (Stassfurt) and z3 (Leine) are large sedimentary Zechstein series. The values for $\dot{\varepsilon}_s$ differ by a factor of more than 100 under the same stress and temperature conditions (Hunsche [1994a, 1995]).*

Table 2.2 Equations for the steady-state creep of rock salt.
Values of constants and details in Hunsche and Schulze [1994] and in the referenced publications.
R = universal gas constant; $\sigma^* = 1$ MPa (normalization); $T^* = 1$ K (normalization)

---

1. BGRa $\dot{\varepsilon}_S = A \cdot \exp[-Q/(RT)] \cdot (\sigma/\sigma^*)^n$
   Wallner *et al.* [1979], Hunsche [1994b].

2. BGRb $\dot{\varepsilon}_S = \{A_1 \cdot \exp[-Q_1/(RT)] + A_2 \cdot \exp[-Q_2/(RT)]\} \cdot (\sigma/\sigma^*)^n$
   Hunsche and Schulze [1994] .

3. Erlangen $\dot{\varepsilon}_S = A \cdot \exp[-Q/(RT)] \cdot (\sigma/\sigma^*)^2 \cdot \sinh(C_1 \cdot C_2^{T/T^*} \cdot \sigma)$
   Vogler [1992].

4. Carter $\dot{\varepsilon}_S = A \cdot \exp[-Q/(RT)] \cdot (\sigma/\sigma^*)^n$
   Horseman *et al.*[1993].

5. Munson $\dot{\varepsilon}_S = \dot{\varepsilon}_{S1} + \dot{\varepsilon}_{S2} + \dot{\varepsilon}_{S3}$

   $\dot{\varepsilon}_{S1} = A_1 \cdot \exp[-Q_1/(RT)] \cdot (\sigma/\mu)^{n_1}$

   $\dot{\varepsilon}_{S2} = A_2 \cdot \exp[-Q_2/(RT)] \cdot (\sigma/\mu)^{n_2}$

   $\dot{\varepsilon}_{S3} = <H(\sigma - \sigma_o)> \cdot \{B_1 \cdot \exp[-Q_1/(RT)]$
   $\qquad + B_2 \cdot \exp[-Q_2/(RT)]\} \cdot \sinh[q \cdot (\sigma - \sigma_o)/\mu]$

   Munson *et al.*[1989].

6. Wawersik $\dot{\varepsilon}_S = D \cdot \exp\{-Q_{CS}/(RT) \cdot [\ln(A) - \ln(\sigma/\mu)]\}$
   $\qquad\qquad = D \cdot [\sigma/\mu \cdot 1/A]^{Q_{CS}/(RT)}$
   Wawersik [1988].

7. LUBBY2 $\dot{\varepsilon}_S = \sigma/[C \cdot \exp(m \cdot \sigma)]$
   Lux and Heusermann [1983].

8. Composite model, Hampel

   $$\dot{\varepsilon}_S = C \cdot \exp\left(\frac{-Q}{R \cdot T}\right) \cdot \left(\frac{\sigma}{E(T)}\right)^2 \cdot \sinh\left(\frac{b \cdot \Delta a \cdot \sigma_{eff}}{M \cdot k \cdot T}\right)$$

   $\sigma_{eff} = \sigma$ = back stress , $E(T)$ = elastic modulus

   Δa = Activation area, C, b, M, k = constants.

   Hampel *et al.* [1996], Section 3.1 of this book.

Table 2.3   Equations for the transient creep of rock salt
Values of constants and details in Hunsche and Schulze [1994] and in the referenced publications
R = universal gas constant, $\sigma^* = 1\,\text{MPa}$ (normalization), $T^* = 1\,\text{K}$ (normalization)
$t_N = 1\,\text{d}$ (time normalization).

---

1. Menzel and Schreiner

$$\varepsilon_{tr} = K \cdot (\sigma/\sigma^*)^n \cdot (t/t_N)^m \quad \text{or} \quad \dot{\varepsilon}_{tr} = A \cdot (\sigma/\sigma^*)^\beta \cdot \varepsilon_{tr}^{-\mu}$$

Menzel and Schreiner [1977], Salzer and Schreiner [1991].

2. LUBBY2

$$\varepsilon = \varepsilon_{tr} + \varepsilon_{st}$$
$$\varepsilon_{tr} = (\sigma/B) \cdot [1 - \exp(-B/A \cdot t/t_N)]$$
$$\varepsilon_{st} = (\sigma/C) \cdot (t/t_N), \qquad t_N = 1\,d$$

Lux and Heusermann [1983].

3. Carter

$$\varepsilon_{tr} = A \cdot (\sigma/\sigma^*)^n \cdot (t/t_N)^m \cdot (T/T^*)^p$$

Carter and Hansen [1983].

4. Cristescu and Hunsche

$$\dot{\varepsilon} = \frac{\dot{\sigma}}{2G} + \left(\frac{1}{3K} - \frac{1}{2G}\right)\dot{\sigma}\mathbf{1} + k_T \left\langle 1 - \frac{W_T(t)}{H(\sigma)}\right\rangle \frac{\partial F}{\partial \sigma} + k_S \frac{\partial S}{\partial \sigma}$$

Cristescu and Hunsche [1992, 1993a,b]. See Chapter 4, this book.

5. Munson

$$\dot{\varepsilon} = F \cdot \dot{\varepsilon}_S$$
$$F = \exp[+\Delta(1 - \zeta/\varepsilon^*)^2] \quad \text{for} \quad \zeta < \varepsilon^*$$
$$F = 1 \quad\quad\quad\quad\quad\quad\quad \text{for} \quad \zeta = \varepsilon^*$$
$$F = \exp[-\delta(1 - \zeta/\varepsilon^*)^2] \quad \text{for} \quad \zeta > \varepsilon^*$$
$$\varepsilon_S = \varepsilon_{S1} + \varepsilon_{S2} + \varepsilon_{S3} \quad (see\ \text{Table}\ 2.2)$$
$$\Delta = \alpha + \beta \cdot \log(\sigma/\mu) \quad\quad \delta = const.$$
$$\varepsilon^* = K_o \cdot \exp(c \cdot T) \cdot (\sigma/\mu)^m \quad \zeta = (F - 1) \cdot \dot{\varepsilon}_S$$

Munson et al. [1989].

6. Composite model, Hampel

$$\dot{\varepsilon} = \frac{v_o}{1 + F \cdot (\sigma/E)} \cdot \exp\left(-\frac{Q}{R \cdot T}\right) \cdot \frac{b}{M} \cdot \rho \cdot \sinh\left(\frac{b \cdot \Delta a \cdot \sigma_{eff}}{M \cdot k \cdot T}\right)$$

$\sigma_{eff} = (\sigma - backstress)$, $E(T) = elastic\ modulus$
$\Delta a = activation\ area$, $\rho = dislocation\ density$.
For details see in Section 3.1, this book.

7. Aubertin

$$\dot{\varepsilon} = \frac{\hat{S}}{2G} + \frac{\hat{I}\mathbf{1}}{9K_b} + A \left\langle \frac{\hat{X}_{ae} - R}{K} \right\rangle^N + g_1 \left\langle 1 - \frac{\varepsilon_e^i}{\varepsilon_L} \right\rangle^{-g_2} \left\langle \frac{\sqrt{\hat{J}_2} - F_o F_*}{F_r} \right\rangle^M \frac{\partial Q}{\partial \hat{\sigma}}$$

Aubertin et al. [1996a] (see also Section 5.4, this book).

---

present book. To be mentioned also are Takeuchi [1989] and Misra and Murrell [1965]. Tables 2.2 and 2.3 give a number of practically used examples for steady-state and transient creep of rock salt. More creep laws have been determined, e.g., by Senseny [1985], Wawersik [1985], Hamami *et al.* [1996], Munson *et al.* [1996]. They can be used as a guide for other crystalline materials as well. Table 2.4, taken from Ranalli [1995] (Table 10.3), gives a collection of parameters for various crustal materials for steady state creep laws like BGRa (see Table 2.2). It shows a wide variety of possible parameters. They are given here without detailed explanation; this can be found in Hunsche and Schulze [1994] or in the reference papers. More explanations will be given in Chapter 3. Especially the composite model for creep given by Weidinger *et al.* [1996a] and Hampel *et al.* [1996] will be described in more detail in Section 3.1 because it is far advanced with respect to the consequent use the physical basis of creep deformation.

The formulas in Tables 2.2 and 2.3 as well as the creep curves in Figure 1.16 may give the impression that the behavior of the same kind of material is very similar. This is not the case. It has to be emphasized that the creep behavior of different types of the same rock can differ considerably as shown in Figure 2.30 for ten specimens taken from successive stratigraphic layers of the same salt dome. This test series also shows that the changing conditions during evaporation and sedimentation can produce systematic change of creep behavior (here older salt is fast creeping, younger salt is slow). Various types of rock salt can exhibit very significant differences, even if they are separated by only a few meters from each other. The tests have been backed by microscopic and geochemical investigations which show that mainly particle hardening is the reason for the change in creep behavior. Microscopic impurities have a particularly important influence on creep. It is well known from materials science that it is not the overall mass-proportion of impurities but the number, distribution and type of particles that affect the ductility. This is easy to comprehend, since every defect in the crystal lattice is an obstacle to the moving dislocations. In addition to the strength of obstacles within the lattice, the distance between these obstacles is also critical (see Section 3.1). Hunsche *et al.* [1996] demonstrated that this is also the case for rock salt (see also Chapter 3). Figure 2.31 shows a compilation of a great number of steady-state creep rates derived from tests at room temperature. One can notice that the results can differ by a factor of more than 100 for identical test conditions for rather pure rock salt (> 95% halite) and, of course, this must be taken into account in reliable model calculations. In addition, it can be observed that, on an average, rock salt from the stratigraphic unit z2 creeps considerably faster than that from the (younger) unit z3. This confirms an old miner's rule for salt domes in Northern Germany.

In a survey it is not possible to determine the microscopic distribution of the impurities and other microscopic parameters in detail. Therefore, it would be practical to observe instead the macroscopic parameters like grain size which correlate indirectly with creep behavior. An attempt has been made by Albrecht *et al.* [1990] and Plischke [1996]. The classification of rocks goes into the same direction.

The observed differences in creep are so large that model calculations must take them into account, because they can lead to very different convergences of the cavities as well as underestimation of stress accumulation in the less ductile zones. The spatial distribution of creep characteristics of the rock must be mapped in as much detail as is possible and useful during the course of the engineering geological survey. The salt deposit can be subdivided into homogeneous zones of approximately similar mechanical behavior. Albrecht *et al.* [1990, 1993] described a method for this, and Hunsche [1995] outlined an application. The degree of

Table 2.4   Creep parameters of crustal materials for equation BGRa in Table 2.2
$(A_D = A,\ E = Q)$ (after Ranalli [1995]).

Material	$A_D\ (MPa^{-n}s^{-1})$	n	$E\ (kJ\ mol^{-1})$
Rock salt	6.3	5.3	102
Quartz	1.0×10^{-3}	2.0	167
Plagioclase (An$_{75}$)	3.3×10^{-4}	3.2	238
Orthopyroxene	3.2×10^{-3}	2.4	293
Clynopyroxene	15.7	2.6	335
Granite	1.8×10^{-9}	3.2	123
Granite (wet)	2.0×10^{-4}	1.9	137
Quartzite	6.7×10^{-6}	2.4	156
Quartzite (wet)	3.2×10^{-4}	2.3	154
Quartz diorite	1.3×10^{-3}	2.4	219
Diabase	2.0×10^{-4}	3.4	260
Anorthosite	3.2×10^{-4}	3.2	238
Felsic granulite	8.0×10^{-3}	3.1	243
Mafic granulite	1.4×10^{4}	4.2	445

spatial detail required is of course dependent on the problem being addressed.

It is helpful for the interpretation of steady-state creep on the basis of physically based micromechanical creep laws to examine a deformation mechanism map. One example for rock salt is given in Figure 2.32. It reflects the various deformation mechanisms dominant under different stress and temperature conditions. More details are given by Frost and Ashby [1982], Albrecht and Hunsche [1980], Munson and Dawson [1984] and Poirier [1985]. Although the "boundaries" between the different dominating mechanisms can vary, there are obviously strict limitations as to how for a single material law, which is determined experimentally for a particular stress and temperature regime, can be extrapolated into other domains. Figure 2.32 also gives the limit resolution for normal creep tests ($\dot{\varepsilon} \approx 10^{-10} s^{-1} \approx 1\,\mu m/d$), creep tests with high precision at BGR ($\dot{\varepsilon} \approx 10^{-12} s^{-1}$) the regime of halokinesis (growth of salt domes), and the regime in a repository for high level radioactive waste.

It is of interest that irradiation of a material also causes point defects in crystals, which have an influence on deformation like impurities. This has been studied by Schulze [1984, 1986].

Again it has to be kept in mind that an extrapolation of creep laws for long times or low strain rates, which cannot be accomplished in laboratory tests, can only reliably be done on the basis of knowledge of the physics of the acting deformation mechanisms on a microscopic basis. It has to be emphasized that the deformation mechanisms acting in rock salt are the same as in metals and ceramics (e.g., Ilschner [1973], Haasen [1974]). Therefore, knowledge of materials sciences can be transferred. The essential difference is that salt and other rocks show in addition the brittle behavior which is expressed by dilatancy and fracture.

It is of interest that a close correlation was found between the creep ductility and the strength of rock salt, as is also known from materials sciences. Salt with high strength does creep slowly and vice versa. Examples are given in Hunsche [1995, 1996]. To our knowledge such studies do not exist for other rocks.

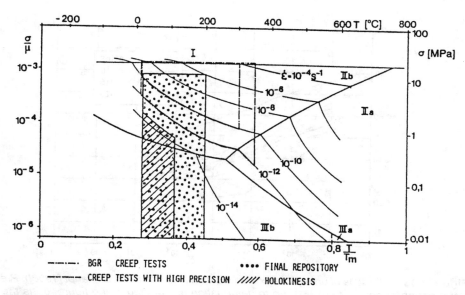

Figure 2.32 *Deformation mechanism map for natural polycrystalline rock salt.*
I, IIa, IIb, IIIa, IIIb: different deformation mechanisms (Hunsche [1984b, 1988]).

This chapter gives practical examples for the time-dependent deformation behavior of rocks together with information about the deformation mechanisms. The following Chapter 3 deals more deeply with the deformation mechanisms for creep.

2.2.2   Relaxation tests and stress drop tests

Relaxation tests are uniaxial or triaxial deformation tests where, after a certain amount of deformation, the relative movement of the end faces of the specimen is suddenly stopped ($\dot{\varepsilon}=0$) and kept so for a long time interval. During this time interval the relaxation of the axial stress is measured. The relaxation is driven by the stored elastic energy and is controlled by the internal creep-deformation (transient creep under decreasing stress). Therefore, the lateral deformation increases during relaxation.

Relaxation tests are sometimes used in rock mechanics for the characterization of rock behavior since stress reduction (and subsequent relaxation) also occurs in nature in the vicinity of new underground openings. Relaxation can be described by the creep —transient and steady state— or by the concept of rheological models. Relaxation tests can also be used to determine a lower creep limit; results of such tests, theoretical considerations (diffusional creep, see Equation (2.33)), and the analysis of the growth of salt domes indicates that a real creep limit does not exist (Hunsche [1978], Albrecht and Hunsche [1980], Hunsche and Schulze [1994]). As in stress drop tests, these tests have the advantage that the microscopic substructure is almost constant during the relaxation phase because practically no deformation-induced recovery takes place. This is favorable for the basic mechanics studies. The stress relaxation tests can be used to determine the activation area (Poirier [1985]).

It is important to state that relaxation tests are difficult to perform since the condition $\dot{\varepsilon}=0$

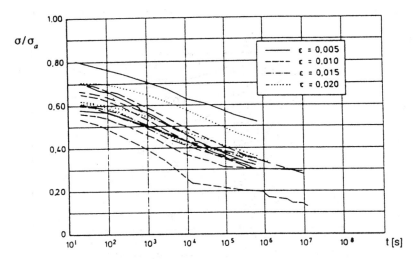

Figure 2.33   *Results of uniaxial relaxation tests on rock salt at 35°C (Haupt [1988]). Previous deformation rates:* $\dot{\varepsilon} = 4 \times 10^{-5} s^{-1}$ *up to deformation ε as noted in the figure. σ_a is the axial stress σ_1 before start of relaxation phase.*

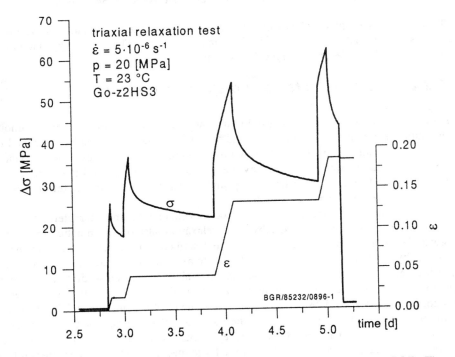

Figure 2.34   *Result of a triaxial relaxation test on a rock salt sample at BGR. The test consists of four loading phases ($\dot{\varepsilon} = const$) with four succeeding relaxation phases, where the stress decreases. Finally the sample is unloaded.*

is not easy to realize. Haupt [1988] has carried out a number of uniaxial relaxation tests on rock salt. Results are shown in Figure 2.33. Figure 2.34 shows the results of a long term triaxial relaxation test with four subsequent loading phases.

Since the creep behavior after a relaxation test is governed mainly by transient creep, one can use the results of such tests also for a crucial validation of transient creep laws in order to check the ability to describe extreme kinds of transient response at decreasing stress.

Stress drop tests (stress dip tests) are usually performed in materials sciences for the determination of the internal long range backstress evolving during creep. In such a test the axial stress is suddenly reduced by a certain amount, preferably after reaching steady-state creep. Drop of the strain rate is possible as well. The internal backstress σ_i is ascertained from tests with different stress reductions and is defined as that reduced stress where no forward or backward creep occurs for some time after stress drop. Usual values are 50 to 70% of the previously applied stress (Blum and Finkel [1982], Blum and Weckert [1987], Hunsche [1988]). At higher reduction a small backward creep can be observed over a longer period of time. This is due to the bending of dislocation segments back to their unloaded position and the rearrangement of dislocations (e.g., in pile up).

As in relaxation tests, the internal structure is not changed during and right after the stress drop. Transient creep laws can incorporate the evolution of backstress as an internal variable and this is necessary for the description of backward creep and inverse transient creep. A number of stress drop tests on rock salt have been performed by Hunsche [1988] with the high-precision device for creep tests shown in Figure 1.17. Figure 2.35 displays some of the results with different amounts of stress reduction after a long period of creep. After a stress drop of about 30% practically no deformation occurs for a longer period of time. Therefore, the long range backstress is about 70% of the previously applied stress in these tests. The dependence of the internal backstress on temperature and stress is addressed by Aubertin *et al.* [1991]. They found a dependence of the ratio σ_i/σ on temperature or stress and gave an equation for this

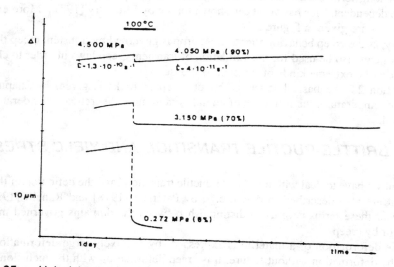

Figure 2.35   *Uniaxial creep experiment on rock salt from the Asse mine (Northern Germany) with a number of stress drops (BGR). Sample length: 250 mm, measuring segment: 170 mm. For rig see Figure 1.17 (Hunsche [1988]).*

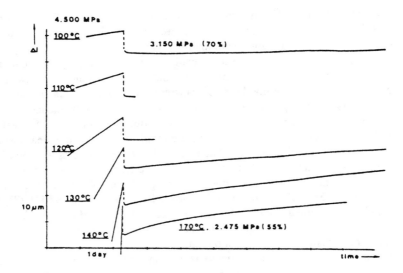

Figure 2.36   *Uniaxial creep experiment on a rock salt specimen from the Asse mine (Northern Germany) with stress drops of 30% at different temperatures (BGR). Sample length: 250 mm, measuring segment: 170 mm. For rig see Figure1.17 (Hunsche [1988]).*

dependence. After a large stress drop one can observe small bachward creep due to the backstress, even for several days and weeks. An example is given by Hunsche [1988]. Due to recovery, creep will finally return to forward creep, however. In Figure 2.36 stress drop tests to 70% of the previously applied stress is shown for several temperatures, showing that recovery of long term backstress and consequently of the properties of transient creep is temperature dependent. This has also been shown for ice by Lliboutry [1987]. More examples for stress drops are given in Figure 2.26.

Anyway, as the creep behavior after a stress drop is governed by transient creep the result of these tests can also be used for the validation of transient creep laws in order to check the ability to describe extreme kinds of transient response.

In Section 2.2 the basic knowledge about creep of rocks is given. In Chapter 3 the influence of temperature, the influence of humidity, and more theoretical considerations will be addressed.

2.3 BRITTLE-DUCTILE TRANSITION AND YIELD STRESS

At this point we have to deal with the "brittle-ductile transition" and the definition of the "yield stress" that are also described in detail, e.g., by Paterson [1978] and Ranalli [1995]. The discussion of these terms help us to distinguish between the domains governed mainly by fracturing or by creep.

Ductile deformation of a material is expressed by relatively large deformation before failure or by deformation without failure. It is creep deformation with the inclusion of some dilatation. Brittle behavior means only little creep deformation until failure. Figure 2.37 shows this difference schematically. The transition from one mode to the other is named the "brittle-ductile transition". Often a value of 3-5% strain to failure is taken as a practical definition

(Paterson [1978]). The transition boundary is dependent on material, temperature, hydrostatic pressure, chemical environment, and loading conditions and is usually drawn in a p-T diagram for a certain material as shown in Figure 2.38 (Figure5.5c in Ranalli). The large influence of the confining pressure on failure strength is shown in Figure 1.8. The transition boundary is also crossed at about $p = 30$ MPa in this case. The influence of temperature is also large and an example is given in Figure 2.39. It must be stressed that this transition is not sharp. Generally, the increase of temperature T or hydrostatic (confining) pressure p increases the ductility. Microfracturing, dilatancy, and under certain conditions cataclastic flow, are

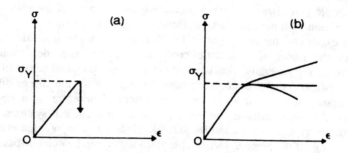

Figure 2.37   *Schematic stress-strain curves for   a) brittle failure and   b) ductile failure. σ_Y = (macroscopic) yield strength (after Ranalli [1995]).*

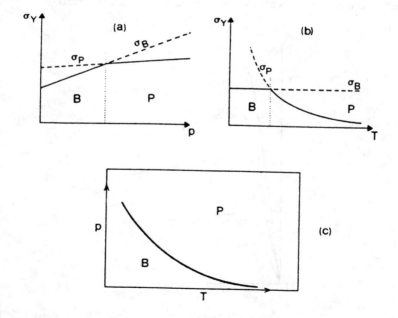

Figure 2.38   *Variation of the brittle yield stress σ_B and ductile yield stress σ_P with a) pressure, and b) temperature. The combined effects of temperature and pressure are shown in c). P denotes region with ductile failure, B brittle failure (after Ranalli [1995]).*

dominating in the brittle field, whereas creep is dominating in the ductile field. The temperature and pressure dependences of deformation and failure are different in both fields as is shown in Figure 2.38 where their effect on the yield stress is discussed. The yield stress is equal to the failure strength in the brittle field, and equal to the flow stress during creep in the ductile field (see also below). In the brittle region practically no dependence on T is observed but is on p, whereas in the ductile region practically no dependence on p exists but does on T. It must be stressed that this transition is not identical with the dilatancy boundary, but there are some relations (see below). The transition can be different for extensional and compressional load geometry, as has been shown for limestone by Heard [1960]. The above definition is the so-called macroscopic yield stress, which is practically the strength of the material (see Figure 2.37) in combination with the associated deformation, the yield strain.

But, there exists also the definition of the microscopic yield stress (also microyielding), which denotes the onset of nonelastic creep deformation (i.e. deviation from elastic deformation). It occurs, generally, much below the macroscopic yield stress. In microphysical terms it is the critical shear stress (Peierls stress) necessary for the slip of dislocations in a certain glide plane, in a certain direction (slip system). Therefore, a number of different microscopic yield stresses exist in one material due to different slip systems. This has been observed in single crystals of rock salt by Skrotzki [1984] and by Wanten *et al.* [1996]. The practical observation of the precise microscopic yield stress can be quite difficult since various deformation mechanisms can superpose and the measurements are difficult to perform (see Section 3.1). The smallest stress causing creep deformation, the flow limit, is not necessarily similar to the yield stress and is strongly dependent on the precision and the duration of the experiments and the question whether diffusion creep can be observed. For several reasons

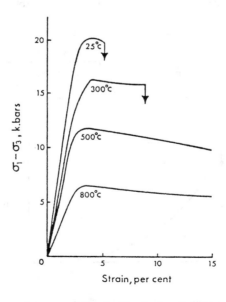

Figure 2.39   *Deformation curves for granite at a confining pressure of 5 kilobars and various temperatures. It illustrates the temperature influence on the brittle ductile transition (after Griggs et al. [1960]).*

given in Chapter 4 the flow limit is practically zero for most materials. It may occur that below a certain temperature, for certain materials only transient creep exists and virtually no steady-state creep. But again this is a question of the resolution of the experiments and their duration, where different models may satisfy the obtained data.

In a homogeneous polycrystal, compatible deformation (without the production of voids) is only theoretically possible if five independent slip systems for dislocations are active, and if glide is the only mechanism. However, rock consists mostly of various minerals exhibiting different deformation behavior that do not possess five independent slip systems — which can be activated— at all temperatures. This fosters microscopic differences into local states of stress, strain, and elastic distortion. This in addition supports grain boundary sliding microcracking, cataclastic flow, and the effect of fluids. Limited deformation can also be produced by twinning of certain minerals and by grain boundary movement (recrystallisation). It seems that the combination of these deformation mechanisms allows compatible deformation below certain effective stresses or above certain hydrostatic pressures, so that dilatation is suppressed. Above these effective stresses dilatation is not suppressed. This would be a qualitative explanation of the existence of the dilatancy boundary.

These considerations show that our picture of the brittle-ductile transition is still a broad outline which needs more research and precision. The inclusion of fracture mechanics for the initiation of growth and closure of microcracks would be very fruitful.

3 Deformation Mechanisms for Creep

3.1 INFLUENCE OF STRESS AND TEMPERATURE

Let us now study the influence of stress and temperature on the creep behavior of rock in more detail on the basis of Section 2.2. This cannot be done and understood without consideration of the deformation mechanisms. If we exclude very low stresses (< 1 Mpa), the dislocations are carrying the deformation being the source of the creep behavior, the fundamental question must be addressed: Which microscopic deformation step determines the speed of creep at low temperatures ($RT<T<T_m/2$; RT = room temperature, T_m = melting temperature) and not very high stresses? The following perspective is derived from Oikawa and Langdon [1985] for some metals and metallic solid solutions: the slowest and therefore the dominant step during recovery of strain hardening is the climb of dislocations pinned by obstacles (i.e. other dislocations or particles). This step is dependent on the diffusion coefficients of the vacancies within the crystal lattice. At moderate temperatures diffusion occurs mainly along the dislocations itself (pipe or core diffusion), whereby in rock salt Cl⁻ vacancies in the NaCl crystal lattice are the slowest to diffuse. Weertman and Weertman [1987] have considered this not only for dislocation climb but also for dislocation cross-slip. The description of the movement of edge and screw dislocations, their visualization and the related theory are given in many textbooks and are not repeated here. Some textbooks are: Cottrell [1953], Ilschner [1973], Haasen [1974], Nicolas and Poirier [1976], Poirier [1985], Ranalli [1995]. It has to be stressed again that these deformation mechanisms are acting during transient as well as during steady-state creep. Therefore a comprehensive constitutive model for creep should consider this close connection. We will come back to the stress dependence later.

Since the rate-controlling step (e.g., diffusion of vacancies, iterative glide or climb of dislocations) is generally a thermally activated rate process, i.e., both thermal agitation and applied stress help overcome a barrier, the temperature dependence of creep is to be described by an exponential Maxwell-Boltzmann term, the Arrhenius function (see, e.g., Ilschner [1973], Haasen [1974], Poirier [1985], Ranalli [1995]):

$$\dot{\varepsilon} \sim \exp\left(-\frac{Q}{RT}\right) . \tag{3.1.1}$$

The activation energy Q describes the influence of temperature T and is characteristic for the acting diffusion process (see Frost and Ashby [1982]), e.g., vacancies in bulk (lattice) diffusion, pipe (core) diffusion, or grain boundary diffusion as is shown schematically in Figure 3.1. R is the universal gas constant.

The activation energy is to be calculated from the measured steady state creep rates by the formula

$$Q = -R \frac{d(\ln \dot{\varepsilon}_s)}{d(1/T)} . \tag{3.1.2}$$

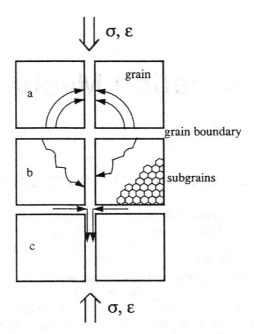

Figure 3.1   *The different kinds of diffusion in a polycrystalline material: a) bulk diffusion (Nabarro-Herring creep); b) pipe or core diffusion, along dislocations; c) grain boundary diffusion (Coble creep).*

This is an effective activation energy if more than one deformation mechanism is important. An example for measured results on rock salt is given in Figure 3.2 (Hunsche and Schulze [1994]). In this case, the slope and subsequently the activation energy obviously change at about 100°C. This indicates a change in the dominating deformation mechanism, which is not yet clearly understood in this special case. It is guessed that there is a change from climb to cross-slip with decreasing temperature. This experimental result showing a change has also been found in a number of publications, e.g., Carter and Hansen [1983], Wawersik [1988], Vogler [1992], Horseman *et al.* [1993]. Albrecht and Hunsche [1980] gave a collection of values for Q for various temperatures from the older literature. It is stated that Q is always increasing with increasing temperature. The diagram of the creep law BGRb for steady-state creep (see Table 2.2) in Figure 3.3 includes this dependence of Q on temperature. The older creep law BGRa also shown does not reflect this fact. Because the Arrhenius therm has a sound physical background, it is used in most of the creep laws for steady-state creep shown in Table 2.2, in which temperature is taken into account, and in others not shown (see also the table in Albrecht and Hunsche [1980]).

Let us now come back to the discussion of the stress dependence. A general and physical based creep law mostly starts from the Orowan equation

$$\dot{\varepsilon} = \frac{b}{M}\rho v , \qquad (3.1.3)$$

where b is the length of the Burgers vector, M the Taylor factor for polycrystals ($M=3$), ρ the average density of (mobile) dislocations, v the average velocity of dislocations. It connects

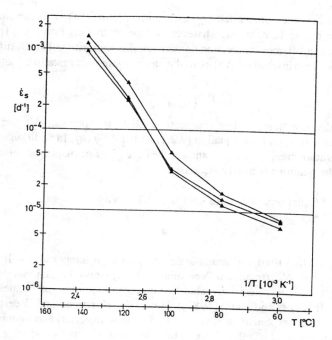

Figure 3.2 *Dependence of steady-state creep rates of three samples on temperature (60 °C to 140 °C) at σ =10 MPa. A bent exists at 100 ° C. Samples from Gorleben salt dome (Germany) (Hunsche and Schulze [1994]).*

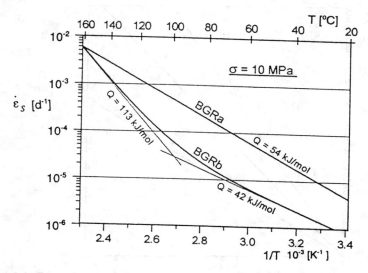

Figure 3.3 *Representation of the two creep laws BGRa and BGRb for steady state creep for σ = 10MPa. Compare with Figure3.2. For details see Table 2.2.*

the macroscopic strain rate (controlled by dislocation movement) with the microphysical deformation process by the dislocations, whatever the rate controlling process is (glide, climb, cross-slip). The relation between dislocation density (all dislocations) and stress differences (in this Chapter 3 we write σ instead of Δσ) is in the steady state for crystalline materials

$$\rho = \left(\frac{\sigma}{\alpha M G b} \right)^2 \tag{3.1.4}$$

with $\alpha \approx 0.1 - 0.4$ (dislocation interaction constant), $\alpha = 0.17$ for rock salt (Vogler [1992], G = shear modulus. This corresponds to Equation (2.2.1) with $L_{sk} = \sqrt{1/\varrho}$. In addition one can show that for high temperature creep ($T > T_m/2$) and for the case of a fine dispersed solid solution the movement of the dislocations is nearly viscous:

$$v \sim \sigma. \tag{3.1.5}$$

The combination of Equations (3.1.3), (3.1.4) and (3.1.5) yields

$$\dot{\varepsilon} \sim \sigma^3. \tag{3.1.6}$$

This relation is often used for steady-state creep at high temperatures. It is called the "natural creep law", and Weertman and Weertman [1983a] stress the fact that this was found without a specific model and has been stated by numerous experimental studies in materials science (see also Biberger and Blum [1989]). However, Weertman and Weertman [1983a] also found that the stress exponent is often $n = 4 - 5$ in pure materials and in impure materials (alloys) without being a solid solution. Several theories exist for this fact which generally base on climb as the predominant mechanism for recovery of strain hardening.

In the regime of low temperature creep ($RT \le T \le T_m/2$) the stress exponent has to be increased by 2, yielding $n = 6 - 7$; experimentally $n = 6 - 9$ is found (Weertman and Weertman

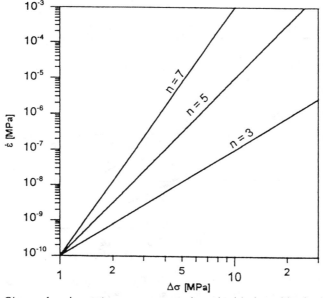

Figure 3.4 *Slope of various stress exponents in a double logarithmic diagram.*

[1983a]). The following reason is given: due to the lower temperature the diffusional path needed for climb of dislocations is restricted under these conditions to the area around the dislocations (pipe diffusion), which again gives a factor of $\varrho \sim \sigma^2$. Figure 3.4 gives a schematic diagram for the different values of n. Therefore, a value of $n = 6 - 9$ should generally be used for steady-state creep of rocks at low temperatures. Oikawa and Langdon [1985] found a stress exponent of $n = 7$ and Weertman and Weertman [1987] found a stress exponent of $n = 5$ for climb and $n = 7$ for cross-slip. The dependence has been investigated extensively for rock salt, which is often used as a model material for rocks as well as for metals and alloys. Many tests have been made on salt single crystals to investigate the active mechanisms and the influence of various glide planes (e.g., Wanten et al. [1996]). The mechanisms are not yet clear, but it seems that glide acts at higher strain rates ($> 10^{-7}$s), and dislocation climb and cross-slip at lower ones.

It is problematic that quite often only a small number of tests are carried out for the determination of a new creep law, and that is true not only for rock salt. Also, it has not always been observed that steady-state creep is really reached. In this case the creep rates are overestimated, especially for small stress differences, and one determines underestimated stress exponents. From a number of test series one receives the impression that a constant value for n is not applicable for stresses above a certain value, above which n seems to increase. For rock salt this value is reported to be about 20 MPa. Therefore, an expression with $\exp(\sigma)$ or $\sinh(\sigma)$ was used for this range in a number of publications. Wallner [1983] and Munson et al. [1989] use the $\sinh(\sigma)$ above a certain level, and Heard [1972] even applies it for the whole stress region. Vogler [1992] uses a mixed expression $\sigma^2 \sinh(\sigma)$, which is used for various substances in the materials sciences (Blum [1991]). This expression is also used for the composite model explained below.

The stress sensitivity of steady-state creep at very low stresses, however, theoretically decreases to $n = 1$. Under these conditions the dislocations are practically unmovable and creep is caused by diffusion alone (see regime IIIb (diffusion along grain boundaries, or most probably along dislocations —pipe or core diffusion) in Figure 2.34). The transition is dependent on temperature and grain size ($\dot{\varepsilon} \sim (d^{-2} - d^{-3})$), and under normal geological and petrographical conditions it is to be expected far below 1 MPa. As a result of this low stress sensitivity, the extrapolation of the material law for steady-state dislocation creep (with $n = 5 - 7$) causes creep rates to be underestimated. These considerations indicate that the stress sensitivity at low stresses or low creep rates must be investigated in more detail for the description of long lasting processes in engineering problems, e.g., for repositories, or in geology. However, since diffusional creep in geological materials occurs only at very small strain rates, which usually cannot be measured in the laboratory, one has to transfer results from the materials sciences or to use natural analogs, e.g., the growth of salt domes (Hunsche [1978]).

In order to have a sound basis for an improved creep law for steady-state and transient creep of rock salt and, because it can act as a model material also for other rocks or metals, an extensive reference test series with a great number of tests (≈ 200) on the same material (z2SP Speisesalz from the Asse mine, Germany) has been performed at BGR in cooperation with the University Erlangen-Nürnberg using uniaxial and triaxial creep tests. An intermediate status and the details of the tests and their evaluation have been reported by Hampel et al. [1996]. The material used is rather clean, has an average grain size of 3 mm and creeps relatively fast. The tests were carried out between 30°C and 250°C and between 1.7 MPa and 35 MPa. For an increased reliability, in each stage of the creep tests the deformation was

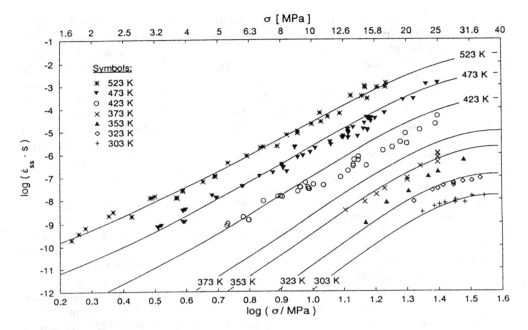

Figure 3.5 *Steady-state creep rate ε̇$_s$ vs. stress difference σ of a series of 197 reference creep tests at different temperatures with a single type of rather clean natural rock salt (z2SP) from the Asse Mine in Northern Germany (symbols). The model curves were calculated with the composite model (after Hampel et al.[1996], improved).*

continued until a steady-state creep rate could be determined accurately. The maximum axial strain in a stage ranged from 1% to more than 40% depending on the test conditions, i.e., mainly the magnitude of the stress change before the stage. Single stages consumed a testing time from 30 minutes to more than one year.

Figure 3.5 shows the result of the extensive reference creep test series in a double logarithmic plot. The lines represent the simultaneously fitted curves of the reference creep law to the data points (see below). The small differences between the data points are due to unavoidable variations in the material itself since the experimental error is only as big as the size of the symbols. The distances between the data sets for different temperatures represent the activation energy (which is not constant). The slope of the sets for a single temperature represents the stress exponent. Up to σ = 25 MPa this exponent is determined by the value $n = 7$, a value which is actually predicted by theory as shown above. The fitted curves are done with the composite model that is a further advanced theoretical approach as shown below.

The basic question arises whether the stress exponent and the activation energy stay the same for another type of salt (z3OSO, Orangesalz, Gorleben saltdome, Germany). Therefore, a still incomplete test series was performed on a slower and more impure salt. The present results are shown in Figure 3.6, where the steady-state creep rates are compared with the fitted curves of Figure 3.5. It results that this salt creep is slower by a factor between 20 (at 250°C) and 100 (at 150°C and 50°C). It is an important result, that the stress dependence appears to be the same for rather pure and the impure material. The effect on the activation energy cannot be deduced from this diagram, but the results given below show that probably it is not changed. From the observations by Hunsche *et al.* [1996] and our knowledge from materials sciences we conclude that the differences in the creep rates of different salt types are related to the intracrystalline impurity distribution. Microscopic particles act as efficient obstacles against the movement of dislocations and therefore have a significant effect on creep. Therefore, it has to be in-corporated into a creep law. Again, it has to be stressed that these results are true also for other crystalline materials like other rocks, metals, alloys, and ceramics.

Now let us come to the formulation of an improved creep law that includes transient and steady-state creep in the desired closely related way, which is able also to model unloading, and which is based on the active microscopic deformation mechanism. Thus, it is possible to use parameters that are independently measurable physical quantities. This gives us the confidence that this law can be extrapolated reliably to long times and to conditions not covered in laboratory tests. The constitutive equation for creep is developed below in more detail because it gives a guideline how the microphysical mechanisms can be incorporated in a creep law.

The basis for this law is the composite model used in materials sciences, as described by Vogler and Blum [1990] and Hofmann and Blum [1993]. This model treats the material as a composite consisting of soft and hard regions, namely the subgrain interiors and the subgrain boundaries, respectively, as described in Section 2.2 and shown in Figure 2.29. Both regions are simultaneously ductile deformable. According to this model, the motion of free

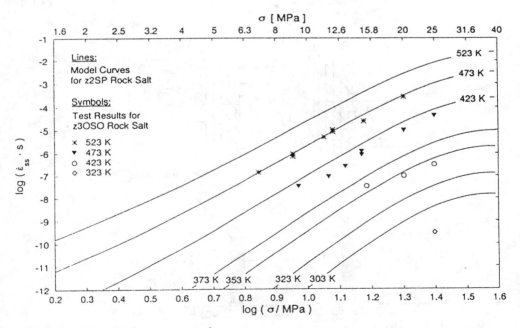

Figure 3.6 *Steady-state creep rate $\dot{\varepsilon}_s$ vs. stress difference σ of a series of 20 creep tests at different temperatures on a single type of impure natural rock salt (z3OSO) from the Gorleben salt dome in North Germany (symbols). The model curves represent the creep law for the rather clean material (Figure 3.5). Note the differences in creep rates for both salt types at the same temperature and stress (Hampel et al.[1996]).*

dislocations through the soft subgrain interiors (d = ductile) and the hard subgrain boundary regions (h = hard) is coupled. This causes an increase of the applied stress in the hard regions due to the resulting additional forward stress, and a decrease in the soft regions due to the resulting additional backstress. This is shown in Figure 3.7, which also gives the definition of some parameters. The stress increase or decrease has been calculated after Sedlacek [1995] for steady-state creep with

$$k_h = \frac{\sigma_h}{\sigma} = 3.4 .$$
(3.1.7a)

This yields (with f_h = 0.07; see below):

$$c_d = \frac{\sigma_d}{\sigma} = 0.819 .$$
(3.1.7b)

The details of the model for salt are given in Weidinger *et al.* [1996b], Hampel *et al.* [1996], Vogler [1992]. Some improvements, which have been made meanwhile by Hampel [1997] are included in the following. For easier explanation, we first formulate the equation for steady-state creep (Equation (3.1.18)) which is then generalized also for transient creep.

First, we write down a few general equations which are deduced from observation. The average subgrain diameter at steady state is for rock salt (see also Equation (3.1.4)):

$$w_S = 33 \frac{Gb}{\sigma} ,$$
(3.1.8)

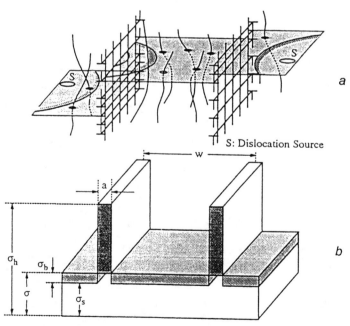

Figuren 3.7 *Diagram of the subgrain structure. a) simplified sketch of dislocation arrangement inside a subgrain and in the subgrain wall; b) stress distribution and some contains some definitions (after Vogler [1992] , Weidinger et al. [1996a]).*

where G is the shear modulus, $b = 3.99 \times 10^{-10}$ m is the Burgers vector, σ is the applied stress difference (Hampel *et al.* [1996]):

$$G = 15\left[1 - 0.73\left(\frac{T - 300}{1070}\right)\right] \qquad (3.1.9)$$

with G in GPa and T in K (Frost and Ashby [1982]). The average distance of free (mobile) dislocations in the subgrain interior in steady-state follows from the real (reduced) stress in the subgrain interior and is

$$d_d = \frac{1}{\sqrt{\rho_d}} = \frac{Gb}{c_d \sigma} \quad , \qquad (3.1.10)$$

where ρ_d is the free dislocation density in subgrains (see Haasen [1974] and (3.1.4)), $c_d = 0.819$ (see Equation (3.1.7)). The relations (3.1.8) and (3.1.10) are obtained from independent microscopic observations on rock salt shown in Figure 3.8. Similar diagrams can be shown for many materials.

The volume fraction of the hard region is

$$f_h = \frac{2a}{w} \qquad (3.1.11)$$

with

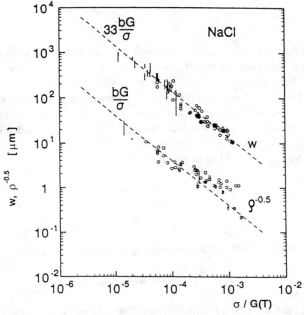

Figure 3.8 *Subgrain diameters w = 33 (b G)/σ and mean dislocation distances ρ⁻⁰·⁵ at steady-state for artificial rock salt determined by various authors. Temperatures from 295 K to 1053 K (after Vogler [1992]).*

$$f_d + f_h = 1 ,$$

(3.1.12)

with f_d, f_h being the volume fractions of hard and soft regions. Sedlacek [1995] has determined a value of $f_h = 0.07$ by theoretical model calculation for rock salt in steady-state creep. Compatible deformation in the two regions means

$$\varepsilon = \sigma_h/E + \varepsilon_h = \sigma_d/E + \varepsilon_d$$

(3.1.13)

where E = elastic modulus = $2(1 + v)G(T)$. From the equilibrium of internal stresses it follows that

$$\sigma = f_d \sigma_d + f_h \sigma_h .$$

(3.1.14)

These formulae have to be considered , together with the observation that dislocations do not only move into and out of the subgrain walls (knitting model, see Blum [1977]) but also subgrains migrate as a whole (Biberger and Blum [1989]). It follows from (3.1.13) that the total creep rate can be expressed by the creep rate for the soft region alone. The connection between the macroscopic quantities $\dot{\varepsilon}$ and $\dot{\varepsilon}_d$ and the quantities describing the microscopic processes yields again the Orowan equation (3.1.3):

$$\dot{\varepsilon} = \frac{b}{M} \rho v .$$

An often used expression for the average velocity v yields

$$v = v_o \exp\left(-\frac{Q}{RT} \right) \sinh\left(\frac{b \, \Delta a \, \sigma_{eff}}{MkT} \right)$$

(3.1.15)

where v_o is a constant velocity, Δa = activation area for thermally activated movement of dislocations in the subgrain interior, σ_{eff} = effective stress, driving the dislocation in the subgrain interior. For the choice of the often used sinh see Heard [1972], Vogler [1992], Barrett and Nix [1965]. The activation energy a is temperature-dependent (see above).

The activation area Δa is a central feature and describes the average area over which the dislocation sweeps in one deformation step. $\Delta a/b$ is equal to the distance of the intersecting dislocations in the soft region $1/\sqrt{\rho_s}$ or of the distance of the hindering impurity particles d_p. The combination yields, if working in parallel:

$$\Delta a = \frac{b}{1/d_p + \sqrt{\rho_s}} .$$

(3.1.16)

Finally the effective stress within the soft region can be defined as

$$\sigma_{eff} = \frac{1 - f_k k_h}{1 - f_h} \sigma - \alpha M G b \sqrt{\rho} - \sigma_p ,$$

(3.1.17)

where $\alpha = 0.17$ is the dislocation interaction constant (Haasen [1974]) (see Equation (3.1.4)). So, d_p resembles the distance (or the distribution) of the impurities, whereas σ_p represents their strength (σ_p must be made stress-dependent at low stresses, since it would otherwise form a creep limit, or pure diffusional creep must be superimposed).

The **final formula for creep** derived from the Orowan equation using the composite model is

$$\dot{\varepsilon} = \frac{1}{1-F'}\frac{b}{M}\rho\,v_o\exp\left(\frac{Q}{RT}\right)\sinh\left(\frac{b\,\Delta a\,\sigma_{\it eff}}{MkT}\right)$$

(3.1.18)

with special relations for the stress dependence of Δa, $\sigma_{\it eff}$, and ρ. Q is temperature-dependent. For steady-state all quantities on the right-hand side are constant and F' is zero. For transient creep F' is obtained by a relation given below. This equation forms a very reliable basis for the description of creep and its extrapolation, because all quantities of the formula have a physical meaning which can be found from other independent observations or experiments. For instance, it is very convincing that the stress-dependent activation areas Δa for the rock salt types z2SP (Figure 3.5) and z3OSO (Figure 3.6) which were found by a fit to the respective steady state creep rates were independently given by stress dip experiments. Moreover it is very convincing that the differences of the two data sets in Figures 3.5 and 3.6 can be related to each other by only changing the average distance of the particles ($d_p = 1\,\mu m$ for z2SP and $d_p = 0.5\,\mu m$ for z3OSO), which is an important finding. The function for activation energy and the values for v_o and σ_p did not have to be changed in this case. Finally it is satisfying that, in general, only three parameters (v_o, d_p, σ_p) have to be fitted for the description of steady-state creep of different types of salt. In addition, these parameters represent a distinct physical meaning, where v_o is related to the measurable average velocity of dislocations (Equation (3.1.18)).

Now we have a **creep law for steady-state creep** based on the microscopic deformation mechanism of dislocation movement, containing only quantities with physical meaning, which

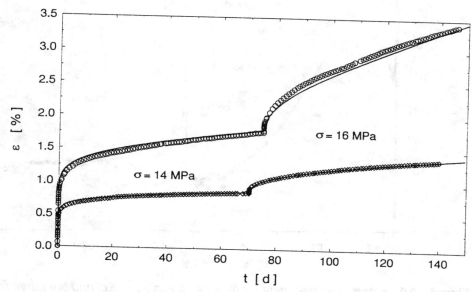

Figure 3.9 *Two measured creep curves for two different types of rock salt and the curves fitted with the composite model. $T = 22°C$.*

describes the reference creep law in Figure 3.5 very well. The values for the parameters are for this salt (z2SP):

$$v_o = 3000 \; m/s \,, \quad d_p = 1 \, \mu m \,, \quad \sigma_p = 0.4 \; \text{MPa}.$$

For the description of **transient creep** Equation (3.1.18) can easily be extended on a physical basis. To do this one has to consider which quantity is changing during transient creep. Again these are physical quantities, namely $w =$ subgrain diameter, $a =$ subgrain wall thickness, $\rho =$ dislocation density inside the subgrain.

It is reasonable to assume that, for instance, the subgrain diameters w change proportional to deformation (deformation hardening) and also the difference between the (new) steady-state value w_s and the actual value w. This yields

$$dw = (w_s - w)\frac{1}{k_w} d\varepsilon \tag{3.1.19}$$

and by integration

$$w = w_s + (w_o - w_s)\exp\left(\frac{\varepsilon_o - \varepsilon}{k_w}\right) \tag{3.1.20}$$

with $w_o =$ starting value of w (deformation history), $w_s =$ steady-state value of w for the new test condition (3.1.8), $\varepsilon_o =$ starting value of ε, $k_w =$ relaxation constant. Thus, w approaches the new value asymptotically with an asymptotic dependence on ε. Similar equations exist

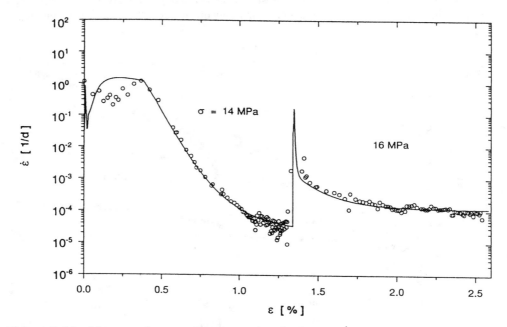

Figure 3.10 *Measured creep data for rock salt given as $\dot{\varepsilon} = f(\varepsilon)$ and the curve fitted with the composite model. $T = 22\,°C$. The loading phase has a duration of 14 min, or 0.4% strain. Same test as in Figure 3.11.*

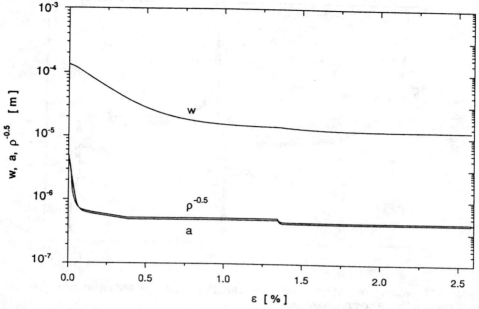

Figure 3.11 *Calculated evolution of the average values for subgrain diameter w, thickness of subgrain boundary a, distance of free dislocations inside the subgrains $\rho^{-0.5}$. Same test as in Figure 3.10.*

for α and ρ. The only three quantities which have to be introduced for transient creep are k_w, k_p, k_a. They have to be determined from measurements. The only relation still to be given is

$$F' = \frac{\sigma}{E} \frac{1-k_h}{(1-f_h)^2} \frac{1}{w} \left(2 \frac{a_s-a}{k_a} - f_h \frac{w_s-w}{k_w} \right).$$

(3.1.21)

F' is zero for steady-state creep.

Now having the formulation for transient creep as well, one can fit complete creep curves. In Figure 3.9 examples for two salt types are shown, and the fit is very good. Figure 3.10 shows the creep rate for a similar test and Figure 3.11 shows the strain-dependent evolution of w, ρ and a for the same test. Finally, Figure 3.12 shows that the model is also able to describe more complex experiments with stress and temperature changes. The model is also suitable to describe stress reductions properly. Let us recall that the model has only 6 parameters to be fitted for transient and steady-state creep of different salt types.

The composite model as described here takes account of isotropic hardening only, despite the fact that kinematic hardening is also present (see Julien *et al.* [1996]). However, since in geosciences and in mines the direction and value of stress differences are always changing slowly, this can be disregarded (condition of collinearity).

The composite model shown above is given for rock salt. However, it can be used for all kinds of crystalline materials as well, because it is based on general physical principles describing the micromechanical deformation mechanism. The use of models of this kind is

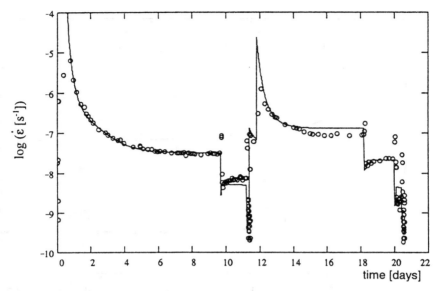

Figure 3.12   *Measured creep curves of a complex test on rock salt given as* $\dot{\varepsilon} = f(\varepsilon)$ *and the fit curve with the composite model.*

an important step forward since it gives an increased reliability for model calculations in geosciences and in materials sciences.

In addition, one must consider that additional deformation mechanisms can also be acting, based on recrystallisation, solution-precipitation, humidity induced creep, micro-cracking, fracturing, chemical reactions, etc. They are partly time-dependent.

3.2 *WATER ENHANCED CREEP*

The mechanical behavior of rock can be strongly affected by water, even by a rather small amount. A great number of publications about this topic exist because it is of great interest, for instance, for studies of crustal and lithospheric deformation. Most experimental studies have probably been performed on quartzite, olivine, calcite, and halite. Results and discussions can be found, e.g., in the books by Knipe and Rutter [1990], Karato and Toriumi [1989], Barber and Meredith [1990]. The effect is often called "hydraulic weakening", first observed by Griggs and Blacic [1965] and Griggs [1967]. A series of tests on quartzite is given in Figure 3.13, as an example. As shown by, e.g., den Brok and Spiers [1991], Jaoul *et al.* [1984], Blacic and Christie [1984], quartzite is considerably weakened by added water ($< 0.5\%$) by a factor of 10 or more at constant strain rates $\dot{\varepsilon}$. The weakening apparently occurs only at high enough confining pressures and above a certain critical temperature which is dependent on the water content. Under these conditions it exhibits a stress exponent of 1.2 - 2, lower at lower strain rates. This value is supported by the theories ($n \approx 1$) and are, apart from deformation mechanisms, carried by dislocations (here $n = 3$). The hydraulic weakening mechanism in hard rocks is explained in different ways. The classical way is that water molecules, or their ionized parts or e.g., SiOH-groups, move through the bulk crystal

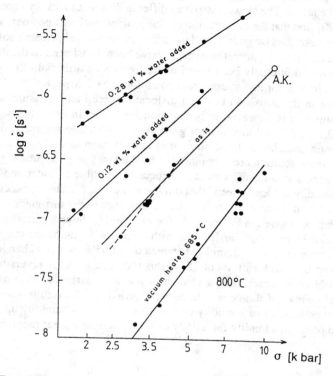

Figure 3.13 *Result of creep tests on Heavitree quartzite at a confining pressure of p = 1.5 GPa on dryed samples, undried samples and samples with water added. One can observe the continuous increase of the steady-state creep rates with increasing amount of water available as well as the decrease of the stress exponent (Jaoul et al. [1984]).*

by diffusion. Climb and glide of dislocations are enhanced by dissolved water defects in the crystal lattice (e.g. Paterson [1989]). However, a number of experiments have caused doubts concerning this explanation (den Brok and Spiers [1991], Carter *et al.* [1990]). Especially the first authors have given good arguments that rather pressure solution (also named precipitation-solution process or solution transfer creep) is the dominant mechanism for hydraulic weakening, continuously producing new and recrystallized material (see also Jaoul [1984]). Perhaps this process acts in combination with the above weakening process. Their experiments on quartz show that material is dissolved in the fluid phase at stressed locations and transported by diffusion in a water-saturated interconnected network of microcracks within the dilatant rock, and finally it is precipitated again in the open microcrack preferably in zones with low stress. Therefore, new developed crystals are found in the experiments. The whole process works only at high enough stresses and under dilatant conditions.

The term "pressure solution" is also used for the diffusional transport on water-bearing grain boundaries (see Rutter [1976, 1983], Spiers *et al.* [1990], Stocker and Ashby [1973]). The theory may be somewhat different. But again, an interconnected network of fluid by a film or by channels is necessary on the grain boundaries, but not dilatancy. Anyway, the process consists of three phases: solution, diffusion, precipitation, and the slowest phase determines the deformation behavior. The slowest step may differ in different rocks. In a meticulous study Spiers *et al.* [1990] find that the deformation of fine grained rock salt powder, saturated with brine, can well be described by grain boundary diffusion -controlled pressure solution and the related formulae. In the same paper, these results have been transferred to the deformation of dense salt and have additionally been based on test results on artificial salt with very small grain sizes. It results that for low stress-strain conditions ($\sigma < 5$ Mpa, $\varepsilon < 10^{-12}$ s^{-1}) as found in a mine, we are in the transition between dislocation creep and pressure solution as the dominant mechanism. It is, however, to be questioned whether in natural rock salt with little water of about 0.1 - 0.2% and much larger grain sizes, interconnected water films are really formed on the grain boundaries. In addition, they stress that in natural rock salt, not flooded by brine, the pressure solution creep is inhibited by dilatancy, e.g., around a gallery.

At least in rock salt, a material which is especially sensible to water and humidity, one observes the "humidity-induced creep" that only works under dilatant stress conditions. The following can be found. In uniaxial creep tests, the influence of the surrounding air has a large influence on creep, as is given in Figure 3.14. The experiment shown , performed by Hunsche and Schulze [1996], has been carried out with stepwise variation of the humidity. The increase and decrease in relative humidity between 0 and 65% causes a change in the rate of steady-state creep by a factor of about 15, in this test. The effect is reversible and normal transient creep always occurs after a change in humidity after decrease as well as after increase. A great number of similar tests have been carried out. The results show that the rock reacts almost immediately to a humidity change. An increase in humidity up to 75%, which is the maximum possible humidity for a body of rock salt over a great range of temperatures

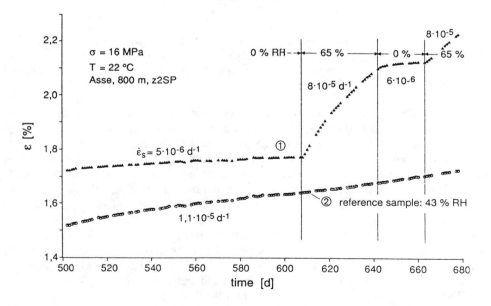

Figure 3.14 *Section of uniaxial creep tests with stepwise varied relative humidity of the surrounding air. The second specimen was tested in the same rig but was subjected to a constant air humidity.*

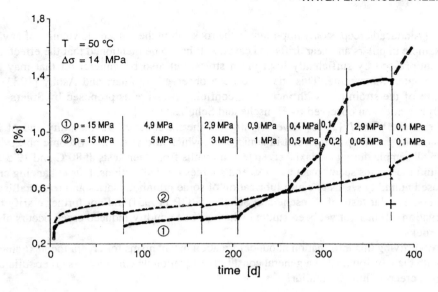

Figure 3.15 *Section of two triaxial creep tests at 50°C with stepwise varied confining pressure p. Specimen no.2 is only slightly affected by the confining pressure since it was impregnated with mineral oil at the beginning of the test. +: specimen was cleaned and the test was started again.*

(3- phase equilibrium), can even result in a factor of 50. Similar tests, published by Varo and Passaris [1977] and Horseman [1988], show the same tendency.

Additional triaxial creep tests on rock salt show that the humidity-induced creep is influenced by confining pressure (or the mean stress). Above about $p = 3$ MPa (with $\Delta\sigma \approx 15$MPa) the humidity-induced creep is completely suppressed and the test behaves as if it was performed in completely dry air. Figure 3.15 illustrates such an experiment with stepwise variations in the confining pressure p. A number of tests show that the rate of steady-state creep changes by a factor of as much as about 35 between $p = 0.1$MPa and 3 MPa. Again, normal transient creep is observed after each change of the confining pressure. Brodsky and Munson [1991] found similar results, when samples were flooded with brine.

On the basis of a great number of tests, Hunsche and Schulze [1996] (see also Schulze [1993] and Hunsche and Schulze [1994]) have developed an empirical equation for the description of the acceleration by humidity-induced creep, giving its nonlinear dependence on relative humidity of the air and the confining pressure. The effect is explained by them as follows: if the rock salt is sufficiently permeable for air as a result of open grain boundaries or microfractures, then water vapor can diffuse through this open inner space towards points of enhanced energy —for example at points with strain hardening due to stress concentrations at the contact between two grains. This hardening will be removed by the adsorbed vapor— acting as a catalyst — by micro-recrystallization causing a repetitive recovery at various sites. Due to this mechanism, humidity-induced creep can only be active in a sufficiently open pore space within the rock. Therefore, it can only be observed under stress conditions, which develop dilatancy. This fact has been proven in Figure 2.14. It is shown there that humidity-induced creep occurs only above the dilatancy boundary. The same effect is obviously present when the material is completely flooded. In practical terms these results mean that the humidity-induced creep is only important in the rock salt in the immediate vicinity of cavities, for example in pillars and near drifts and caverns. It has to be mentioned that the effect is not only suppressed by sufficiently high mean stress, but also by mineral oil that may have contaminated the sample. This has also been observed by Pharr and Ashby [1983]. An increase of the strain rate with increasing confining pressure as proposed by Spiers *et al* [1990] has not been observed by Hunsche and Schulze [1996].

Le Cleac'h *et al.* [1996] have investigated the influence of air humidity and of fluid inclusions on the creep of rock salt in a microcell, which allows the microscopic observation of a small sample during auniaxial creep test. It results from their tests at 80°C and 12.5 MPa that fluid inclusions seem to act as sinks and sources for dislocations, thus enhancing creep. Increased humidity seems to cause lubrication of some grainboundaries and recrystallisation at others. Similar tests of Hunsche and Schulze (BGR [1995]) did not further clarify these observations. It has not yet been studied whether the humidity-induced creep occurs also in other rocks.

After giving this information about deformation mechanisms for creep, the next chapters deal with the development at a general constitutive equation which invol ves creep, dilatancy, damage, creep failure, and failure.

4 Rheological Constitutive Equations for Rocks

4.1 INTRODUCTION

The history of the formulation of general constitutive equations for rocks, which describe various possible time effects and irreversible volumetric changes in rocks, is recent. Most authors have started from uniaxial tests describing either creep or loading-rate effects and have tried afterwards to match the data with some uniaxial empirical mathematical expressions. Some others have started from physically based arguments. The irreversible deformation of the volume, of great significance for rock mechanics (see previous chapters), was generally disregarded. For a survey of the literature and of the experimental foundation of various uniaxial, first formulated empirical constitutive equations and various variants of constitutive equations, see Vutukuri *et al.* [1974], Lama and Vutukuri [1978 vol.III], Ladanyi and Gill [1983], Ranalli [1995], Langer [1988], Cristescu [1989a,e, 1993c, 1994a, 1996b], Karato and Toriumi [1989], Di Benedetto and Hameury [1991], Haupt [1991], Dusseault and Fordham [1993], Hudson [1993b], Ladanyi [1993], Fakhimi and Fairhurst [1994], Jeremic [1994], Durup and Xu [1996], Munson [1997], and many more papers mentioned in Chapters 1, 2, 3.

By analogy with metal rheology, most often elementary uniaxial rheological models were used in the past for geomaterials, metals, polymers, and organic matter. No attempt was done to incorporate into the model the volumetric behavior. For stationary creep, the power function

$$\dot{\varepsilon} = A\,\sigma^n \tag{4.1.1}$$

is very common (Obert and Duvall [1967]) and used by very many authors for a variety of rocks. An Arrhenius term $\exp(-Q/RT)$ as

$$\dot{\varepsilon} = A\,\sigma^n \exp\left(-\frac{Q}{RT}\right) \tag{4.1.2}$$

is often incorporated to take into account the temperature effect in geomaterials, as well as in metals. For more examples and explanation see Sections 2.2 and 3.1, and Takeuchi [1989], Misra and Murrell [1965], Munson *et al.* [1996], Hamami *et al.* [1996]. Another variant of the form

$$\dot{\varepsilon} = B\,\sigma^m \exp\left[-\frac{Q}{RT}\left(1 - \frac{\sigma}{\sigma_o}\right)\right] \tag{4.1.3}$$

has been suggested for rock salt (see Wanten *et al.* [1996]), with σ_o the reference flow stress at 0 K. Here Q is the activation energy for a mole, R is the universal gas constant and T is the absolute temperature. Humidity may also be included in the creep laws (as shown in Section 3.2).

An important step forward toward a correct formulation of general constitutive equation for rocks was the experimental finding of the irreversible volumetric deformation of rocks. As already mentioned, Bauschinger (see Bell [1973]) and Bridgman [1949] have pioneered the description of volumetric changes in several rocks. Further pioneering experimental works are due to Brace [1965], Walsh [1965], Brace et al. [1966], Bieniawski [1967], Rice [1975], Schock [1976, 1977], Holcomb [1981], van Mier [1986], Dudley et al. [1994] (see the literature mentioned by Cristescu [1989a] and in several chapters of Hudson [1993a], Flavigny and Nova [1989], and Chapter 1).

Constitutive equations describing time effects but not specifically compressible/dilatancy volumetric changes, will not be described here (Wan [1996]). Also time-dependent constitutive equation for crushed rock will not be described (see Cristescu and Duda [1989], Liedtke and Bleich [1988], Zeuch [1990], Spiers and Brzesowsky [1993], BGR [1995], Cristescu [1996a] for broken coal, Hansen et al.[1996], Korthaus [1996], Zhang et al. [1996], etc.).

4.2   HOW TO CHOOSE THE MOST APPROPRIATE CONSTITUTIVE EQUATION

From the previous section one may have the feeling that the formulation of a constitutive equation is based on the researcher's intuition and that there is no general procedure to be followed to formulate a constitutive equation. In the present chapter we try to show that the procedure to determine a constitutive equation is logical, systematic and follows very precisely established successive steps. We have to start by making a few **diagnostic tests** that have to reveal the basic mechanical properties of the rock considered, and to show which are the successive steps to be followed, what systematic tests are necessary, what information we try to obtain from these tests, how this information is then analyzed to determine step by step the various coefficients or functions involved in the constitutive equation, how to check the constitutive equation against the data, how to improve it if the matching is not satisfactory, etc.

Before discussing the necessary diagnostic tests let us first observe that a preliminarily established uniaxial constitutive equation does not suffice for the formulation of a triaxial one. The reason is the volumetric behavior of all geomaterials: their volume is either irreversible compressible or irreversible dilatant. This is shown in Figure 4.1 (Cristescu [1986]) for sandstone in a uniaxial test. For small values of axial stress the volume is compressible, while for higher values the volume is dilatant. Close to failure the volume increases very rapidly and, generally, surpasses its initial value.

Another example is shown in Figure 4.2 for a very soft artificial rock, the filler used to fill underground excavated caverns (Mateescu et al.[1983]). It is made of broken rock mixed with a very small quantity of cement. The uniaxial compressive strength is σ_c= 6.86 MPa and initial porosity is 25%. The test shown in this figure was obtained with a loading rate $\dot{\sigma}_1$=0.147 MPa s^{-1}. For much lower loading rates the obtained compressibility reaches ε_v = 0.25 %, i.e., much higher values. In other words, in creep tests the compressibility of this artificial rock is much more pronounced than the one obtained with uniaxial traditional quasistatic constant loading rate deformation tests. The last point on the σ_1-ε_1 curve corresponds to failure. Failure was not recorded on the σ_1-ε_2 curve. K=5.04 GPa, shown in Figure 4.2, was determined in unloading tests following the procedure presented in conjunction with Figure 4.15.

The volumetric behavior is the main difference between mechanical behavior of metals and that of geomaterials. With most metals we know in advance that the volume response is

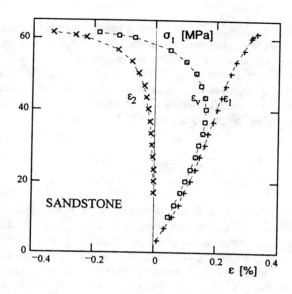

Figure 4.1 *Uniaxial stress-strain curves for sandstone showing compressibility at small stress levels and dilatancy at higher stresses.*

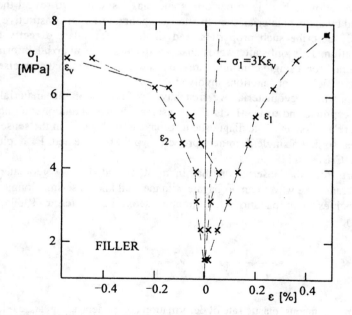

Figure 4.2 *Uniaxial stress-strain curves for filler.*

essentially elastic (with the exception maybe of states close to failure, or of the highly porous metals); therefore, if we establish a uniaxial constitutive equation (called in plasticity theory the "universal stress-strain curve"), we can rewrite it easily in invariants, the determination of plastic potential follows from here with the normality assumption (assuming an associated constitutive equation, which seems a reasonable assumption for metals), and thus via this procedure we obtain the constitutive equation for three dimensions. This procedure cannot be used for geomaterials. For instance, if a uniaxial stress-strain curve is obtained for a certain particular geomaterial, the uniaxial strain ε_1 can be expressed in terms of invariants as $\varepsilon_1 = \bar{\varepsilon} + \varepsilon_V/3$, with $\bar{\varepsilon}$ being the equivalent strain

$$\bar{\varepsilon} = \left(\frac{2}{3} \varepsilon' \cdot \varepsilon' \right)^{\frac{1}{2}}$$

(4.2.1)

and ε_V the volumetric strain. Thus, from the recording of ε_1 in a uniaxial test we cannot distinguish the contribution of $\bar{\varepsilon}$ from that of ε_V in the total magnitude of ε_1. Moreover, ε_V can increase and decrease during a monotonic increase of ε_1. In conclusion, uniaxial tests must be considered as preliminary informative tests only. If we make the (strong, i.e., physically incorrect) assumption that the mechanisms producing the change in shape and those producing volumetric changes are not coupled, then stress-strain relations obtained either in uniaxial tests or in triaxial tests for a single confining pressure may be useful to establish a triaxial constitutive equation. However, this information is not **sufficient** for the formulation of a general constitutive equation.

The approach followed here is to perform some diagnostic tests to reveal the kind of property exhibited by the geomaterial, then to choose a general triaxial constitutive equation that can probably describe such properties, and finally to formulate **directly** a triaxial constitutive equation. It is only afterwards that we discuss how uniaxial information or information obtained for a single confining pressure could be used to make precise some of the constitutive coefficients or functions involved.

When dealing with a geomaterial we first have to find out if this material exhibits **instantaneous response** and **time effects** as for instance, transient and/or stationary creep, rate effect, stress relaxation, etc. The diagnostic tests have to be "clean" in the sense that they must be imagined so that a single dominant property would be revealed. Examples will be given below (see Cristescu [1989a]).

First we start from the observation that in most extended body geomaterials both longitudinal and transverse waves can propagate with the well known seismic longitudinal and shear wave velocities of propagation (for literature see, e.g., Cristescu [1989a, Chap.5], Siggings [1993]):

$$v_P^2 = \frac{3K}{\rho} \frac{1-\nu}{1+\nu} \qquad v_S^2 = \frac{G}{\rho} \ .$$

(4.2.2)

In this case the instantaneous, elastic rate of deformation component $\dot{\varepsilon}^E$ must be related to the stress fluxes (or rates) $\dot{\sigma}$ by

$$\dot{\varepsilon}^E = \frac{\dot{\sigma}}{2G} + \left(\frac{1}{3K} - \frac{1}{2G} \right) \dot{\sigma}\mathbf{1} ,$$

(4.2.3)

where the shear and bulk moduli G and K may depend on stress and/or strain invariants, or on some damage parameter (history of damage evolution); σ is the Cauchy stress, σ is the mean stress and **1** is the unit tensor. Equation (4.2.3) defines the **instantaneous response.** How the two moduli can be determined from tests will be described further on.

The next step is to see if the considered material possesses time-dependent properties such as creep and relaxation, phenomena that have the same physical explanation. As described in the previous chapters, geomaterials exhibit time-dependent properties. For instance Figure 4.3 shows three stress-strain curves for schist obtained with three distinct loading rates (interrupted lines). Thus, when the loading rate is increased the **whole stress-strain curve rises,** starting from the smallest values of the stress and strain. The same Figure 4.3 shows the result of a creep test, the stress being held constant at various levels for three days, then for four days, etc., as shown; the total is of 35 days. We have to point out that creep occurs starting from the smallest values of the loading stresses. Similar results have been reported for rock salt by Hansen et al. [1984] and Hunsche and Schulze [1994]; see also Section 3.1. Since the stress-strain curves are loading-rate-dependent from the smallest applied stresses and since the creep is to be observed for the smallest applied stresses, we must choose a constitutive equation with **practically zero initial yield stress** (see also Allemandou and Dusseault [1996]). The stars at the ends of various curves mark failure. Thus, the model must describe the loading-rate dependence or the loading-history dependence of failure: **with increased of the loading rate, the stress at failure increases** (see also Section 2.1), while

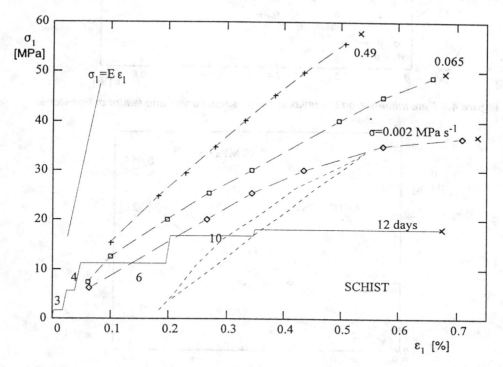

Figure 4.3 Uniaxial stress-strain curves for schist for various loading rates, showing time influence on the entire stress-strain curves, including failure (Cristescu [1986]).

the strain at failure changes too. Also, in creep tests (or very long term loading tests) failure is expected at a much lower stress level than in uniaxial standard tests. The creep at the first four loading levels in Figure 4.3 is a transient creep; it ultimately ends at a stabilized state, i.e., transient creep. However, the last and highest loading certainly results in steady-state and tertiary creep as well. All these are various time-effects which have to be described by the constitutive equation. A time-independent elasto-plastic constitutive equation will not do it. Irreversible deformation during unloading is also shown in Figure 4.3 (short-dashed line). The elastic slope shown in Figure 4.3 as a full line was obtained by unloading procedure as

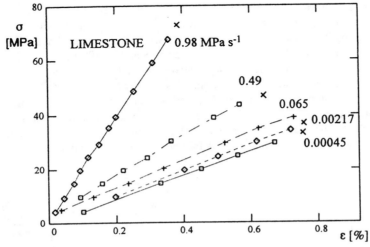

Figure 4.4 *Rate influence on the uniaxial stress-strain curves and failure of limestone.*

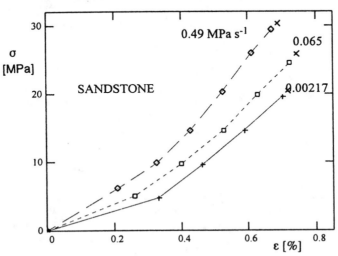

Figure 4.5 *Loading rate influence on the uniaxial stress-strain curves and failure of sandstone (Cristescu [1984]).*

described below in conjunction with Figures 4.15 and 4.16.

The results shown in Figure 4.4 were obtained for limestone in uniaxial tests (Cristescu [1984]). Five loading rates were considered. Even though these stress-strain curves are nearly linear, their slopes change starting from the smallest stresses, and consequently there is no question of using linear elasticity once the initial slope is loading-rate-dependent. The stars at the end of these curves marks failure. Again, with rocks, stress and strain at failure depend on the loading history. Similar results are shown in Figure 4.5 for sandstone. This time the stress-strain curves are nonlinear, but for each loading rate we get another curve starting from the origin. Failure is again loading-rate-dependent. Again, if we make an analogy with the influence of the loading rate on metals, where this influence is felt above the yield stress only, we can conclude once more that the yield stress for most rocks is close to zero; see also Section 2.3.

Sometimes the loading rate or strain rate influence on the uniaxial stress-strain curve is not so obvious. For instance Figure 4.6a shows four uniaxial stress-strain curves for granite plotted with the data published by Sano et al.[1981]. The $\sigma_1 - \varepsilon_1$ curves are only slightly influenced by the strain rate. However, the $\sigma_1 - \varepsilon_2$ curves, where ε_2 is the transverse (diameter) strain, are influenced by the strain rate. This influence as shown at the upper part of the curves, corresponds to dilatancy. Concerning the influence of the strain rate on the volumetric deformation, the main conclusion is the following: with lower strain rates the granite is

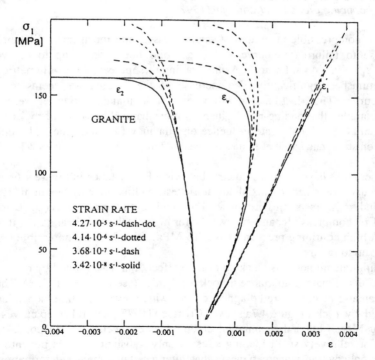

Figure 4.6a Influence of the strain rate on unconfined stress-strain curves for granite (using the data by Sano et al. [1981]) showing significant volumetric dilatancy.

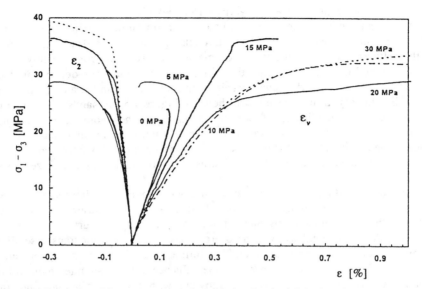

Figure 4.6b *Volumetric and transverse strain variation in triaxial deviatoric tests on limestone, showing compressibility followed by dilatancy at small confining pressures, but compressibility only at high confining pressures (Maranini [1997]).*

slightly more compressible at relatively small stresses, and much more dilatant at higher stresses close to failure (the curve shown as solid line extends up to the volumetric deformation $\varepsilon_v = -0.0644$ at failure). As the loading rate increases, the volumetric dilatancy at failure diminishes significantly. The increase of the loading rate seems to straighten the $\sigma_1 - \varepsilon_v$ curves. Thus, fast loading will produce little dilatancy up to failure. Figure 4.6a shows how complex the loading-rate influence on mechanical behavior of rocks can be (see also Cristescu [1982] for the rate influence on dilatancy for sandstone). The influence of loading rate on stress and strain at failure is described in Section 2.1 (Figures 2.1, 2.3, 2.4 and 2.5).

Some other rocks have other volumetric behavior. For instance in Figure 4.6b are shown stress-strain curves obtained in triaxial deviatoric tests on limestone by Maranini [1997]. Only volumetric and transverse strains are shown. For small confining pressures (i.e., 0 and 5 MPa) this rock is first compressible, and afterwards, for high octahedral shear stress, it is dilatant. However, at high confining pressures (over 15 MPa) the rock is compressible only, for all stress states, up to failure.

The loading-rate influence is one kind of time effect that is exhibited by rocks. **Creep** is also a time effect of major importance in rock mechanics (see Chapters 1, 2, 3). The uniaxial creep tests can also be considered diagnostic tests, which revealsthe time-dependent properties exhibited by rocks. Figure 4.7a shows (Cristescu [1975]) several creep curves for schist from Palazu Mare obtained in uniaxial creep tests. Each curve corresponds to another specimen. For relatively small loading stresses only transient creep is put into evidence. Deformation only by transient creep means that, after a certain time, the deformation by creep practically stops. We say that a "stabilization for the transient creep" has taken place. The corresponding stress and strain state are points on what will be called the "**stabilization**

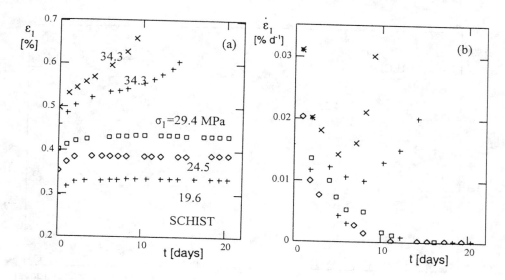

Figure 4.7 *Uniaxial creep curves for schist obtained with various loading stresses. Variation in time of $\dot{\varepsilon}$ shows, if and when, stabilization takes place ($\dot{\varepsilon}_1$ in %/day).*

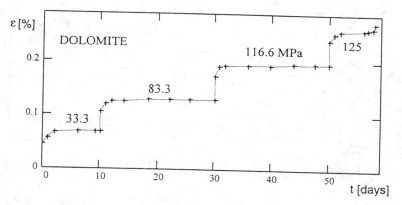

Figure 4.8 *Typical uniaxial creep curves for dolomite showing transient creep for small loading stresses, and stationary and tertiary creep for high stresses.*

boundary". Obviously for each loading stress we obtain another point on this stabilization boundary. When stabilization takes place is decided by plotting the variation in time of the axial rate of strain. Figure 4.7b shows this variation for the creep curves shown in Figure 4.7a. For higher loading stresses, however, the strain becomes more significant, and perhaps steady-state creep, followed by tertiary creep leading to failure, is also shown. Let us recall that steady-state creep means creep taking place with constant strain rate, while tertiary creep to a creep taking place with increasing $\dot{\varepsilon}$. The last two upper curves correspond to two distinct specimens loaded with the same stress, but exhibiting slightly different behavior. The last point on these curves corresponds to failure. Occasionally a period of tertiary creep

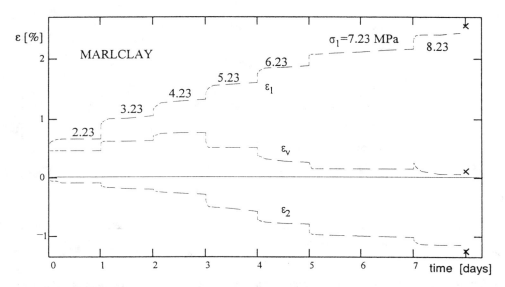

Figure 4.9a *Creep curves for marlclay in uniaxial stress tests showing volumetric compressibility at smaller stresses and dilatancy at higher stresses.*

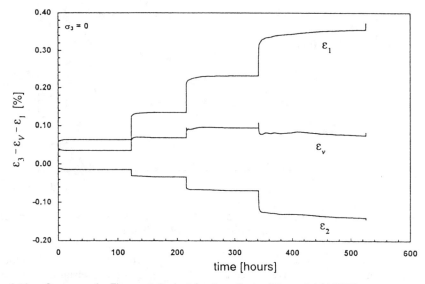

Figure 4.9b *Same as in Figure 4.9a but for limestone (Maranini [1997]).*

(increasing strain rate) is also possible before failure. Another example of uniaxial creep tests is shown in Figure 4.8 for dolomite (Cristescu [1985e]). The curve shown corresponds to a rock with uniaxial compression strength $\sigma_c = 166.6\,\text{MPa}$. It is only for loading surpassing $0.7\,\sigma_c$ that the creep becomes stationary and ultimately tertiary creep produces failure. For small loading stresses very long time intervals are necessary to possibly reveal a stationary creep.

If during creep tests the volumetric strain is also recorded, interesting aspects are revealed. For instance Figure 4.9a shows the creep curves for marlclay (Cristescu [1993c]) subjected to loading stress increased in successive steps. For small stresses the volume of marlclay decreases by creep (compressibility). However, as the loading stress is increased the volume becomes dilatant, i.e., dilatancy is exhibited during creep. More precisely, at each loading the volume first deformed elastically (compressible) and afterwards during creep the volume is dilatant. Failure occurred during the last reloading. Similar results have been obtained for rock salt by Hunsche (see Figure 2.13). In Figure 4.9b are shown similar results for limestone obtained by Maranini [1997]: in the first three loading steps, volumetric compressibility is obtained. At the fourth reloading, after an instantaneous elastic compressibility, there follows a creep producing dilatancy.

The irreversible volumetric changes, such as compressibility and dilatancy, are not only loading-rate-dependent, but also temperature-dependent. To give an example we have chosen a man-made rock-like material, bituminous concrete, whose mechanical behavior is highly temperature-dependent. The bitumen content ranges between 7.5 and 8.5 %. Figure 4.10 shows the volumetric behavior of bituminous concrete in uniaxial tests (Cristescu and Florea [1992]) at various temperatures shown and loading rate 0.01 MPa s^{-1}. Obviously, the temperature has a significant influence not only on the strength, which varies between 1.5 MPa at 50°C and 8.5 MPa at -15° C, but also on the compressibility and dilatancy characteristics.

Time effects such as volumetric creep, for instance, have been recorded at very high confining pressures as well. Figure 4.11 shows two uniaxial compression curves for schist in confined tests (no lateral displacement) (Cristescu [1979]; see also Cristescu and Suliciu [1982]). Two loading rates were used in these two tests: 392 MPa min^{-1} and 98 MPa min^{-1}. Even in such tests a small change in the loading rate has a sensitive influence on the schist response. Moreover, if at $\sigma_1 = 370$ MPa or $\sigma_1 = 750$ MPa we keep the stress constant for a

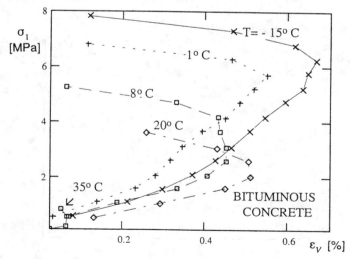

Figure 4.10 *Dependence of compressibility and dilatancy of bituminous concrete on temperature.*

few minutes only, an obvious volumetric creep producing compressibility is recorded. Thus, even at very high pressures compressibility by creep is possible for this hard rock.

Another example of volumetric compressibility by creep in uniaxial confined tests (oedometer kind of tests - see Cristescu [1989a] Section 3.2) is that of cement concrete (mortar) shown in Figure 4.12 (Constantinescu [1981]; see also Cristescu [1984]). For the three tests shown, three rates of strain were used ($6.4\times10^{-4}s^{-1}$, $1.67\times10^{-3}s^{-1}$ and 2.55×10^{-3} s^{-1}). Again, a strain-rate influence is obvious. At $\sigma_1=1.166$ GPa the stress was held constant for 102 min, 87 min, and 70 min, respectively. The volume compressibility by creep was recorded. Most of the volumetric strain is irreversible.

When we try to find a model to describe creep of geomaterials we can choose a nonlinear $\Delta\sigma_1$ and keep it constant until a stabilization of the transient creep (i.e., $\dot{\varepsilon}\approx0$) is obtained. We now increase the loading stress up to $2\Delta\sigma_1$, and determine another point belonging to the stabilization boundary, and so on. By this procedure one can determine the whole stabilization boundary. If we repeat the procedure, using this time from the beginning at each loading step a stress of double magnitude $2\Delta\sigma_1$, for instance, we will obtain another stabilization boundary

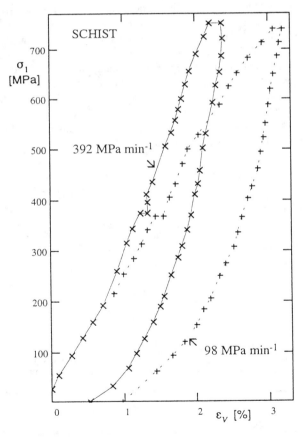

Figure 4.11 *Loading-rate influence on volumetric compressibility in uniaxial confined tests. Two plateaux of creep compressibility are obtained within a few minutes.*

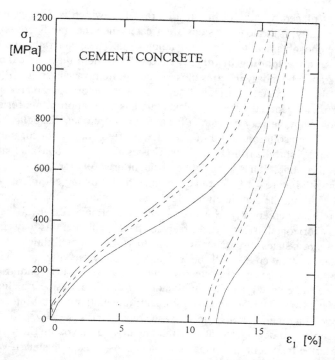

Figure 4.12 *Volumetric compressibility in uniaxial confined tests (oedometer type) of cement concrete (mortar) for three loading rates $\dot{\varepsilon}_v = 6.4 \times 10^4 \, s^{-1}$ (full line), $1.67 \times 10^3 \, s^{-1}$ (dotted line), $2.55 \times 10^3 \, s^{-1}$ (interrupted line), showing also volumetric creep.*

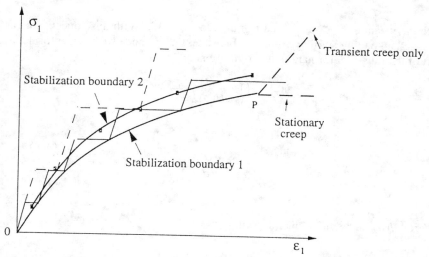

Figure 4.13 *History dependence of the stabilization boundary and possible description of steady-state creep at high stresses.*

(shown by squares placed on a solid line in Figure 4.13), which is distinct from the first obtained one. Generally, if the loading steps are greater, then the stabilization boundary is situated above the one obtained with smaller loading steps. Thus, generally, the stabilization boundary depends on the **loading history**. It follows from here that one cannot write the equation of stabilization boundary in terms of stress and strain invariants only, $F(\sigma, \varepsilon) = 0$, but the equation of this boundary must also depend on a history parameter κ, say, e.g., $G(\sigma, \varepsilon, \kappa) = 0$. Therefore, generally, the viscoelastic models are inappropriate for geomaterials and one has to look from the very beginning for a viscoplastic one. Viscoelastic models can be used for approximate solutions valid in a restricted stress domain, or for methodological reasons. The concept of **stabilization boundary** (Cristescu [1979, 1989a]) is fundamental for the formulation of the model. It has to be defined straightforwardly for the triaxial stress- state.

In conjunction with Figure 4.13 one can also make the following comment. If we find for a certain geomaterial that the stabilization boundary is a strictly increasing curve, then that material exhibits transient creep only. However, if above a certain stress level the stabilization boundary becomes horizontal, steady-state creep can be described in this way (i.e., with such kind of model) for stress states surpassing this limit stress. Other possibilities to describe both transient and steady-state creep will be discussed below, having in mind that most geomaterials exhibit both transient and steady-state creep, even at low loading levels, and it is only the time interval elapsed from the application of the load, besides the load itself, that decides which of the two stages of creep is dominant: for relatively short intervals, transient creep, and for very long time-intervals, steady-state creep. Thus generally, both **transient** and **steady-state creep** are to be incorporated in the model, and maybe **tertiary creep** as well.

In order to describe **transient creep** that ultimately ends in a stable state if very long time intervals are involved, one usually uses the bracket

$$\langle A \rangle = \frac{1}{2}(A + |A|) = A^{+} \tag{4.2.4}$$

with the meaning of "positive part of the function A". With this concept we can describe "elastic" unloading starting from a stable state. For instance we can use for the irreversible part of the rate of deformation due to transient creep

$$\dot{\varepsilon}_{T}^{I} = k_{T} \left\langle 1 - \frac{W(t)}{H(\sigma)} \right\rangle \frac{\partial F}{\partial \sigma} , \tag{4.2.5}$$

or

$$\dot{\varepsilon}_{T}^{I} = k_{T} \left\langle 1 - \frac{W(t)}{H(\sigma)} \right\rangle N(\sigma) \tag{4.2.5'}$$

where $H(\sigma)$ is the yield function, with

$$H(\sigma(t)) = W(t) \tag{4.2.6}$$

the equation of the stabilization boundary (which is the locus of the stress states at the end of transient creep when stabilization takes place, i.e., when $\dot{\varepsilon}^I = 0$, $\dot{\sigma} = 0$; this boundary depends on the loading history), with

$$W(T) = \int_0^T \sigma(t) \cdot \dot{\varepsilon}^I(t) \, dt = \int_0^T \sigma(t) \dot{\varepsilon}_v^I(t) \, dt + \int_0^T \sigma'(t) \cdot \dot{\varepsilon}^{I'}(t) \, dt = W_v(T) + W_D(T)$$

(4.2.7)

the irreversible stress power per unit volume at time T, used as a work-hardening parameter or as internal state variable. Thus the "history" is involved in $W(T)$. Further, in (4.2.5), $F(\sigma)$, which may not exist or cannot be determined, is a viscoplastic potential that controls the orientation of $\dot{\varepsilon}_T^I$. If F coincides with H we say that the constitutive equation is "associated" to a prescribed yield function H. Otherwise the constitutive equation is said to be "non-associated," as often happens for most geomaterials.

Let us recall that if a yield condition is written as $H(\sigma) = W(t)$, an "associated" (i.e., associated to the yield condition) constitutive law assumes proportionality between $\dot{\varepsilon}^I$ and $\partial H / \partial \sigma$. In particular, for Mises type of yield condition, i.e., $II_{\sigma'} \equiv H(\sigma)$, the proportionality is between $\dot{\varepsilon}^I$ and σ', the stress deviator (see (4.2.15)). For rocks, soils, and particulate materials, this assumption is not in agreement with experiment. The orientation of $\dot{\varepsilon}^I$ is not in the direction of $\partial H / \partial \sigma$, but in some other one which has to be found from experiment (see next chapter). If $\dot{\varepsilon}^I$ is proportional to $\partial F / \partial \sigma$, then $F(\sigma)$ is called the viscoplastic potential.

Alternatively one can use the form (4.2.5') of the constitutive equation, which does not involve the concept of viscoplastic potential, which may not exist or cannot be determined, but that of the "viscoplastic strain rate orientation tensor" $N(\sigma)$ (Cazacu [1995], Cazacu et al. [1997]). k_T is some kind of "viscosity coefficient": it may depend slightly on stress and strain invariants, and maybe on a damage parameter describing the history of microcracking and/or history of pore collapse to which the rock was subjected. The whole factor $k_T(\partial F / \partial \sigma)$ (or $k_T N(\sigma)$) can be considered to be some kind of variable viscosity coefficient.

In order to describe **steady-state creep** one can adapt accordingly either the function H (see, for example, the above comments related to Figure 4.2), or one can add to (4.2.5) an additional term, as for instance

$$\dot{\varepsilon}_S^I = k_S \frac{\partial S}{\partial \sigma} \, ,$$

(4.2.8)

where $S(\sigma)$ is a viscoplastic potential for steady-state creep and k_S is a viscosity coefficient for steady-state creep, which may possibly depend on stress invariants and on damage, if necessary. Since the bracket $\langle \rangle$ is absent in (4.2.8), the creep described by such a term will last so long as stress is applied. If the geomaterial can also be compressible, the function $S(\sigma)$ must satisfy some restrictive conditions ensuring that the volumetric creep during compressibility is of transient nature only. Let us recall that steady-state creep means creep taking place with constant $\dot{\varepsilon}$ under constant stress state. Transient and steady-state creep are quite often difficult to distinguish. One can describe them either by a single term (as mentioned above) or by two additive terms. This last procedure is easier to handle in both formulations of the model and when the model is used to solve some practical problems. The main reason to describe transient creep and steady-state creep by two additive terms, instead of a single one, is due to distinct behavior of the volume in the two types of creep. If we would like to describe with the constitutive equation both compressibility and/or dilatancy, then the transient creep term must describe compressibility in the compressibility domain (see Figure 4.14) but dilatancy in the dilatancy domain. However, the steady-state term has to describe either dilatancy or incompressibility. It is quite difficult to formulate a model which would describe distinct volumetric behavior for the two kinds of creep, if both transient and steady state- creep are described by a single term in the constitutive equation, though from a physical point of view it might be desirable to have a single term in the constitutitve equation.

A few remarks are necessary. First we have assumed that the bracket $\langle\rangle$ is involved in (4.2.5) in a linear form. This assumption has been made to simplify the computations when the constitutive equation is used in mining or civil engineering applications. However, it may be sometimes involved in a nonlinear form, as for instance

$$\dot{\varepsilon}_T^I = \frac{k_T}{E}[1 - \exp(\lambda \langle H(\sigma) - W(t)\rangle)]\frac{\partial F}{\partial \sigma} \tag{4.2.9}$$

or

$$\dot{\varepsilon}_T^I = \frac{k_T}{E}\left(\frac{H(\sigma) - W(t)}{a}\right)^n \frac{\partial F}{\partial \sigma} \tag{4.2.10}$$

if necessary (see Cristescu and Suliciu [1982], Chapter 2). Here $\lambda > 0$, $a > 0$ and $n > 0$ are constants. Secondly, let us mention also that we could use another parameter than the irreversible work per unit volume, to describe the irreversible isotropic hardening. For instance, we could use the irreversible equivalent strain

$$\bar{\varepsilon}^I(t) = \sqrt{\frac{2}{3}\varepsilon^I(t) \cdot \varepsilon^I(t)} \tag{4.2.11}$$

or the irreversible equivalent integral of the rate of deformation tensor

$$\bar{\varepsilon}^I(T) = \sqrt{\frac{2}{3}} \int_0^T \sqrt{\dot{\varepsilon}^I(t) \cdot \dot{\varepsilon}^I(t)}\, dt . \tag{4.2.12}$$

However, both these last expressions cannot distinguish between irreversibility produced by compressibility and irreversibility produced by dilatancy. These expressions can be used to describe irreversible isotropic hardening for those materials that are either compressible only, or dilatant only.That is the reason why we will use from now on the work-hardening parameter $W(t)$ only. Let us mention also that it is the stress work that is involved in thermodynamic considerations,in formulation of work-hardening laws in plasticity, etc.,while other parameters as (4.2.11) and (4.2.12) can be used in classical plasticity for special loading cases only. In addition to the arguments of "classical plasticity" here we must describe the irreversible behavior of the volume as well. Finally we must mention that some of the constitutive coefficients or functions may depend on other parameters as temperature and humidity.

For isotropic materials all the functions involved in constitutive equation must satisfy the invariance requirement which, for instance for $N(\sigma)$, is written as

$$N(Q\sigma Q^T) = QN(\sigma)Q^T$$

for any orthogonal transformation Q; here the superscript T means "transpose".

To formulate a **general constitutive equation** we make several main assumptions (Cristescu [1987, 1989a, 1993a, 1994a, b]):

- For the time being, we consider only homogeneous and isotropic rocks. Thus the constitutive functions will depend on stress and strain invariants only, besides, maybe, on an isotropic damage parameter. From all possible stress invariants we assume that of greater significance for geomaterials are the mean stress

$$\sigma = \frac{1}{3}(\sigma_1 + \sigma_2 + \sigma_3) \tag{4.2.13}$$

and the equivalent stress $\bar{\sigma}$ or the octahedral shear stress τ

$$\bar{\sigma}^2 = \sigma_1^2 + \sigma_2^2 + \sigma_3^2 - \sigma_1\sigma_2 - \sigma_2\sigma_3 - \sigma_3\sigma_1 \tag{4.2.14}$$

and

$$\tau = \frac{\sqrt{2}}{3}\bar{\sigma} = \left(\frac{2}{3}II_{\sigma'}\right)^{1/2} \tag{4.2.15}$$

with $II_{\sigma'} = (1/2)\sigma' \bullet \sigma'$ the second invariant of the stress deviator $\sigma' = \sigma - \sigma\mathbf{1}$.

- We assume that the material displacements and rotations are small (i.e., nonlinear terms are negligible with respect to the linear ones) so that the rate-of-deformation components are additive

$$\dot{\varepsilon} = \dot{\varepsilon}^E + \dot{\varepsilon}^I . \tag{4.2.16}$$

- The elastic rate-of-deformation component $\dot{\varepsilon}^E$ is given by (4.2.3).
- The irreversible rate-of-deformation component $\dot{\varepsilon}^I$ satisfies either (4.2.5) or (4.2.8), or is the sum of the two expressions, depending on the need to describe either transient creep or steady-state creep only, or, finally, both kinds of creep occurring simultaneously.
- The initial yield stress of the rock can be assumed to be zero or very close to it.
- The constitutive equation is valid in a certain constitutive domain bounded by a short-term failure surface, which will be incorporated in the constitutive equation.

Thus the **constitutive equation** will be written in the form

$$\dot{\varepsilon} = \frac{\dot{\sigma}}{2G} + \left(\frac{1}{3K} - \frac{1}{2G}\right)\dot{\sigma}\mathbf{1} + k_T\left\langle 1 - \frac{W(t)}{H(\sigma)}\right\rangle\frac{\partial F}{\partial \sigma} + k_S\frac{\partial S}{\partial \sigma} \ . \tag{4.2.17}$$

Let us examine what types of properties can describe such a constitutive equation. For the time being let us disregard the last term so that the irreversibility will solely be due to the transient creep. From (4.2.5) we obtain for the irreversible volumetric rate of deformation

$$\left(\dot{\varepsilon}_V^I\right)_T = k_T\left\langle 1 - \frac{W(t)}{H(\sigma)}\right\rangle\frac{\partial F}{\partial \sigma}\cdot\mathbf{1} \ . \tag{4.2.18}$$

Let us assume that at the initial moment t_o (moment when an excavation is started, say), the initial stress state (the so called "primary" stress) $\sigma^P = \sigma(t_o)$ is an equilibrium stress state, i.e., $H(\sigma(t_o)) = W^P$, with $W^P = W(t_o)$ for the primary stress state (see Cristescu [1989a]). A stress variation from $\sigma(t_o)$ to $\sigma(t) \neq \sigma(t_o)$ with $t > t_o$, due to an excavation, for instance, will be called **loading** if

$$H(\sigma(t)) > H(\sigma(t_o)) \ , \tag{4.2.19}$$

and three cases are possible depending on which of the following inequalities is satisfied by the new stress state:

$$\frac{\partial F}{\partial \sigma}\cdot\mathbf{1} > 0 \quad \text{or} \quad \frac{\partial F}{\partial \sigma} > 0 \quad \text{compressibility} \tag{4.2.20}$$

$$\frac{\partial F}{\partial \sigma}\cdot\mathbf{1} = 0 \quad \text{or} \quad \frac{\partial F}{\partial \sigma} = 0 \quad \text{compressibility/dilatancy boundary} \tag{4.2.21}$$

$$\frac{\partial F}{\partial \sigma}\cdot\mathbf{1} < 0 \quad \text{or} \quad \frac{\partial F}{\partial \sigma} < 0 \quad \text{dilatancy} \ . \tag{4.2.22}$$

We observe that $(\partial F/\partial\sigma)\cdot\mathbf{1} = \partial F/\partial\sigma$ if we take into account that F depends on stress invariants. Therefore the behavior of the volume is governed by the orientation of the normal to the surface $F(\sigma) = $ constant at the point representing the actual stress state (see Figure 4.14): if the projection of this normal $(\partial F/\partial\sigma)\cdot\mathbf{1}$ on the σ-axis is pointing towards the positive orientation of this axis, that stress state produces irreversible compressibility, otherwise - dilatancy. There where this normal is orthogonal to the σ-axis, there are no irreversible volumetric changes; the volumetric behavior is elastic.

If instead of (4.2.19) the new stress state satisfies

$$H(\sigma(t)) < W(t_o) \tag{4.2.23}$$

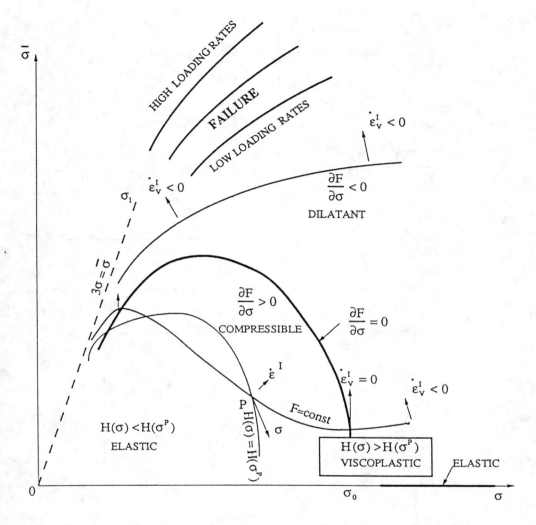

Figure 4.14 *Domains of compressibility, dilatancy, and elasticity in the constitutive domain: thick line is the compressibility/dilatancy boundary $\partial F/\partial \sigma = 0$; failure depends on the loading rate.*

then an **unloading** takes place, and the response of the geomaterial is elastic, according to (4.2.3). Obviously, the "unloading" concept has a meaning for the transient term, only; since a decrease of the stress means for the steady state term a "slower" creep.

A very important remark is necessary concerning the correct determination of the **elastic parameters**. Their correct determination influences the whole procedure for the determination of the constitutive equation since elastic parameters are involved in the estimation of elastic components of strain, of inelastic components, of irreversible stress work, etc. The elastic parameters can be determined either by dynamic procedures or by static ones. By the dynamic

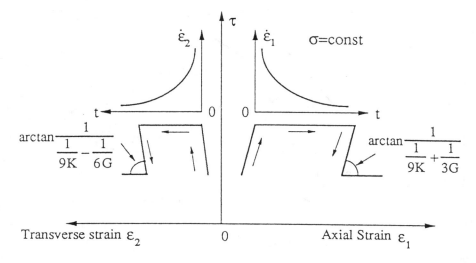

Figure 4.15 *Static procedure to determine the elastic parameters in unloading processes following short creep periods.*

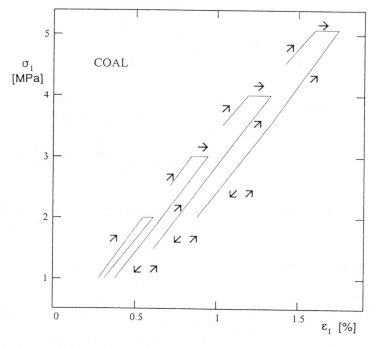

Figure 4.16 *Procedure to determine elastic parameters from the beginning of unloading slopes obtained after short period (10 minutes) of creep. Unloading and reloading follow the same path (Cristescu [1989a]).*

procedure one measures the travel times of two seismic waves (see (4.2.2)) and afterwards the elastic parameters follow from

$$K = \rho\left(v_P^2 - \frac{4}{3}v_S^2 \right) \quad , \quad G = \rho v_S^2 \ .$$ (4.2.24)

By the static procedure one has to determine them at various stress and strain levels. Since the initial shape (at small stresses) of the stress strain curves is dependent on the loading rate (see Figures 4.4 and 4.5) or the strain rate (Figures 4.6a and 2.2), and since most geomaterials are creeping even for very small applied stresses, these parameters are to be determined after a certain time-interval elapsed from the last loading, during which the rock is allowed to deform by creep (see Cristescu [1989a]). This is shown in Figure 4.15, where for a chosen loading stress level the rock is allowed to creep until the rate-of-deformation components become (in absolute value) quite small, ensuring that during the subsequent unloading performed in a comparatively much shorter time interval, no significant interference between creep and unloading phenomena will take place. For this purpose, frequently, a preliminary creep of the geomaterial lasting half an hour or one hour might be quite sufficient. The length of this time interval greatly depends on the kind of geomaterial under consideration, and occasionally even shorter creep intervals will suffice. An example is shown in Figure 4.16 for coal (Cristescu [1989a]). The axial stress was kept constant for ten minutes only. If only a partial unloading is performed (one third of the total stress or even less), the unloading and reloading follow quite closely straight lines that practically coincide. The reason to perform only a partial unloading is that the specimen is quite "thick" and as such the stress state in the specimen is not really uniaxial. Thus, during complete unloading additional phenomena due to the "thickness" of the specimen will be involved, including, e.g., kinematic hardening in the opposite direction (see Sections 2.2.2 and 3.1). Thus, according to the "static" procedure the elastic parameters are to be determined from the first portions of the unloading slopes obtained in triaxial tests, after a short-term creep test. In another example in Figure 4.17 are shown some

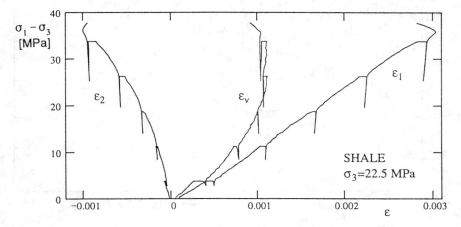

Figure 4.17 *Stress-strain curves obtained in triaxial test on shale; the unloadings shown follow a period of creep of several minutes.*

of the results obtained by Nawrocki *et al.* [1997] for shale in triaxial tests. For each confining pressure at various levels of the axial stress, this stress component was held constant for several minutes. Small plateaux of creep have been obtained. After this creep period, small unloading/reloading cycles have revealed the magnitude of the elastic constants for that particular stress state for both compressibility and dilatancy domains. Tests have been performed for several confining pressures. Figure 4.17 shows the results obtained for $\sigma_3 = 22.5$ MPa. It has been found that the bulk modulus increases with the mean stress. With the increase of the stress difference $\sigma_1 - \sigma_3$ and constant confining pressure ($\sigma_3 =$ constant) the elastic parameters first increase (in the compressibility domain) and then decrease (in the dilatancy domain). The same procedure to determine the elastic parameters has been applied for Tournemire shale by Niandou *et al.* [1997].

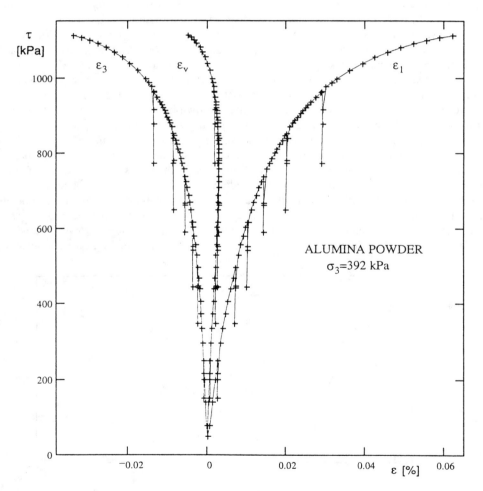

Figure 4.18 *Stress-strain curves for alumina powder in triaxial test at confining pressure $\sigma_3 =$ 392 kPa, showing nearly linear partial unloading/reloading following a stress relaxation of 15 min.*

If the testing device is controlling strain, one can use a similar procedure by keeping a certain time interval the strain constant and allowing the stress to relax. Afterwards, after a certain time interval, when the stress rate is already quite small, an unloading/reloading is applied for the determination of the elastic parameters. An example of such stress-strain curves obtained in a triaxial test is given in Figure 4.18 for alumina powder. Though it is not a rock, the characteristics are the same as for most geomaterials (Jin and Cristescu [1996]). At various loading levels the strain was kept constant for 15 minutes during which a significant stress relaxation was recorded. Afterwards an unloading and reloading followed the same straight line, and from the corresponding slopes the elastic parameters were determined. In Figure 4.18 about half of the stress decreases shown are due to relaxation, and the other half to unloading. This procedure is proposed and described by Cristescu [1989a]. No complete unloading is to be performed since the lower portion of the unloading curves are influenced by other phenomena as well (as the thickness of the specimen, etc., as already mentioned). Finally let us observe that the values of the elastic parameters determined by static procedures as described above, compare very well with those determined by dynamic procedures.

These parameters are not constant. For instance, Figure 4.19 shows the variation of the Young's modulus E_1 for shale after Cristescu and Cazacu [1995]. Since shale is an anisotropic rock, in Figure 4.19 the result for the Young's modulus is shown in the direction of the axis of symmetry only. Using the data for shale obtained by Niandou [1994] the following stress dependence was found:

$$E_1(\sigma, \tau) := E_{1\infty} - \left[E_{1a}\tau^2 + \frac{E_{1b}}{\tau + E_{1c}} \right] \exp(-E_{1d}\sigma) \qquad (4.2.25)$$

with τ the octahedral shear stress defined by (4.2.15) and σ the mean stress (see (4.2.13)). The other coefficients are constants: $E_{1\infty} = 1.59 \times 10^4$, $E_{1a} = 0.09$, $E_{1b} = 2.501 \times 10^5$, $E_{1c} = 16$, and $E_{1d} = 0.0313$. Generally it is found that with increasing pressure the elastic parameters tend towards a constant asymptotic value, which is the value of that parameter for the geomaterial in which all the pores and microcracks have been closed by a high applied pressure. Since

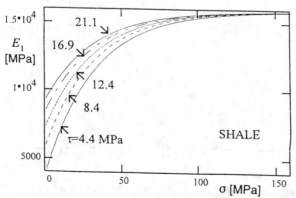

Figure 4.19 *The variation with stress of Young's modulus for shale.*

the data available for the determination of E_1 were obtained in Kármán-type triaxial tests, i.e., for a constant confining pressure, the successive readings are done along the lines $\sqrt{2}\,\sigma - \tau =$ constant, or $3\sigma - \bar{\sigma} =$ constant (see Figure 4.20). In order to determine the individual dependence on σ and τ one reads the data not along the lines $\sqrt{2}\,\sigma - \tau =$ constant, but along the interrupted horizontal lines $\tau =$ constant. With this procedure one can determine the dependency of the elastic parameters on σ first, and afterwards the coefficients thus found are assumed to depend on τ. In this way one can determine the dependence on τ and σ of any constitutive function from data obtained in the Kármán type of triaxial tests. As an example, Figure 4.21 shows how the Young's modulus E_1 was determined for shale from the data obtained in Kármán triaxial tests (Cazacu [1995]). The exact location of the horizontal asymptote shown in Figure 4.19 is difficult to find since it requires tests performed at very high confining pressures. Thus, in most cases the asymptote is guessed from the overall behavior of that particular constitutive parameter at smaller confining pressures. Generally, for most geo-materials the elastic parameters increase smoothly with σ. For small variations of σ, this variation can, sometimes, be neglected. The dependence on τ may depend on that particular geomaterial. The shale considered in Figure 4.19 is a compressible rock only, with no dilatancy exhibited up to failure. That is why for this rock the Young's modulus E_1 is increasing with τ. For other rocks which are dilatant for stress states in the neighborhood of failure, the dependence on τ of the elastic moduli can be more involved (i.e., when τ is increasing and is approaching failure, the elastic parameters may decrease). For high values of τ the elastic parameters can vary during creep tests (constant stresses) since progressive damage is taking place (see Chapter 6).

To give another example, Figure 4.22 shows for alumina powder the dependence of the

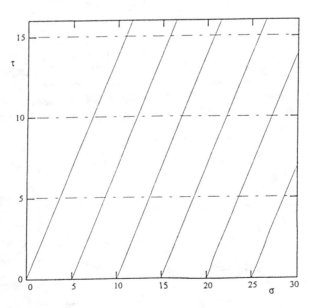

Figure 4.20 *Grid of lines* $\tau =$ *constant (dash-dot lines) and* $\sqrt{2}\,\sigma - \tau =$ *constant (solid lines) used to determine the dependence of any constitutive function on* τ *and* σ*, from data obtained in standard triaxial compression tests.*

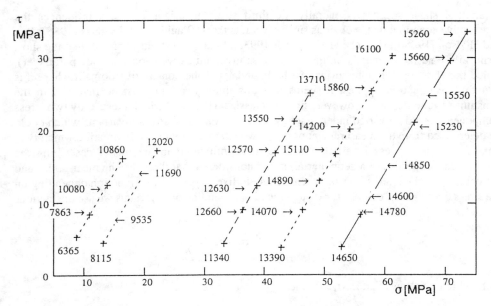

Figure 4.21 *Lines $\sqrt{2}\sigma - \tau = constant$ along which experimental values of E_1 were obtained for shale and the dependency of E_1 on σ and τ was formulated.*

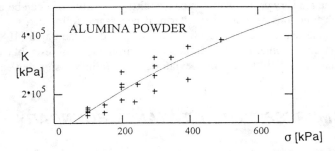

Figure 4.22 *Variation of the bulk modulus of alumina powder with pressure.*

bulk modulus on pressure. Generally, the bulk modulus is strictly increasing with increasing pressure, and tending asymptotically towards a constant value as pressure is increased very much. However, in the performed tests no such high pressures were applied.

Generally, most measurements of the dependence of the elastic parameters of geomaterials on stresses have considered the dependence on the mean stress only (Volarovich *et al.* [1974], King [1966, 1970], Vincké [1994]). The variation of the elastic parameters with the stress state is quite smooth when stresses and/or strains are varying. Occasionally one can consider constant values for these parameters around a certain location, where an excavation will be made.

Let us give an explanation why we have assumed in the following chapters that the constitutive functions depend primarily on two stress invariants: the mean stress σ and the octahedral shear stress τ. Generally the third stress invariant III_σ also influences the constitutive equation; that is obvious since several authors (Paul [1968], Demiris [1985], Pan and Hudson [1988b], Hunsche [1990, 1991, 1993a], Lade [1993]) have shown that the short term failure surface depends on the third stress invariant as well (on the Lode parameter). Since the short term failure surface is the boundary of the constitutive domain where this constitutive equation applies, it seems that this third invariant must be involved in the constitutive functions too. However, since in classical (Kármán) triaxial tests only two stress components could be varied independently, there are only two stress invariants which could be determined from the measurement of these two stress components. Estimating which two, out of the three stress invariants, are the most influential in the model describing the mechanical properties of a geomaterial, it was considered that these are the mean stress and the octahedral shear stress. Thus, when we would like to pass from a function depending on the stress components σ_1 and $\sigma_2 = \sigma_3$ to a function depending on invariants we use the obvious formulae

$$\sigma = \frac{1}{3}(\sigma_1 + 2\sigma_2) \quad , \quad \bar{\sigma} = |\sigma_1 - \sigma_2| = \frac{3}{\sqrt{2}}\tau \ , \tag{4.2.26}$$

from which it follows that if $\sigma_1 > \sigma_2$ then

$$\sigma_1 = \sigma + \frac{2}{3}\bar{\sigma} = \sigma + \sqrt{2}\tau \quad , \quad \sigma_2 = \sigma - \frac{\bar{\sigma}}{3} = \sigma - \frac{\tau}{\sqrt{2}} \ . \tag{4.2.27}$$

The above formulae are very useful. Afterwards the third stress invariant can be incorporated in the constitutive functions as a corrective term, if necessary, following, for instance the procedure used by Desai and Zhang [1987] and Desai and Varadarajan [1987]. For this purpose one has to make either true triaxial tests, or Kármán type of tests, but with loading performed in at least two directions in the deviatoric plane (compression and extension).

4.3   COMPRESSIBILITY/DILATANCY BOUNDARY

A very important concept which may characterize a certain type of rock is the volumetric behavior: dilatancy and/or compressibility. Most rocks are compressible for small octahedral shear stress but dilatant for higher values of the shear stress. The boundary in a stress space (or plane), between the domain where the rock is compressible and the domain where the rock is dilatant, is called the compressibility/dilatancy boundary (Cristescu [1985b,e]). This boundary is the geometrical locus in the σ-$\bar{\sigma}$ plane, or σ-τ plane, where the rock, or generally the geomaterial, passes from compressibility to dilatancy. Let us first assume that this boundary is a curve in the σ-$\bar{\sigma}$ plane.

The compressibility/dilatancy boundary is obtained quite easily by finding where (i.e. for what stress state) along the $\sigma_1^R - \varepsilon_v^R$ curves obtained in classical triaxial tests performed with various confining pressures $\sigma_2 =$ constant (i.e., $\sqrt{2}\sigma - \tau =$ constant) the slope of the tangent to these curves is equal to the elastic one ($\dot{\sigma}_1^R = K\dot{\varepsilon}_v^{ER}$, where $\dot{\varepsilon}_v^I = 0$; we recall that in the classical triaxial test mean stress varies when $\sigma_1^R = \sigma_1 - \sigma_2 = \bar{\sigma}$ is increased, and the confining pressure $\sigma_2 =$ constant; thus the elastic part of the volumetric strain varies too). Here the

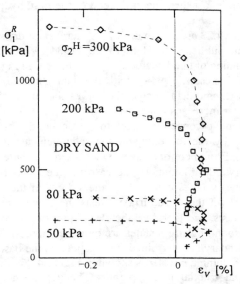

Fig. 4.23 *Dependency of relative volumetric strain on relative equivalent stress for dry sand.*

superscript R means "relative", i.e. the stress or strain measured with respect to the state at the end of the hydrostatic part of the test; these relative components are just the stresses and strains during the second stage of the triaxial test (see Cristescu [1989a]) for which the

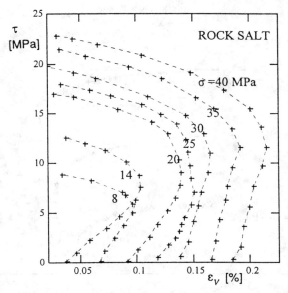

Figure 4.24 *Dependency of volumetric strain on octahedral shear stress for rock salt in true triaxial tests (same results as Figure 2.11).*

reference configuration is just the state at the end of the first, hydrostatic, stage of the test (when $t = T_H$).

For instance, in Figure 4.23 are shown the $\sigma_1^R - \varepsilon_v^R$ curves for dry sand (Cristescu [1991a]), plotted by using the data published by Hettler *et al.* [1984]. Volumetric strains are relative, i.e., referred to the beginning of the deviatoric stage of the test. In Fig. 4.24 several curves for rock salt are plotted (Cristescu [1994a]) with the data obtained in true triaxial tests by Hunsche (see Sections 1.3 and 2.1). Thus, the curves shown are obtained for various tests with σ = constant when τ increases.

A considerable volume compaction of 0.5 to 1% occurs during the hydrostatic loading phase, partially caused by the loosening of the specimen during coring and machining (Hunsche [1991, 1992]). From here one can see that when τ is increased the rock salt is first compressible and afterwards dilatant. In the case of true triaxial tests, the passing from compressibility to dilatancy takes place where the slopes of the various curves shown in Figure 4.24 are vertical (in these true triaxial tests the mean stress does not vary in the deviatoric stage of the tests, and therefore the elastic part of the volumetric strain does not vary either). The corresponding stress states are on the compressibility/dilatancy boundary, and that is how this boundary is determined (in practice, on the basis of many tests).

Another example for alumina powder is shown in Figure 4.25 (Jin and Cristescu [1996]), obtained with a classical triaxial test apparatus. The dependence of the compressibility/

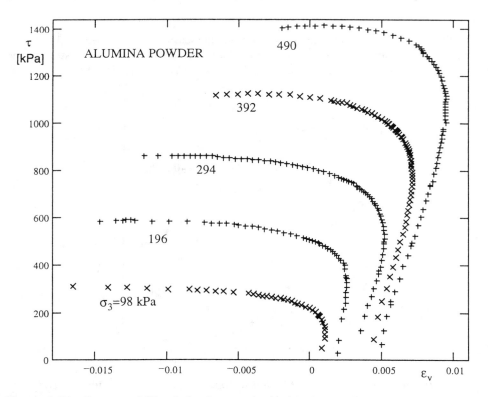

Figure 4.25 *Compressibility of alumina powder in classical triaxial tests.*

dilatancy boundary on confining pressure is obvious. From this figure follows an important conclusion for the consolidation (compaction) of powders: maximum consolidation is obtained if a shearing stress is superposed over a hydrostatic pressure. Also with increasing pressure the dilatancy at failure decreases, as for other materials.

After determining the stress states at the compressibility/dilatancy boundary, for classical triaxial tests one uses the formulae

$$\bar{\sigma} = \sigma_1^R = \sigma_1 - \sigma_2 \quad , \quad \sigma = \frac{\bar{\sigma}}{3} + \sigma_2 \tag{4.3.1}$$

to represent the above locus in the σ-$\bar{\sigma}$ plane. In the case of true triaxial tests this locus is obtained straightforwardly from the data. Let

$$X(\sigma,\bar{\sigma}) = 0 \tag{4.3.2}$$

be the equation of the compressibility/dilatancy boundary which can be obtained by approximating the data with a simple empirical expression. Sometimes the equation of this locus is written in the form

$$X(\sigma,\bar{\sigma}) := -\left(\frac{\bar{\sigma}}{\sigma_\star}\right)^{\frac{m}{n}} + f\frac{\sigma}{\sigma_\star} = 0 \tag{4.3.3}$$

with m, n and f positive constants, and $\sigma_\star = 1$ (the unit stress, here $\sigma_\star = 1\,\text{MPa}$).

For example, for dry sand this relation is linear up to $\sigma = 700$ kPa:

$$X(\sigma,\bar{\sigma}) := -\frac{\bar{\sigma}}{\sigma_\star} + 2f\frac{\sigma}{\sigma_\star} = 0 \tag{4.3.4}$$

with $f = 0.696$ (Cristescu [1991a] on the basis of the data by Hettler $et\ al.$ [1984]), whereas for rock salt we write the equation of this boundary in the form

$$X(\sigma,\tau) := -\frac{\tau}{\sigma_\star} + f_1\left(\frac{\sigma}{\sigma_\star}\right)^2 + f_2\frac{\sigma}{\sigma_\star} \tag{4.3.5}$$

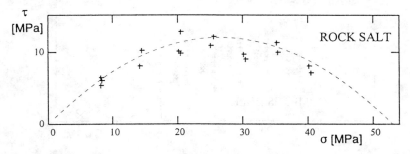

Figure 4.26 *Compressibility/dilatancy boundary for rock salt.*

with $f_1 = -0.01697$ and $f_2 = 0.8996$, on the basis of the data given in Section 2.1 (shown as dotted line in Figure 4.26). Therefore for dry sand the constitutive equation, where (4.3.4) is incorporated, can be used up to about $\sigma = 700$ kPa.

For alumina powder it is

$$X(\sigma,\tau) = 0.555\,\sigma - \tau. \tag{4.3.6}$$

In principle, the compressibility/dilatancy boundary must pass through the point $\sigma = \sigma_o$, $\tau = 0$, i.e., $X(\sigma_o, 0) = 0$, and by definition we will accept $X(\sigma, 0) = 0$ for $\sigma \geq \sigma_o$. Here σ_o is the smallest pressure closing all pores and microcracks. This requirement is necessary since when the hydrostatic pressure is increased we have to accept that the irreversible compressibility stops at that pressure which closes all microcracks and pores, and that no additional irreversible compressibility is possible for higher pressures. It may be that this limit pressure is difficult to be determined by tests or that we are not interested to extend the constitutive equation up to such high pressures. Thus if the experimental data available do not cover the whole pressure interval up to $\sigma = \sigma_o$, then we will be unable to determine the compressibility/dilatancy ("C/D" for short) boundary up to σ_o, but only a portion of it. Nevertheless this portion of the boundary may still be useful for the formulation of a model of limited validity (see Cristescu [1989a] for such partially determined boundary for a variety of rocks, Cristescu [1991a] for dry and saturated sand; see Figure 2.14 for various examples for rock salt). This limited validity means that the model can be used in a very precise delimited pressure interval only. Since for the sake of a coherent model, the irreversible compressibility must end somewhere at high pressures, the models considering unrestricted compressibility (for instance "cap" models) have a limited validity, i.e. for a certain limited interval of variation of σ, only (not for large values of σ).

In conjunction with the C/D boundary one can make further remarks. If we examine in the neighborhood of the C/D boundary (along which $\dot{\varepsilon}_V^I = 0$) the width of the domain where $\dot{\varepsilon}_V^I$ has a small (negligible) variation, we come to the conclusion shown in Figure 4.27 for rock

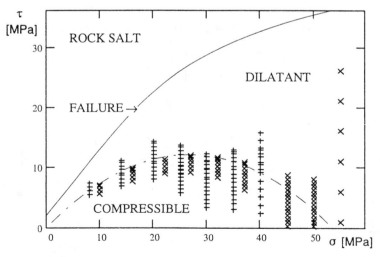

Figure 4.27 *Domains of compressibility and dilatancy, separated by a strip of incompressibility (+ + +: experimental data; ×× ×: model prediction).*

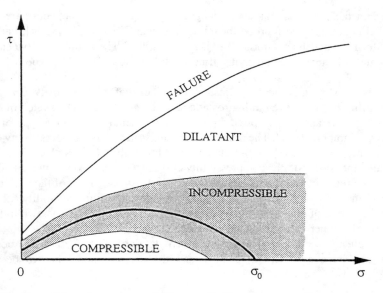

Figure 4.28 *Domains of compressibility and dilatancy, separated by a strip of incompressibility, convenient to be used in models.*

salt. Here by +++ lines mark the measured width of the domain where the irreversible volumetric strain has a negligible variation (the volumetric strain variation is smaller than the maximum reached on the boundary minus 0.4×10^{-4}). Also by xxx lines the same intervals are shown as predicted by the model described in the next chapter. Therefore we can consider that in quite a wide domain in the neighborhood of the C/D boundary the volumetric behavior is practically elastic only (no irreversible volumetric strain variation). This domain of incompressibility enlarges very much at very high pressures. Thus the C/D boundary seems not to be a surface but rather a domain of incompressibility (Figures 4.28 and 2.12). For convenience and in order to keep the model simple, one can represent this boundary as a surface in a stress space (or a curve in the σ-$\bar{\sigma}$ plane). Since the domain of incompressibility enlarges much towards high pressures, and since there the irreversible volumetric changes are small, it is quite difficult to determine experimentally the C/D boundary as a surface at high pressure.

There are rocks which are considered only slightly compressible. In this case the compressibility is sometimes disregarded and in the σ-$\bar{\sigma}$ plane only two domains are considered: incompressible and dilatant (see Chan *et al.* [1992], Aubertin *et al.* [1996a]). However, there are some other rocks, generally with high initial porosity, which are practically compressible only, up to (or very close to) the failure surface. In this case it is the domain of compressibility which covers practically the whole constitutive domain (domain in the σ-$\bar{\sigma}$ plane of all possible stress states with $\sigma < \sigma_o$). The "progressive increase in plasticity with confining stress" was reported for cemented shale by Leddra *et al.* [1992].

Sometimes, for rocks of high initial porosity, the pores may collapse under hydrostatic loading (Mowar *et al.* [1994]), sometimes showing significant time effects (Andersen *et al.* [1992]) with important consequences for petroleum geomechanics (Roegiers [1994]). As mentioned in the previous chapters, changes in pore and crack networks are related to electrical conductivity and fluid permeability (see also Glover *et al.* [1994], Fan and Jones [1994], Stormont and Fuenkajorn [1994], Peach [1996]).The coupling between pore fluid diffusion and dilatancy accompanying friction sliding was studied by Rudnicki and Chen [1988].

There is also another possible case. Some rocks of high porosity can change the compressibility behavior even in hydrostatic tests, by a collapse. The sketch in Figure 4.29 shows such a case and was inspired from the experimental data, for a type of limestone, obtained by Maranini [1997]. The dilatancy domain is small. The compressibility domain can be divided into three areas. In C_1 a volumetric compressibility takes place; in C_2 the pore collapse and the slope of the compressibility curves ($\sigma \sim \varepsilon_V$ slope) is much smaller, and the creep rates increase in creep tests, i.e., the viscoplastic deformation of the rock takes place easier in domain C_2. In C_3, after the collapse of pores, the compressibility continues with an increasing slope of the $\sigma \sim \varepsilon_V$ curves. How the various boundaries shown in Figure 4.29 extend towards higher values of τ and σ is not known.

The conclusion would be that for various rocks the C/D boundary and pore collapse boundary can have a variety of shapes, and the desired model has to be formulated accordingly.

4.4   FAILURE SURFACE

As already mentioned in conjunction with Figure 4.14, failure of geomaterials is a strongly time-dependent phenomenon. For many geomaterials failure is expected earlier if the loading is faster and will take place for much higher stresses and smaller strains (see Figures 4.3 - 4.6).

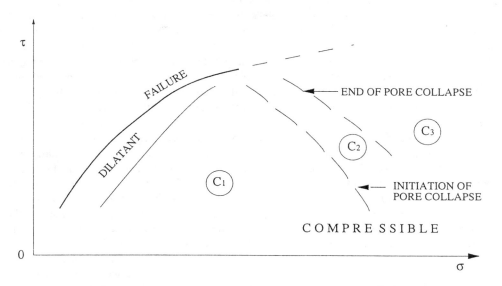

Figure 4.29   *Domains of compressibility, dilatancy and pore collapse for rocks of high porosities.*

Vice versa, very slow loadings, such as those taking place around underground excavations, say, may produce a failure after months or even many years as creep failure, for stresses much smaller than the one determined in conventional triaxial tests. It seems therefore that failure conditions which disregard the loading history would not have much meaning altogether. However, "short term" failure conditions, considered as a "limiting" condition, i.e., failure obtained in fast quasi-static loadings, have great importance in characterizing the property of the rock and for the formulation of any model. These are the failure conditions determined with standard triaxial tests on rocks in relative short time intervals as shown in Sections 1.3 and 2.1 (see also Paterson [1978]). How these failure conditions are supplemented with "creep failure" is a topic which will be discussed in Chapter 6.

When using standard triaxial tests one determines the short-term failure surface from the maxima of the stress-strain curves (see Figures 1.7, 1.8, 2.1, 2.2, 2.11b, 2.18). As an additional example Figure 4.30 shows for alumina powder the stress-strain curves obtained in classical triaxial tests with various confining pressures shown. Thus the **equation of the short term failure surfaces** is written in terms of stress invariants, if the rock is isotropic,

$$Y(\sigma, \tau) = 0 \ , \tag{4.4.1}$$

assuming that only these two stress invariants are involved. In the case of strong dependence on the third stress invariant, it can be incorporated in (4.4.1) as well. We have found it

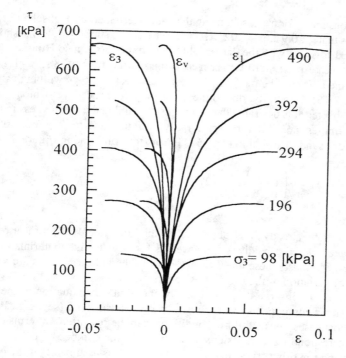

Figure 4.30 *The maxima on the stress-strain curves for alumina powder show where failure is taking place.*

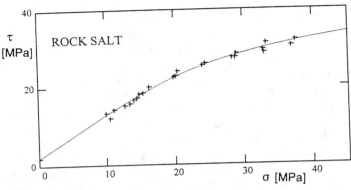

Figure 4.31 *Short term failure boundary for rock salt (the solid line is the prediction of (4.4.2))*

sometimes convenient to write this equation in the form

$$Y(\sigma,\tau) := -r\frac{\tau}{\sigma_*} - s\left(\frac{\tau}{\sigma_*}\right)^6 + \tau_o + \frac{\sigma}{\sigma_*} = 0 \qquad (4.4.2)$$

with r, s, τ_o positive constants. For rock salt, using the data determined by Hunsche (Sections 1.3 and 2.1), it follows that $r = 0.91$, $s = 1.025 \times 10^8$, $\tau_o = 1.82$; the surface (4.4.2) is shown as solid line in Figure 4.31 together with the data used to determine it (see also Hunsche [1992, 1993a] improved). From recent experimental data by Hunsche the improved values $r = 0.88$, $s = 1.1 \times 10^8$, $\tau_o = 2$ have been obtained. It must be mentioned that this is a rock salt type with high strength (S8 and S9 in Figure 2.15). Generally, due to the mathematical procedure which will be presented in the next chapter, it is desirable that the terms containing the mean stress in (4.4.2) be expressed in the simplest possible form (preferably even in linear form, if possible). In another example for dry sand the relation (4.4.1) is linear (the solid line in Figure 5.14):

$$Y(\sigma,\bar{\sigma}) := -\left(1 + \frac{\alpha}{3}\right)\frac{\bar{\sigma}}{\sigma_*} + (2f + \alpha)\frac{\sigma}{\sigma_*} = 0 \qquad (4.4.3)$$

with $\alpha = 0.984$ and $f = 0.696$ already given above, following Equation (4.3.4). Linear relations are found for many geomaterials, mainly soils, as well as for powder-like materials for small to moderate pressures. Obviously these linear expressions for the failure surfaces are valid for a restricted interval of variation of the mean stress only.

Finally we would like to observe that occasionally one writes the equation of the failure surface in terms of stress components even though the rock is isotropic (see, e.g., Paterson [1978]). In fact in most cases these failure conditions can be rewritten in terms of stress invariants, but the corresponding expressions may sometimes be cumbersome.

Remark concerning the third stress invariant. Above it was assumed that the equation of the short term failure surface depends on first and second stress invariants. Generally, the question arises if the third stress invariant has a significant influence on all the constitutive functions. As shown by many authors (see Hunsche [1992] and Sections 1.3 and 2.1) for geomaterials the failure surface depends on the third stress invariant as well. Moreover, Hunsche [1996] has found from tests that for rock salt the C/D boundary practically does not depend on the third stress invariant, even for various types of rock salt. Thus it seems that the third stress invariant influences the mechanical properties in the neighborhood of the failure surface only. Therefore, in most of the constitutive domain the influence of the third invariant can be disregarded, as long as the stress state is not close to the failure surface. If for some geomaterials the influence of the third invariant is felt in most of the constitutive domain, and the experimental data are accurate and complete, the influence of this invariant can and has to be considered in the determination of the constitutive functions.

4.5   GENERALIZATION OF THE MODEL FOR FINITE STRAINS

The generalization of the model (4.2.17) for large deformations is due to Cleja-Tigoiu [1991]. The formulation is with respect to the configuration at time t. It is possible that the rock has previously been deformed under possible unknown large deformation. The small relative deformation with respect to a fixed configuration at time t may be superposed on a finite deformation (for these concepts see Truesdell and Noll [1965], Malvern [1969]).

It is assumed that the deformations are finite at any time and that the elasto-viscoplastic behavior is written with respect to the configuration at an arbitrary fixed time t, as reference configuration. The generalized constitutive equation is written as

$$D = \frac{1}{2G}\check{T} + \frac{1}{3}\left(\frac{1}{3K} - \frac{1}{2G}\right) tr\check{T}\,\mathbf{1} +$$

$$\left\langle 1 - \frac{W(t)}{H(j(T))}\right\rangle \left[\psi_o(j(T))\,\mathbf{1} + \psi_1(j(T))T' + \psi_2(j(T))(T')^2\right],$$

(4.5.1)

where the constitutive parameters and functions are defined with respect to the configuration at time t, thus they have a "relative" meaning (see Section 7.2). Here D is the rate of deformation tensor $D_{ij} = (1/2)(v_{i,j} + v_{j,i})$, T is the relative Cauchy tensor, and $\check{T}$ is a stress flux. In Cleja-Tigoiu [1991] this stress flux is taken in the Truesdell sense, i.e., $\check{T} = (div\,v)T + \dot{T} + LT - TL^T$, with L the spatial gradient of the velocity $L_{ij} = v_{i,j}$. Further in (4.5.1) ψ_i are three constitutive functions defining the orientation of D^i and which depend on the set of invariants of T, denoted $j(T)$. $\dot{W}(t)$ is the irreversible stress power, i.e., $\dot{W}(t) = T(t) \cdot D^i(t)$. More details and several general variants of the constitutive equation can be found in the mentioned paper by Cleja-Tigoiu [1991].

Generalization of the steady-state creep term is straightforward.

4.6   CONCLUSIONS AND HISTORICAL NOTES

The main concepts incorporated in the constitutive equation (4.2.17) are:

a) The **compressibility/dilatancy boundary**, which is the locus in the $\sigma - \tau$ plane

$$X(\sigma,\tau) = 0 \qquad\qquad (4.6.1)$$

making precise if for a certain stress state considered the rock becomes compressible $X(\sigma,\tau) > 0$ or dilatant $X(\sigma,\tau) < 0$ (see the dash-dot line in Figure 4.32).

b) The short term **failure surface**

$$Y(\sigma,\tau) = 0 \qquad\qquad (4.6.2)$$

which is obtained straightforwardly from tests as the maxima on the stress-strain curves obtained in triaxial tests (solid line in Figure 4.32).

c) The initial **yield surface** (shown as an interrupted line in Figure 4.32) is obtained assuming that the initial primary stress state σ^P (shown as a rhombus in Figure 4.32) is a point located on the initial stabilization boundary $H(\sigma^P) = W^P$ for transient creep. However, the constitutive equation also describes steady state creep, which may be due either to a stress variation produced by an excavation, or to a slow tectonic motion produced by the primary stress state. For this reason steady-state creep in Figure 4.32 produces volumetric dilatancy for stress states corresponding to points belonging to the domain labelled [DE], while from the point of view of the transient creep in the same domain an "unloading" is taking place. Similarly [CE] is an "elastic" domain for transient creep, but an "incompressible" one for steady-state creep. If no excavation is performed, a primary stress state represented by a point located in a domain [C] or [CE] will produce a steady-state creep (a slow tectonic motion, or land slides, for example) with no irreversible volumetric changes, while a primary stress state

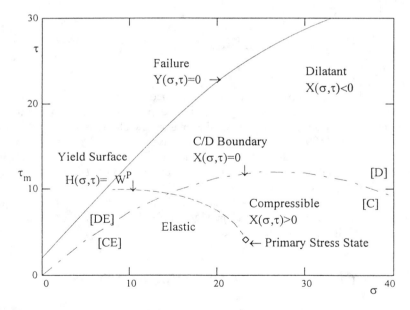

Figure 4.32 *Possible stress states around a cavern. The interrupted line represents the initial yield surface and the rhombus a possible primary stress belonging to this surface. The dash-dot line is the compressibility/dilatancy boundary.*

located in the domains [D] or [DE] will result in a steady-state creep producing irreversible dilatancy besides a change in shape.

d) The response of the rock to any sudden loading or unloading is **elastic**. As already mentioned "loading" means a stress change $\sigma(t_o) \rightarrow \sigma(t)$, with $\sigma(t) \neq \sigma(t_o)$ for which $H(\sigma(t)) > H(\sigma(t_o))$, while "unloading" means $H(\sigma(t)) < H(\sigma(t_o))$. "Neutral" loading, i.e., $H(\sigma(t)) = H(\sigma(t_o))$, does not play an important role. If the yield function $H(\sigma)$ is determined from triaxial tests done in the laboratory, W^P follows from $H(\sigma^P) = W^P$ if the initial primary stress state is a state on the stabilization boundary.

For the rock considered and for the corresponding depth (i.e., primary stresses) one is plotting the "map" of Figure 4.32. If the elastic stress state just after an excavation can be found in analytical form, then this stress state can be superposed on this map to find out where around the excavation the rock becomes dilatant, where compressible, where failure will take place and where evolutive damage is to be expected. Afterwards, on the same map one can follow the stress evolution during creep, while the elastic (instantaneous) solution will be used as "initial" data for the creep process.

A general associated constitutive equation of the form (4.2.17) describing both irreversible dilatancy and compressibility during transient creep was first presented at the International Symposium "Plasticity Today" held in Udine in 1983. A summary was published by Cristescu [1985d, 1985e] and an extended version by Cristescu [1987]. Full details can be found also in the book by Cristescu [1989a], where the procedures to be followed in order to determine the constitutive function H were also given (see also Cristescu [1994b]). In these papers the concept of compressibility/dilatancy boundary was introduced. This concept plays a fundamental role in the model. In the book by Cristescu [1989a] one can find examples for a variety of rocks: granite, andesite, several kinds of coal, cement concrete (mortar). The generalization of the model for the non-associated case and how the viscoplastic potential has to be determined from tests was given in Cristescu [1991a], where examples for dry and saturated sand were given. Additional details can be found in Cristescu [1994b]. The generalization of the model for finite strains is due to Cleja-Tigoiu [1991]. Finally, the procedure to determine the potential for the steady-state creep so that dilatancy could also be described was given by Cristescu [1993a]. Examples have been given for rocksalt (Cristescu [1991b, 1992, 1994b], Cristescu and Hunsche [1992, 1993a,b, 1996]), for bituminous concrete (Cristescu and Florea [1992]), for chalk (Dahou [1994], Dahou et al. [1995]), for limestone and sand (Brignoli and Sartori [1993]), for limestone (Maranini [1997]), for anisotropic shale (Cristescu and Cazacu [1995]), and for ceramic powder (Jin et al. [1996a, 1997], Cristescu et al. [1997], Cristescu and Cazacu [1997], Cazacu et al. [1997]). The model (4.2.5'), where instead of the concept of viscoplastic potential the concept of "viscoplastic strain rate orientation tensor" is used, is due to Cazacu [1995], Cazacu et al. [1996] (see also Cristescu [1996b]).

5 General Constitutive Equation

5.1 GENERAL FORMULATION

In the previous chapter we presented the experimental evidence which leads to the formulation of the three-dimensional constitutive equation in the form (4.2.17)

$$\dot{\varepsilon} = \frac{\dot{\sigma}}{2G} + \left(\frac{1}{3K} - \frac{1}{2G} \right) \dot{\sigma} 1 + k_T \left\langle 1 - \frac{W(t)}{H(\sigma,\tau)} \right\rangle \frac{\partial F}{\partial \sigma} + k_S \frac{\partial S}{\partial \sigma} . \qquad (5.1.1)$$

Here G and K are shear and bulk moduli respectively (generally not constant), $H(\sigma,\tau)$ is the yield function, $F(\sigma,\tau)$ is the viscoplastic potential, $S(\sigma,\tau)$ is the potential for the steady-state creep, k_T and k_S are the two corresponding viscosity coefficients, and $W(t)$ is the irreversible stress work per unit volume.

In order to determine explicitly all the constitutive functions and coefficients involved in (5.1.1), one starts with the elastic parameters G and K. The procedure to determine them from experimental data was described in the previous chapter. Generally both G and K are not constant. For most geomaterials they vary smoothly with the stress state or with the damage evolution of the rock. Afterwards one has to determine the yield function $H(\sigma,\tau)$ by computing from triaxial tests $W(t)$, since $H(\sigma,\tau) = W(t)$ is the equation of stabilization boundary. If at this stage one assumes that the constitutive equation is associated (i.e. associated to a certain yield function), then the next parameter to be determined is k_T. This parameter can be obtained from any kind of creep test or even triaxial test in which time is also recorded. If the constitutive law is nonassociated, or we expect it to be so, the next step is to determine the viscoplastic potential $F(\sigma,\tau)$ from tests. If $F(\sigma,\tau)$ cannot be determined, one can replace the term $\partial F/\partial \sigma$ by an irreversible strain rate orientation tensor $N(\sigma,\tau)$ that is somehow easier to determine from data. Thus all the constitutive functions and parameters involved in the transient creep term would be determined.

If we intend to use the constitutive equation to describe creep lasting a very long time interval, then we have to determine the potential $S(\sigma,\tau)$, basically from creep tests.

Transient creep and steady-state creep are described in (5.1.1) by two additive terms, though from a physical point of view, a single term would be preferred (see Chapter 3). The reason is that in this way it is easier to determine the constitutive coefficients from the data. Moreover, the main reason is that while in transient creep we have to describe both volumetric compressibility and/or dilatancy, in steady-state creep one has to describe either incompressibility and/or dilatancy. Thus, from the point of view of volumetric creep, transient creep and steady-state creep have to describe in a distinct way the volumetric creep. That is why it is quite difficult to describe transient and steady-state creep with a single term in the constitutive equation.

In this chapter it will be shown how the constitutive functions involved in (5.1.1) can be determined from the data. The procedures that have been used are presented. These procedures are obviously not unique. The authors' intention is to suggest some procedures, and to encourage other researchers to find out other possible variants. The diversity of mechanical properties exhibited by geomaterials imposes a variety of procedures to determine the constitutive functions from the data, and, maybe, even a variety of constitutive equations.

In brief, that is what follows in this chapter (see also Cristescu [1987, 1989a, 1991a, 1994b, 1996b]).

5.2 TRANSIENT CREEP

In the constitutive equation (5.1.1), the irreversible rate-of-strain component describing transient creep is (see (4.2.5))

$$\dot{\varepsilon}_T^I = k_T \left\langle 1 - \frac{W(t)}{H(\sigma,\tau)} \right\rangle \frac{\partial F}{\partial \sigma} \quad . \tag{5.2.1}$$

Thus, in order to formulate such a constitutive equation for a specific geomaterial, one has to determine from tests all the constitutive functions involved. After determining the elastic parameters the first to be determined from the data is the yield function $H(\sigma,\tau)$ and afterwards the viscoplastic potential $F(\sigma,\tau)$. The constitutive law (5.2.1) describes both transient volumetric creep and changes in shape. Volumetric changes could be either irreversible compressibility or dilatancy.

5.2.1 The yield function

When a rock is loaded (for instance, an excavation is performed at time t_o) then, after excavation, at time $t > t_o$, in most domains around the excavation it follows that $H(\sigma(t),\tau(t)) > W(t_o)$ and the irreversible strain increases according to (5.2.1), since the

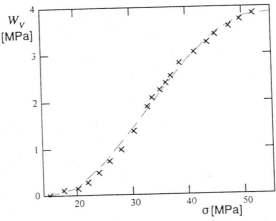

Figure 5.1 Variation of W_V with σ for chalk in hydrostatic test.

bracket $\langle\rangle \neq 0$. If stress is kept constant as in creep tests, $W(t)$ is increasing in time and $W(t) \to H(\sigma,\tau)$. Thus equilibrium is reached when $W(t) = H(\sigma,\tau)$. If the stress is slightly varying the story is practically the same. This suggests the procedure to determine $H(\sigma,\tau)$ in creep tests of very long duration, performed in triaxial test devices, for various confining pressures and octahedral shear stresses. Though the time intervals considered are long, they are not so long as to involve steady-state creep as well; even if steady-state creep starts being involved, we consider here long time intervals during which steady-state creep is still negligible with respect to the transient one. If tests of long duration are not available, any triaxial tests will suffice, but the model will, more or less, underestimate the magnitude of the transient strain, but will still describe the main features: dilatancy, compressibility, creep failure, short-term failure, etc.

In triaxial tests, as a rule, one subjects the specimen to a hydrostatic loading followed by a deviatoric one, either in the Kármán sense, or in the true triaxial sense. Let us recall that in the Kármán type of triaxial tests, the so-called "deviatoric" part is usually performed in compression tests along straight lines $\sqrt{2}\,\sigma - \tau = $ constant (see Figure 4.21), and thus they are not really proper deviatoric tests. For this reason, and for the physical reasons related to the finding of the behavior of rock subjected to a hydrostatic pressure or/and shear stress, we look for a yield function of the form

$$H(\sigma,\tau) := H_H(\sigma) + H_D(\sigma,\tau) \qquad (5.2.2)$$

where the two subscripts come from "hydrostatic" and "deviatoric". Obviously H_D must satisfy the condition $H_D(\sigma,0) = 0$, if the rock is isotropic (i.e. no change in shape is possible under a hydrostatic stress state). The function $H(\sigma)$ is obviously not unique. The determination of H may depend on the loading path. Here we follow the "triaxial" loading path, which is very close to the loading path followed when a borehole is excavated (see Chapters 8 and 9).

$H_H(\sigma)$ is determined in hydrostatic tests, where $\sigma_1 = \sigma_2 = \sigma_3$ are increased in a certain chosen time interval $t \in (0, T_H]$. During this stage of the test we determine from triaxial data

$$W_V(T_H) \equiv W_H(T_H) = \int_0^{T_H} \sigma(t)\dot{\varepsilon}_V(t)\,dt - \frac{\sigma^2(T_H) - \sigma^2(0)}{2K}, \qquad (5.2.3)$$

where the last term corresponds to the elastic component of the volumetric strain. The results are used to determine the stabilization boundary

$$H_H(\sigma) = W_H(t) \qquad (5.2.4)$$

for hydrostatic stress states when time increases very much. For instance Figure 5.1 shows the variation of W_V with σ for chalk, obtained using the data by Shao and Henry [1990]. As a general trend, initial concavity is always directed upwards and afterwards it changes for higher pressures σ. If the pressure is increased very much all the microcracks and pores are closed and no further irreversible volumetric compaction is possible. Let us denote by σ_o the level of the pressure that closes all microcracks and pores (see Figure 4.14, where this limit pressure is marked). For $\sigma > \sigma_o$ the slope of the $W_V - \sigma$ curve is horizontal since the irreversible volumetric strain is no longer varying. Quite often, for practical purposes, the model we would like to formulate is needed for much smaller pressures, and thus this limit pressure remains undetermined. The model can be formulated even if this limit pressure is not determined.

Another example for a very hard artificial rock (mortar) in uniaxial strain confined tests is shown in Figure 5.2 (Cristescu [1989a] using the data obtained by Constantinescu [1981]). This time the pressures are very high, but the trend is the same.

For the powder-like materials, including soils, the trend is again the same. As an example Figure 5.3 shows the W_v-σ curve for alumina paste (powder mixed with 50% per volume with water) (Jin and Cristescu [1996], Jin et al. [1996a], Cazacu et al. [1997]). For several other examples see Cristescu [1989a, 1994a].

The data shown in Figures 5.1 - 5.3, or similar ones, can be approximated by simple empirical functions. For instance, for rock salt one has obtained (Cristescu [1994a,b])

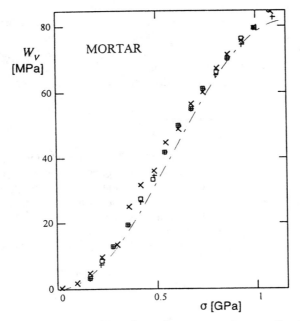

Figure 5.2 *Variation of irreversible volumetric stress work per unit volume with pressure for mortar.*

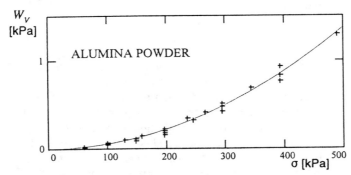

Figure 5.3 *Variation of W_v with σ for alumina powder.*

$$H_H(\sigma) := \begin{cases} h_o \sin\left(\omega \dfrac{\sigma}{\sigma_\star} + \varphi\right) + h_1 & if \quad \sigma \leq \sigma_o \ , \\ h_o + h_1 & if \quad \sigma \geq \sigma_o \ , \end{cases} \qquad (5.2.5)$$

with $h_o = 0.116$ MPa, $h_1 = 0.103$ MPa, $\omega = 2.91°$, $\varphi = -64°$, $\sigma_o = 53$ MPa, $\sigma_\star = 1$ MPa. A similar expression has been used for mortar:

$$H_H(\sigma) := c\left(1 - \cos\omega \dfrac{\sigma}{\sigma_\star}\right) \ , \qquad (5.2.6)$$

with $c = 0.04$ GPa, $\omega = 156.06°$, $\sigma_\star = 1$ GPa.

For chalk the expression found to approximate the data reasonably well is

$$H_H(\sigma) := \begin{cases} h_o \sin[\omega(\sigma - s_1) + \varphi] + h_1 & if \quad \sigma < 55 MPa \ , \\ h_o + h_1 & if \quad \sigma \geq 55 MPa \ , \end{cases} \qquad (5.2.7)$$

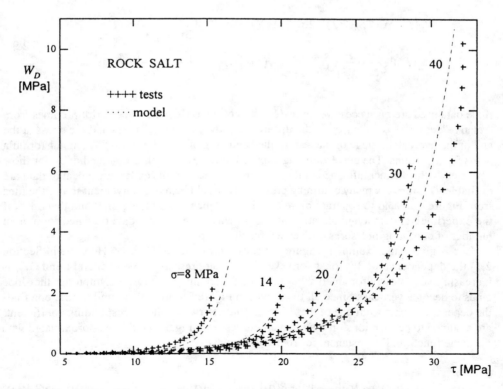

Figure 5.4 *Irreversible stress work per unit volume as function of* τ *during deviatoric true triaxial test, for various constant pressures.*

with $h_o = 2.31$ Mpa, $h_1 = 2.35$ MPa, $\omega = 4.641$ MPa^{-1}, $s_1 = 16$ MPa, $\varphi = -90°$.

These functions could also well be approximated with a polynomial (Dahou [1994] for shale, Jin and Cristescu [1996] for alumina, etc.) with the additional condition to be imposed that for pressures surpassing σ_o, the slope must be horizontal.

The function $H_D(\sigma, \tau)$ is to be determined in the second stage of the triaxial tests, i.e. in the deviatoric one. If the test is a true triaxial test, then during the deviatoric phase one usually has $\sigma =$ constant and it is τ which increases starting from zero. If the triaxial test is a conventional compressional (Kármán) test, then during this test the confining pressure $\sigma_2 = \sigma - \tau/\sqrt{2} =$ constant; though σ is varying during this test, for convenience, such a test will still be called a "deviatoric" test. If the test is performed with a true triaxial apparatus, then the formula to be used for the determination of W_D when τ is steadily increased is

$$W_D(T) = \int_{T_H}^T \sigma'(t) \cdot \dot{\varepsilon}\,(t)\,dt - \frac{\sigma'(T) \cdot \sigma'(T)}{4G} + \sigma(T_H) \int_{T_H}^T \dot{\varepsilon}_v(t)\,dt \tag{5.2.8}$$

where the last term corresponds to the volumetric part (we recall that in true triaxial deviatoric tests the mean stress is constant, but the volumetric strain is not). If the deviatoric test is performed with a classic apparatus where $\sigma_2 = \sigma_3 =$ constant $= \sigma - \tau/\sqrt{2}$, then the formula is more involved (see Cristescu [1989a], where all the details are given):

$$W_D(T) = \sigma_1(T_H)\varepsilon_1^R(T) + 2\sigma_2(T_H)\varepsilon_2^R(T) - \sigma_1(T_H)\left[\frac{1}{3G} + \frac{1}{9K}\right]\sigma_1^R(T)$$

$$- 2\sigma_2(T_H)\left[\frac{1}{9K} - \frac{1}{6G}\right]\sigma_1^R(T) + \int_{T_H}^T \sigma_1^R(T)\dot{\varepsilon}_1^R(t)\,dt - \frac{1}{2}\left[\frac{1}{3G} + \frac{1}{9K}\right][\sigma_1^R(T)]^2 \tag{5.2.9}$$

if the data used are on successive lines $\sqrt{2}\sigma - \tau =$ constant. Here the superscript R comes from "relative": $\sigma^R = \sigma - \sigma^H$, i.e., σ^R is the difference between the total stress and the stress at the end of the hydrostatic stage of the test (or the beginning of deviatoric test). A similar formula holds for the strains. The corresponding expression for $H_D(\sigma, \tau)$ thus obtained does not show very clearly the distinct influence of the pressure and that of τ in H_D. If a great deal of data are available, i.e., for very many confining pressures, it would be more convenient to use the idea from Figure 4.20 and to separate more clearly the dependence of H_D on σ and that on τ, if the experimental data available allow it. Very many experimental data obtained for a great number of confining pressures are needed for this purpose.

Let us give some examples. Figure 5.4 shows for rock salt (tests by Hunsche in Section 2.1) the dependence of W_D on τ for several constant pressures. As a general trend W_D is increasing very slowly for small τ but towards failure this increase is asymptotic (the slope tends to become vertical). Since all the curves in Figure 5.4 are rather similar, first one finds the dependence on τ from one of these curves, and afterwards the corresponding coefficients are assumed to be functions on σ so that all curves from Fig.5.4 could be approximated with the same function. For instance, for rock salt we have

$$H_D(\sigma, \tau) := A(\sigma)\left(\frac{\tau}{\sigma_*}\right)^{14} + B(\sigma)\left(\frac{\tau}{\sigma_*}\right)^3 + C(\sigma)\left(\frac{\tau}{\sigma_*}\right), \tag{5.2.10a}$$

$$A(\sigma):=a_1 + \cfrac{a_2}{\left(\dfrac{\sigma}{\sigma_\star}\right)^6} \quad , \quad B(\sigma):=b_1\dfrac{\sigma}{\sigma_\star}+b_2 \quad , \quad C(\sigma):=\cfrac{c_1}{\left(\dfrac{\sigma}{\sigma_\star}\right)^3 + c_3}+c_2 \quad , \tag{5.2.10b}$$

with $a_1 = 7\times10^{-21}\,\text{MPa}$, $a_2 = 6.73\times10^{-12}\,\text{MPa}$, $b_1 = 1.57\times10^{-6}\,\text{MPa}$, $b_2 = 1.7\times10^{-5}$ MPa, $c_1 = 26.12\,\text{MPa}$, $c_2 = -0.00159$ MPa, $c_3 = 3134$ and $\sigma_\star = 1\,\text{MPa}$. The prediction of (5.2.10a,b) is shown as dotted lines.

Another example is shown in Figure 5.5 for dry sand (Cristescu [1991a]) using the data by Hettler *et al.* [1984]. The tests were performed with a classical device; thus the curves shown correspond to tests with different constant confining pressures, while the variation of W_D is taken with respect to the stress-difference $\sigma_1 - \sigma_2$, with σ_2 the confining pressure. The expression found for H_D is

$$H_D(\sigma,\bar{\sigma}):=\left[a_o + a_1\left(\sigma - \dfrac{\bar{\sigma}}{3}\right)\right]\left(\cfrac{\bar{\sigma}}{\sigma - \dfrac{\bar{\sigma}}{3}}\right)^6 + b_o\bar{\sigma} \tag{5.2.11}$$

where $a_o = 3\times10^{-4}$ kPa, $a_1 = 2.1\times10^{-5}$, $b_o = 3.4\times10^{-3}$. The predictions of (5.2.11) are shown in Figure 5.5 as dashed lines.

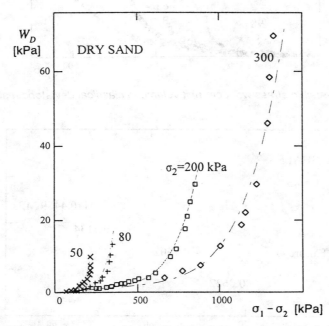

Figure 5.5 *Irreversible stress work per unit volume W_D as function of stress difference in Kármán triaxial tests, for several constant confining pressures.*

A similar behavior is obtained for ceramic powder. Figure 5.6 shows the irreversible stress work (see (4.2.7)) per unit volume for alumina powder (powder mixed with 50% water per volume) obtained in classical triaxial tests, for various confining pressures. This result suggests that the same type of general 3-dimensional models can be used for ceramic powder as for geomaterials.

It is revealing to represent the surfaces H = constant in a σ-τ plane. For instance, Figure 5.7 shows several surfaces H = constant for shale (Cristescu and Cazacu [1995]). Since this rock is anisotropic, these surfaces are depending on the orientation of the isotropy axes with respect to the specimen axes, or to the stress field applied (or *in situ*). Figure 5.7 shows the shape of these surfaces for the orientation of this axis of $\theta = 45°$ only. For more details concerning the shape of the yield surfaces for this orthotropic rock see Cristescu and Cazacu [1995].

The shape of the surfaces H = constant for various rocks can be quite different, mainly if

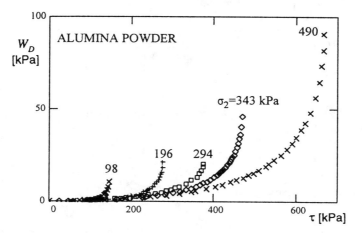

Figure 5.6 *Irreversible stress work per unit volume in classical deviatoric tests.*

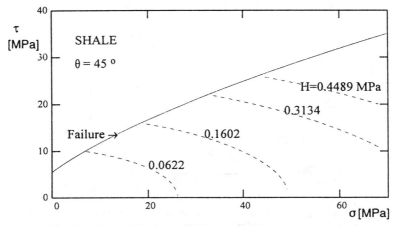

Figure 5.7 *Yield surfaces H = constant for shale for the direction* $\theta = 45°$.

the rock is compressible only, as in Figure 5.7, or is also dilatant (see Figures 5.8 and 5.9). For instance, Figure 5.8 shows the yield surfaces for alumina powder (alumina powder with 50% water per volume). The expression for the function $H(\sigma,\tau)$ will be given below. The C/D boundary was also plotted using the data. Since the C/D boundary does not coincide with the $\partial H/\partial\sigma = 0$, it is obvious that one cannot make the assumption of associativeness and that, for such a material only a nonassociated equation is appropriate. Let us recall that $\partial H/\partial\sigma = 0$ would be the C/D boundary if the constitutive law would be associated. The failure surface (solid line) was also plotted using the data. A conclusion of significant practical importance for compaction of powder is that by adding a shear stress to a hydrostatic pressure one can obtain **an additional compaction**. Comparing Figures 5.8 and 5.7 one can see the difference in the shape of these surfaces for materials which are compressible only (Figure 5.7) and those which are compressible/dilatant (Figure 5.8). From Figure 5.8 one cannot even guess where the limit pressure σ_o which closes all pores would be, since in the experiments performed quite small pressures only were used.

Figure 5.9 shows the yield surfaces for rock salt. This time these surfaces were determined up to high pressures, where all microcracks and pores are closed. Comparing these surfaces with those given in Figure 5.8 it is interesting to note a change in curvature at high pressures: these surfaces become concave. This seems to be the case for all materials that are compressible for some stress states, but dilatant for some other stress states. Also the surface $\partial H/\partial\sigma = 0$ is quite far from the C/D boundary shown as a dash-dot line and plotted straightforward from the data. The assumption of associativeness will be discussed below after determining also the

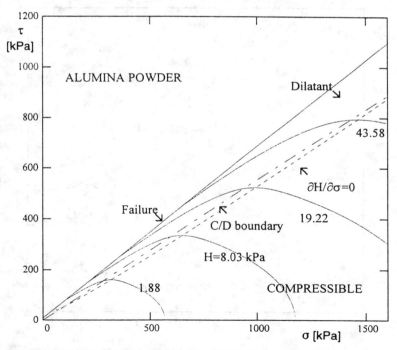

Figure 5.8 *Several yield surfaces for alumina powder. The C/D boundary (dash-dot line) is plotted according to the data and does not coincide with $\partial H/\partial\sigma = 0$, since the material shows a nonassociated behavior.*

expression of the function $F(\sigma,\tau)$, but comparing the boundary $\partial H/\partial\sigma = 0$ with the C/D boundary determined from tests (dash-dot line) one can conclude that, generally, the nonassociated model is to be used for rock salt (see also the experiments on chalk reported by Loe *et al.* [1992]).

Figures 5.7 - 5.9 show how different the yield surfaces can be for various materials.

The same kind of model (5.2.1) has also been used by Brignoli and Sartori [1993]. However, instead of finding elementary empirical formulae for $H(\sigma,\tau)$, they have stored the triaxial test data in a computer and have approximated locally the yield function by spline functions. The results obtained are very good, though they have been applied to a highly porous rock (porous limestone with 34% initial porosity, which is compressible/dilatant for small confining pressures, but compressible only for high confining pressures).

The procedure described above to determine the yield function $H(\sigma,\tau)$ has assumed that H depends on two stress invariants only: the mean stress σ and the octahedral shear stress τ. However, very many experimental data have shown that the failure surface depends also on the Lode parameter

$$m = \frac{2\sigma_2 - \sigma_1 - \sigma_3}{\sigma_1 - \sigma_3} = \frac{3S_2}{S_1 - S_3}\ , \qquad \sigma_1 \ge \sigma_2 \ge \sigma_3 \qquad (5.2.12)$$

(see Section 2.1, Hambley and Fordham [1989], Hunsche [1990, 1991, 1992, 1993a, 1994a],

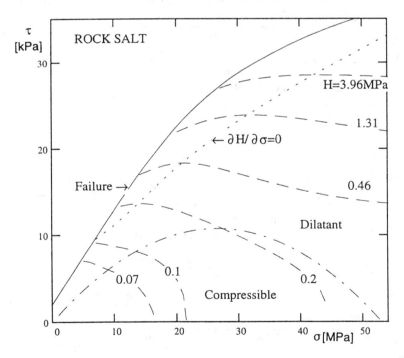

Figure 5.9 *Several yield surfaces for rock salt. The C/D boundary (dash-dot line) is quite distinct from the $\partial H/\partial\sigma = 0$ boundary, showing that a nonassociated constitutive equation is to be formulated.*

Chia and Desai [1994]), i.e., on the third invariant of the stress deviator, or more precisely on the ratio

$$\sqrt[3]{III_T} / \sqrt{II_T} .$$ (5.2.13)

If the failure surface depends on this ratio, one can expect that the yield function also depends on this ratio. After finding the expression of the yield function from compression data only, say, one can find out the generalized form of this function for tension tests as well, by using for instance the idea suggested by Desai and Zhang [1987] to multiply the yield function by a function depending on the expression (5.2.13), and on mean stress. An extensive study concerning the shape of the failure surface for soils and its dependence of the third stress invariant is due to Lade [1993].

5.2.2   The viscoplastic potential

After determining the yield function one can write down the complete constitutive equation (5.2.1) if one assumes that the constitutive law is associated, i.e. if $H = F$. That is a good assumption for most metals, but geomaterials may be expected to behave in some other way. If the law is associated, the only parameter still to be determined is the viscosity coefficient k_T for transient creep. This coefficient can be determined very easily from creep tests, or from any other kind of tests in which time is also recorded. For instance, if a hydrostatic compression test is performed and if the time interval between two successive readings is $t - t_o$, then k_T can be determined using a formula of the form (which is obtained straightforward from (5.2.1))

$$k_T = \frac{\Delta \varepsilon_v - \dfrac{\Delta \sigma}{K}}{(t - t_o)\left(1 - \dfrac{W(t_o)}{H(\sigma(t))}\right) \dfrac{\partial H_H}{\partial \sigma}}$$ (5.2.14)

in conjunction with the data. Here Δ is the increment of the corresponding variable during the time interval $t - t_o$ and $H_H(\sigma) = H(\sigma, 0)$. If any kind of creep tests is available, then one can use a formula of the form

$$k_T = -\ln\left|1 - \frac{W(t_i)}{H(\sigma)}\right|\Bigg|_{t_i}^{t_f} \frac{H(\sigma)}{\dfrac{\partial F}{\partial \sigma} \cdot \sigma} \frac{1}{t_f - t_i} ,$$ (5.2.15)

where t_i is the "initial" moment of creep test, t_f the (arbitrary) "final" moment, $\sigma(t_i) =$ Constant $= \sigma(t_f)$ is constant during the time interval $t \in (t_i, t_f]$ and F is to be replaced by H if the associativeness assumption is accepted. Formula (5.2.15) has a more general validity since it can be used even if F does not coincide with H.

The parameter k_T is measured in s^{-1}. A proper "viscosity coefficient" could be considered to be E/k_T, which is measured in Poise. Also, a correct determination of k_T can be done in long-term creep tests only. Generally, k_T is to be expected to be somewhat variable (possibly smoothly depending on damage) but this subject will not be discussed further since k_T can be determined simultaneously together with F.

It is quite difficult to decide *a priori* if we have to make the associativeness assumption for a certain material. One can compare how far apart are the boundary $\partial H/\partial \sigma = 0$ and the C/D boundary obtained from tests, as we have done above. It may be that these two boundaries are not be too far apart for some rocks in some areas, but that somewhere in the constitutive domain the surfaces H=constant and F=constant would still be quite distinct. That is the reason why we will use a straightforward method to determine the viscoplastic potential. It is only afterwards that we will compare the two functions F and H in order to see if, and where, they are far apart or close together. Also we will compare with the data the prediction of the associated and of the nonassociated constitutive equations in order to decide if for a certain kind of geomaterial, there are circumstances when the associativeness assumption is reasonable (see also Drucker [1988], Lade and Kim [1988], Reed and Cassie [1988], Lade and Pradel [1990]).

If we accept the concept of the viscoplastic potential, the irreversible rate-of-strain component for transient creep is

$$\dot{\varepsilon}_T^I = k_T \left\langle 1 - \frac{W(t)}{H(\sigma)} \right\rangle \frac{\partial F}{\partial \sigma} \quad . \tag{5.2.16}$$

In (5.2.16) the function H depends on the stress invariants σ and $\bar{\sigma}$ (or τ), only. To keep the formulae simpler we have written $H(\sigma)$ instead of $H(\sigma,\bar{\sigma})$. Further, if F depends on the two invariants σ and $\bar{\sigma}$ only, then

$$\frac{\partial F}{\partial \sigma} = \frac{\partial F}{\partial \sigma} \frac{1}{3} \mathbf{1} + \frac{\partial F}{\partial \bar{\sigma}} \frac{\partial \bar{\sigma}}{\partial \sigma} \quad . \tag{5.2.17}$$

Thus, in order to determine the viscoplastic potential from the data, we have to determine the two derivatives $\partial F/\partial \sigma$ and $\partial F/\partial \bar{\sigma}$.

For $H(\sigma,\tau) > W(t)$ one can obtain from (5.2.16) the following fundamental formula

$$k_T \frac{\partial F}{\partial \sigma} = \frac{\dot{\varepsilon}_V^I}{\left\langle 1 - \frac{W(t)}{H(\sigma)} \right\rangle} \quad , \tag{5.2.18}$$

$$k_T \frac{\partial F}{\partial \bar{\sigma}} = \frac{\sqrt{2}}{3} \frac{1}{\left\langle 1 - \frac{W(t)}{H(\sigma)} \right\rangle} \sqrt{\left(\dot{\varepsilon}_1^I - \dot{\varepsilon}_2^I\right)^2 + \left(\dot{\varepsilon}_2^I - \dot{\varepsilon}_3^I\right)^2 + \left(\dot{\varepsilon}_3^I - \dot{\varepsilon}_1^I\right)^2}$$

$$\left(= k_T \frac{\sqrt{2}}{3} \frac{\partial F}{\partial \tau} \right) \quad , \tag{5.2.19}$$

where for simplicity the subscript T was dropped from the rate-of-strain components. In the case of classical triaxial tests the last formula is replaced by

$$k_T \frac{\partial F}{\partial \bar{\sigma}} = \frac{2}{3} \frac{|\dot{\varepsilon}_1^I - \dot{\varepsilon}_2^I|}{\left\langle 1 - \frac{W(t)}{H(\sigma)} \right\rangle}.$$ (5.2.20)

The determination of $\partial F/\partial \sigma$ from the data is done in three stages. First we determine $\partial F/\partial \sigma$ for $\bar{\sigma} = 0$ using formula (5.2.18) and the data for the estimation of $\dot{\varepsilon}_V^I$. We obtain

$$k_T \frac{\partial F}{\partial \sigma}\bigg|_{\bar{\sigma}=0} = k_T \frac{\partial F_H}{\partial \sigma} = \varphi(\sigma)$$ (5.2.21)

for example. The function $\varphi(\sigma)$ is determined in the stress interval $0 \le \sigma \le \sigma_o$ and must possess the property $\varphi(\sigma_o) = 0$. By definition we take $\varphi(\sigma) = 0$ also for $\sigma \ge \sigma_o$, since for $\sigma \ge \sigma_o$ we have $\dot{\varepsilon}_V^I = 0$ (irreversible compressibility is no longer possible for $\sigma \ge \sigma_o$ since all microcracks are already closed for these pressures).

For instance, the results shown in Figure 5.10 are obtained for rock salt (Cristescu and Hunsche [1992]). It was found that the function

$$\varphi(\sigma) := \begin{cases} q\,\sigma(\sigma - q_2)^2 & \text{if} \quad \sigma \le \sigma_o \\ 0 & \text{if} \quad \sigma \ge \sigma_o \end{cases}$$ (5.2.22)

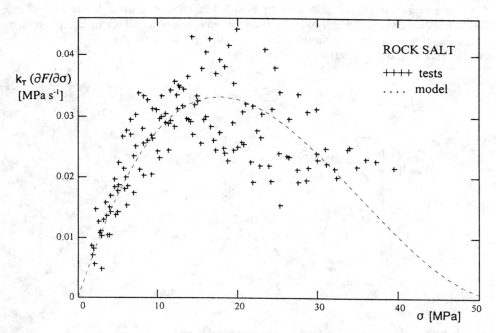

Figure 5.10 *Determination of the derivative $\partial F/\partial \sigma$ from hydrostatic tests.*

can determine the function $\varphi(\sigma)$ on a smaller interval if the whole constitutive equation is to be used for that interval only. For other geomaterials the function $\varphi(\sigma)$ determined for such a smaller interval has a simpler form. As an example, for saturated sand (Cristescu [1991a])

$$\varphi(\sigma) = h\sqrt{\sigma} ,$$

(5.2.23)

with $h = 1.07 \times 10^{-3} (kPa)^{-\frac{1}{2}}$ and σ smaller than 0.1 MPa.

Thus in principle, for a complete constitutive equation, one has to determine all the constitutive functions up to pressures surpassing σ_o, when all microcracks are closed. For practical purposes however, one can determine the constitutive equation on a much smaller interval of pressures. Such a constitutive equation is often quite satisfactory for most engineering applications. The example for sand given here is just such a constitutive equation (see below). One can give also another example for ceramic powder. For alumina powder tested hydrostatically in classical triaxial test the results are shown in Figure 5.11 (Cazacu et al. [1997]) for quite small confining pressures. One can see that at least for these pressures the behavior is nearly linear.

Secondly, we must determine $\partial F/\partial \sigma$ for $\bar{\sigma} \neq 0$. We recall that this function must be positive in the compressibility domain and negative in the dilatancy domain. In other words $\partial F/\partial \sigma$ must change sign when X (from (4.3.2)) changes sign. Also $\partial F/\partial \sigma$ must tend towards minus infinity when Y (from (4.4.1)) tends towards zero, i.e. when the stress state is approaching the short-term failure surface. From here follows that $\partial F/\partial \sigma$ would depend on some combinations involving the stress invariants (namely X and Y), besides depending maybe separately on σ and $\bar{\sigma}$. If the tests are performed with classical test devices, we assume that $\partial F/\partial \sigma$ depends on X, Y, σ, $\bar{\sigma}$ and $\sigma_2 = \sigma - \bar{\sigma}/3$ (=constant during the second stage of the test), if the idea related in conjunction with the Figure 4.20 cannot be used. In the case of true triaxial tests we assume that $\partial F/\partial \sigma$ depends on the same expressions with the exception of σ_2 (since this time σ = constant during the second stage of the test). Thus in the case of true triaxial tests we have to determine a function

$$k_T \frac{\partial F}{\partial \sigma} = G(X, Y, \bar{\sigma}, \sigma)$$

(5.2.24)

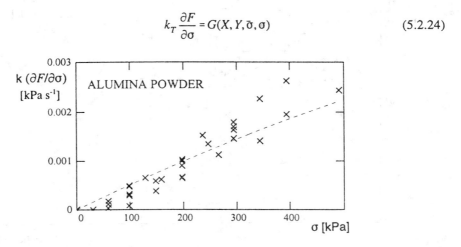

Figure 5.11 *Same as in Fig.5.10 but for ceramic powder.*

which must possess the following properties:

$$k_T \frac{\partial F}{\partial \sigma}\bigg|_{\bar{\sigma}=0} = G(X,Y,0,\sigma) = \varphi(\sigma) \quad (hydrostatic\ loading)$$

$$\left.\begin{array}{l}
G(X,Y,\bar{\sigma},\sigma) > 0 \ \leftrightarrow\ X > 0 \quad (compressibility) \\
G(X,Y,\bar{\sigma},\sigma) = 0 \ \leftrightarrow\ X = 0 \quad (compress/dilatancy\ boundary) \\
G(X,Y,\bar{\sigma},\sigma) < 0 \ \leftrightarrow\ X < 0 \quad (dilatancy) \\
G(X,Y,\bar{\sigma},\sigma) \rightarrow -\infty \ \leftrightarrow\ Y \rightarrow 0 \quad (failure) \ .
\end{array}\right\} \bar{\sigma} > 0 \qquad (5.2.25)$$

Functions with such properties can be found. For instance, the simplest function possessing these properties is

$$k_T \frac{\partial F}{\partial \sigma} = \frac{X(\sigma,\bar{\sigma})\,\Psi(\sigma)}{Y(\sigma,\bar{\sigma})} \qquad (5.2.26)$$

or

$$k_T \frac{\partial F}{\partial \sigma} = [sign\,X(\sigma,\bar{\sigma})] \frac{\Psi_1(\sigma)}{Y(\sigma,\bar{\sigma})}\ . \qquad (5.2.26')$$

In order to match the data better one can use an additional constitutive coefficient $\bar{G}(\bar{\sigma})$ so that instead of (5.2.26) we have

$$k_T \frac{\partial F}{\partial \sigma} = \frac{X(\sigma,\bar{\sigma})\,\Psi(\sigma)}{Y(\sigma,\bar{\sigma})} [\bar{\sigma}\,\bar{G}(\bar{\sigma}) + \sigma] \qquad (5.2.27)$$

or

$$k_T \frac{\partial F}{\partial \sigma} = [sign\,X(\sigma,\bar{\sigma})] \frac{\Psi_1(\sigma)}{Y(\sigma,\bar{\sigma})} [\bar{\sigma}\,\bar{G}_2(\bar{\sigma}) + \sigma]\ . \qquad (5.2.27')$$

In the last formulae Ψ or Ψ_1 are related to the hydrostatic behavior and can be determined from

$$\sigma \frac{X(\sigma,0)\,\Psi(\sigma)}{Y(\sigma,0)} = \varphi(\sigma) \quad or \quad [sign\,X(\sigma,0)] \frac{\Psi_1(\sigma)\,\sigma}{Y(\sigma,0)} = \varphi(\sigma)\ . \qquad (5.2.28)$$

All the coefficient functions involved in (5.2.26) and (5.2.27) are already determined with the exception of $\bar{G}(\bar{\sigma})$, which can be determined using the formula (see (5.2.18))

$$\frac{\dot{\varepsilon}_v - \dfrac{\dot{\sigma}}{K}}{\left\langle 1 - \dfrac{W(t)}{H(\sigma)} \right\rangle} = \frac{X(\sigma,\bar{\sigma})\,\Psi(\sigma)}{Y(\sigma,\bar{\sigma})}[\bar{\sigma}\,\bar{G}(\bar{\sigma}) + \sigma]\left(= [sign\, X(\sigma,\bar{\sigma})]\,\frac{\Psi_1(\sigma)}{Y(\sigma,\bar{\sigma})}\left[\bar{\sigma}\,\bar{G}_2(\bar{\sigma}) + \sigma\right]\right) \qquad (5.2.29)$$

where $\dot{\varepsilon}_v^I = \dot{\varepsilon}_v - \dot{\sigma}/K$ from the left-hand side is to be obtained from the data, and all other terms besides $\bar{G}$ are already known.

For rock salt in most cases the coefficient $\bar{G}(\tau)$ has been disregarded.

In another example, for saturated sand (Cristescu [1991a]) we have

$$k_T \frac{\partial F}{\partial \sigma} = h_1 \frac{(-\bar{\sigma} + 2f\sigma)\sqrt{\sigma}}{(2f + \alpha)\sigma - \left(1 + \dfrac{\alpha}{3}\right)\bar{\sigma}} \qquad (5.2.30)$$

with $h_1 = 2.3 \times 10^{-3}$ (kPa)$^{-1/2}$ s^1; the other coefficients were given following the formulae (4.3.4) and (4.4.3).

In the third and last stage we must determine $\partial F/\partial\bar{\sigma}$. For this purpose we integrate (5.2.27) or (5.2.26) with respect to σ to get

$$k_T F(\sigma,\bar{\sigma}) = \int G(X,Y,\bar{\sigma},\sigma)\,d\sigma + g(\bar{\sigma}) = F_1(\sigma,\bar{\sigma}) + g(\bar{\sigma}) \quad . \qquad (5.2.31)$$

We now differentiate (5.2.31) with respect to $\bar{\sigma}$, and combine the result with formula (5.2.19) to obtain for the determination of $g'(\bar{\sigma})$ the formula

$$g'(\bar{\sigma}) = \frac{\sqrt{2}}{3}\frac{1}{\left\langle 1 - \dfrac{W(t)}{H(\sigma)} \right\rangle}\sqrt{\left(\dot{\varepsilon}_1^I - \dot{\varepsilon}_2^I\right)^2 + \left(\dot{\varepsilon}_2^I - \dot{\varepsilon}_3^I\right)^2 + \left(\dot{\varepsilon}_3^I - \dot{\varepsilon}_1^I\right)^2} - \frac{\partial F_1}{\partial\bar{\sigma}} \qquad (5.2.32)$$

if true triaxial tests are available. In the case of classical triaxial tests this formula reads

$$g'(\bar{\sigma}) = \frac{2}{3}\frac{\dot{\varepsilon}_1 - \dot{\varepsilon}_2 - \dfrac{\dot{\sigma}_1 - \dot{\sigma}_2}{2G}}{\left\langle 1 - \dfrac{W(t)}{H(\sigma)} \right\rangle} - \frac{\partial F_1}{\partial\bar{\sigma}} \quad , \qquad (5.2.33)$$

where in most cases $\dot{\sigma}_2 \approx 0$. In these formulae F_1 and H are known functions, while the rates-of-deformation components $\dot{\varepsilon}_1$, $\dot{\varepsilon}_2$ and $W(t)$ are determined from the data.

Let us give some examples. Figure 5.12 shows (Cristescu [1994b]) the results obtained with formula (5.2.32) using the data for rock salt obtained by Hunsche (see Section 2.1); there are a number of tests each obtained with some other constant value of σ. These data can be approximated by the empirical formula

$$g(\tau) = g_o \frac{\tau}{\sigma_\star} + \frac{g_1}{2} \left(\frac{\tau}{\sigma_\star} \right)^2 + \frac{g_2}{4} \left(\frac{\tau}{\sigma_\star} \right)^4 ,$$

(5.2.34)

with $g_o = 0.01\,\text{MPa s}^{-1}$, $g_1 = 6.5 \times 10^{-5}\,\text{MPa s}^{-1}$, $g_2 = 5.9 \times 10^{-6}\,\text{Mpa s}^{-1}$. In Figure 5.12 the prediction of this formula is shown as a dotted line.

A second example is shown in Figure 5.13 for dry sand (Cristescu [1991a]). The result is obtained this time with formula (5.2.33) and the data by Hettler *et al.* [1984]. The data were approximated by

$$g(\bar{\sigma}) = g_o \bar{\sigma} + \frac{g_1}{3} \bar{\sigma}^3$$

(5.2.35)

with $g_o = 0.003\,\text{s}^{-1}$ and $g_1 = 1.7 \times 10^{-8}\,(\text{kPa})^{-2}\,\text{s}^{-1}$.

It is important to mention that the coefficient function $g(\bar{\sigma})$ plays an important role in the constitutive equation in the sense that in the prediction of creep, for instance, this coefficient

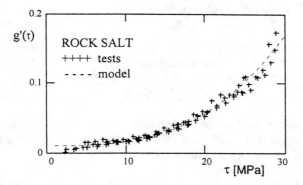

Figure 5.12   *Determination of the function $g'(\tau)$. The dotted line is the prediction of (5.2.34).*

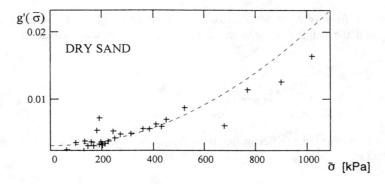

Figure 5.13   *Determination of the function $g'(\tau)$ from data for dry sand.*

is dominant. That is why it has to be determined quite accurately.

5.2.3   The irreversible strain rate orientation tensor N

If one would not like to (or cannot) use the concept of viscoplastic potential, or if the data available are not sufficient for the determination of the function F, we can find an alternative variant of nonassociated constitutive equation.

Let us consider again the constitutive equation (5.2.1). For the nonassociated case the irreversible component of the transient creep can be expressed as

$$\dot{\varepsilon}_T^I = k_T \left\langle 1 - \frac{W(t)}{H(\sigma)} \right\rangle N(\sigma) \ , \tag{5.2.36}$$

where the tensorial function $N(\sigma)$ is governing the orientation of $\dot{\varepsilon}_T^I$ and will be called the **irreversible strain rate orientation tensor** (Cazacu [1995], Cristescu and Cazacu [1995], Cazacu et al.[1997]). As the material is isotropic, N must satisfy the invariance requirement

$$N(Q\sigma Q^T) = Q N(\sigma) Q^T \tag{5.2.37}$$

for any orthogonal transformation Q. From classical theorems of representation of isotropic tensor functions, it follows that N can be represented as

$$N(\sigma) = N_1 1 + N_2 \sigma + N_3 \sigma^2 \ , \tag{5.2.38}$$

where N_i are scalar-valued functions of the stress invariants. In order to reduce the complexity of the problem we assume that $N_3 \equiv 0$, and also we disregard the influence of the third stress invariant. Thus $N(\sigma)$ is considered in the form

$$N(\sigma) = N_1 1 + N_2 \frac{\sigma'}{\tau} \ , \tag{5.2.39}$$

where the coefficients N_i depend on two invariants only $N_i(\sigma, \tau)$ (compare (5.2.39) with (5.2.17)), and the prime indicates the deviator. The constitutive equation (5.2.36) now becomes

$$\dot{\varepsilon}_T^I = k_T \left\langle 1 - \frac{W(t)}{H(\sigma, \tau)} \right\rangle \left[N_1 1 + N_2 \frac{\sigma'}{\tau} \right] . \tag{5.2.40}$$

In order to determine the coefficient functions N_1 and N_2, the formulae (5.2.18) and (5.2.20) are replaced by

$$k_T N_1(\sigma,\tau) = \frac{\dot{\varepsilon}^I_v}{3\left(1 - \dfrac{W(t)}{H(\sigma,\tau)}\right)} \qquad (5.2.41)$$

and

$$k_T N_2(\sigma,\tau) = \frac{3}{\sqrt{2}}\frac{|\dot{\varepsilon}_1 - \dot{\varepsilon}_2|}{\left(1 - \dfrac{W(t)}{H(\sigma,\tau)}\right)} . \qquad (5.2.42)$$

Using (5.2.41) and the data obtained during the hydrostatic part of the test we determine $k_T N_1\big|_{\tau=0}$. We use the notation $k_T N_1\big|_{\tau=0} = \varphi(\sigma)$. For alumina powder a second-order polynomial in σ seems to fit the data well. Thus

$$k_T N_1\big|_{\tau=0} = a\left(\frac{\sigma}{\sigma_*}\right)^2 + b\left(\frac{\sigma}{\sigma_*}\right) , \qquad (5.2.43)$$

with $a = -8\times10^{-11}$ and $b = 1.641\times10^{-6}$. Next $N_1(\sigma,\tau)$ must be determined for $\tau \neq 0$. From (5.2.41) it follows that this function must be positive in the compressibility domain and negative in the dilatancy one. The simplest possible function satisfying these properties is

$$k_T N_1 = \varphi(\sigma) + \frac{\tau}{\sigma_*}\, sign\, X(\sigma,\tau)\, \psi(\sigma,\tau) , \qquad (5.2.44)$$

where $X(\sigma,\tau)$ is the function involved in the definition of the C/D boundary (see (4.3.6)) and

$$sign\, X(\sigma,\tau) = \begin{cases} 1 & for\ X(\sigma,\tau) \geq 0 \\ -1 & otherwise . \end{cases} \qquad (5.2.45)$$

In (5.2.44) all the functions, except $\psi(\sigma,\tau)$, have been already determined. By making use of (5.2.41), from (5.2.44) where results

$$\psi(\sigma,\tau) = \left\{\frac{\dot{\varepsilon}^I_v}{3\left(1 - \dfrac{W(t)}{H(\sigma,\tau)}\right)} - \varphi(\sigma)\right\}\left\{\frac{\tau}{\sigma_*}\, sign\, X(\sigma,\tau)\right\}^{-1} . \qquad (5.2.46)$$

Therefore, from the data obtained in the deviatoric part of the compression tests and

(5.2.46) one can determine $\psi(\sigma,\tau)$. For alumina powder a possible expression for $\psi(\sigma,\tau)$ is

$$\psi(\sigma,\tau) = z_c \left[\frac{\tau}{\sigma_*} - y_a \left(\frac{\sigma - \dfrac{\tau}{\sqrt{2}}}{\sigma_*} \right) - y_c \right]^5 \exp\left(z_b \frac{\sigma - \dfrac{\tau}{\sqrt{2}}}{\sigma_*} \right), \tag{5.2.47}$$

where $y_a = 0.842$, $y_c = -56.06$, $z_b = -0.01$ and $z_c = 1.127 \times 10^{-15}$.

We now use the formula (5.2.42) in conjunction with the data obtained in the deviatoric part of the triaxial tests. For alumina powder and the confining pressures $\sigma_3 = 294$; 392, and 490 kPa, it was found that $k_T N_2$ can be approximated by

$$k_T N_2 = m_1 \left(\frac{\tau}{\sigma_*} \right)^{14} \exp\left[m_2 \left(\frac{\sigma - \dfrac{\tau}{\sqrt{2}}}{\sigma_*} \right) \right] + n_1 \left(\frac{\tau}{\sigma_*} \right)^2 \exp\left[n_2 \left(\frac{\sigma - \dfrac{\tau}{\sqrt{2}}}{\sigma_*} \right) \right] + r, \tag{5.2.48}$$

where $m_1 = 2.38 \times 10^{-33}$, $m_2 = -0.038$, $n_1 = 4 \times 10^{-7}$, $n_2 = -0.005$, $r = 8 \times 10^{-4}$.

5.2.4 Examples

Let us give a few examples of yield surfaces and viscoplastic potentials for various materials.

The constitutive equation for **dry sand** has been obtained by Cristescu [1991a] using the data by Hettler et al. [1984]. The yield function was obtained as

$$H(\sigma,\bar{\sigma}) := \left[a_o + a_1 \left(\sigma - \frac{\bar{\sigma}}{3} \right) \right] \left(\frac{\bar{\sigma}}{\sigma - \dfrac{\bar{\sigma}}{3}} \right)^6 + b_o \bar{\sigma} + c_1 \sigma^2 + c_2 \sigma, \tag{5.2.49}$$

with $a_o = 3 \times 10^4$ kPa, $a_1 = 2.1 \times 10^{-5}$, $b_o = 3.4 \times 10^{-3}$, $c_1 = 1.5 \times 10^{-6} (\text{kPa})^{-1}$, $c_2 = 1.8 \times 10^{-3}$. The viscoplastic potential has a more involved expression:

$$F(\sigma,\bar{\sigma}) := \left[-\frac{\varphi_o \bar{\sigma}}{2f + \alpha} + \frac{(2f\varphi_o - \varphi_1 \bar{\sigma})\left(1 + \dfrac{\alpha}{3} \right)\bar{\sigma}}{(2f + \alpha)^2} + \frac{2f\varphi_1}{(2f + \alpha)^3}\left(1 + \dfrac{\alpha}{3} \right)^2 \bar{\sigma}^2 \right]$$

$$\times \ln\left[(2f + \alpha)\sigma - \left(1 + \frac{\alpha}{3} \right)\bar{\sigma} \right] + \frac{2f\varphi_o - \varphi_1 \bar{\sigma}}{(2f + \alpha)}\sigma + \frac{2f\varphi_1}{(2f + \alpha)^3}\sigma \tag{5.2.50}$$

$$\times \left\{ \frac{1}{2}\left[(2f + \alpha)\sigma - \left(1 + \frac{\alpha}{3} \right)\bar{\sigma} \right]^2 + 2\left(1 + \frac{\alpha}{3} \right)\bar{\sigma}\left[(2f + \alpha)\sigma - \left(1 + \frac{\alpha}{3} \right)\bar{\sigma} \right] \right\} + g_o \bar{\sigma} + \frac{g_1}{3}\bar{\sigma}^3.$$

with $\varphi_o = 3.1 \times 10^{-3}$, $\varphi_1 = 5.3 \times 10^{-6} (\text{kPa})^{-1}$, $\alpha = 0.984$, $f = 0.696$, $g_o = 0.003$, $g_1 = 1.7 \times 10^{-8}$

$(kPa)^{-2}$.

The shapes of several yield surfaces H = constant (interrupted lines) and several surfaces F = constant (short-dashed lines) are shown in Figure 5.14. Everywhere in the constitutive domain shown these two surfaces are quite distinct. The domain of compressibility has a wide extent towards high values of $\bar{\sigma}$, at least up to the pressures shown. The domain of dilatancy is found in the neighborhood of the failure surface (solid line) and extends up to the compressibility/dilatancy boundary $\partial F/\partial \sigma = 0$ (dash-dot line). Since the line $\partial H/\partial \sigma = 0$ is quite distinct from the line $\partial F/\partial \sigma = 0$ and since the surfaces F = constant and H = constant are quite distinct, the associativeness assumption cannot be made anywhere in the constitutive domain. The limit pressure σ_o was not determined from the existing data, so that the validity of the model is limited to the stress state shown in Figure 5.14. Tests carried out with much higher pressures would be necessary, but for sand and civil engineering such high pressures have no practical importance anyway. The elastic parameters as reported by the authors are $E = K$ =700 MPa and G = 210 MPa. Their possible dependence on the stress state was not determined.

The surface $\partial H/\partial \sigma = 0$ uniting all the maxima of the interrupted curves H=constant does however have some deeper meaning discussed by Cristescu [1991a] in conjunction with the problem of sand stability (sand liquefaction), if the sand is saturated and either undrained

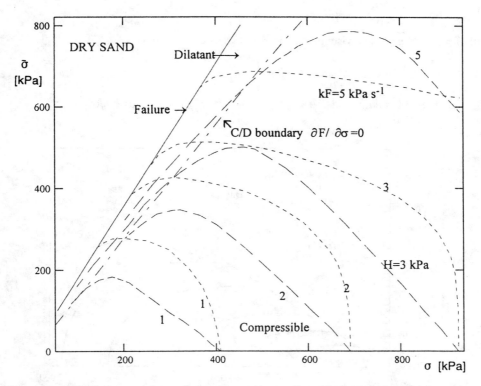

Figure 5.14 *Shapes of the surfaces H = constant (long-dashed lines) and F = constant (short-dashed lines) for dry sand; the dash-dot line is the compressibility/-dilatancy boundary while the failure surface is shown as a solid line.*

or dynamically loaded. Indeed, in the domain $\partial F/\partial\sigma>0$, $\partial H/\partial\sigma<0$, between the maxima on the surfaces $H=$ const and the C/D boundary $\partial F/\partial\sigma=0$, any loading pulse starting from a stress state on one surface $H=$ constant, will produce compressibility of the skeleton. This in turn will increase the pressure in pores and thus decrease the stress in the skeleton. Thus if the loading is dynamic, and mainly repeated (oscillations), and if the sand is not drained, any loading pulse will significantly decrease the stress in the skeleton, down to a stress state unable to carry the load applied. For experiments discussing such a kind of instability see Lade *et al.* [1993], Lade and Yamamuro [1993] and Lade [1994].

A similar model has been determined for **saturated sand** (Cristescu [1991a]). The two sets of surfaces $H=$constant and $F=$constant are also quite distinct and the assumption of associativeness cannot be made.

The model formulated for drained **alumina powder** (powder mixed with 50 % water per volume and grain size of the powder ranging between 40 μm to 200 μm) is of the same kind. In fact, two variants of the model were worked out: a model using the concept of viscoplastic potential and another one which does not use this concept (variant (5.2.36) of the model). For the first variant of the model the yield function $H(\sigma,\tau)$ was determined as

$$H(\sigma,\tau):=\frac{0.27\,\tau^9}{\left\{\sigma-\dfrac{\tau}{\sqrt{2}}+37\right\}^{m/n}}+4.6\times10^{-5}\tau^2+5.833\times10^{-6}\sigma^2+10^{-6}\sigma\ , \qquad (5.2.51)$$

with $m=17$ and $n=2$. The viscoplastic potential was determined as

$$k_T F(\sigma,\tau):=a_1\tau^2\sigma+a_2\frac{\tau^2}{b_1+b_2\sigma}+a_3\tau^2\ln(b_1+b_2\sigma)$$
$$+\exp(b_3\sigma+b_4\tau)\!\left(a_4\tau^2+a_5\tau^2\sigma+a_6\tau^3\right)+a_7\ln(c_1\tau+c_2+c_3\sigma)+a_8\sigma^2+a_9\sigma^3+a_{10}\tau^2 \qquad (5.2.52)$$

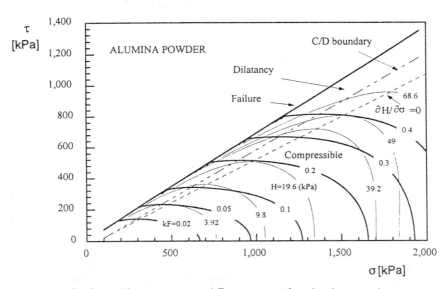

Figure 5.15 *Surfaces H = constant and F = constant for alumina powder.*

with

$a_1 = -1.368\times10^{-6}$, $a_2 = 3.2475$, $a_3 = -1.654\times10^{-5}$, $a_4 = -3.216\times10^{-5}$, $a_5 = -1.0525\times10^{-4}$, $a_6 = 1.691\times10^{-4}$, $a_7 = -1.461\times10^{-6}$, $a_8 = 2.85\times10^{-6}$, $a_9 = 1.767\times10^{-7}$, $a_{10} = 3.15\times10^{-4}$, $b_1 = -207742$, $b_2 = 311153$, $b_3 = -1.0275$, $b_4 = 0.7266$, $c_1 = -1$, $c_2 = 6.575\times10^{-2}$, $c_3 = 0.6842$.

In Figure 5.15 are shown the shapes of several surfaces H=constant and F=constant for alumina powder. These are considerably distinct, though the C/D boundary and the $\partial H/\partial\sigma = 0$ boundary are not too far apart, at least for the small pressures shown. The domain of compressibility extends up to very high values of τ, the dilatancy region being a strip close to the failure surface (see also Cristescu and Cazacu [1997], Cristescu *et al.* [1997]).

The third example for **rock salt** is shown in Figure 5.16: the long-dashed lines represent the surfaces H=constant, while the short-dashed lines are the surfaces $k_T F$= constant, plotted according to

$$
\begin{aligned}
k_T F(\sigma,\tau) := \sigma_* \Bigg\{ & \frac{f_1 p_1}{4}[Y(\sigma,\tau)]^4 + \left[-\frac{4}{3}f_1 p_1 Z(\tau) + \frac{f_2 p_1 + f_1 p_2}{3} \right][Y(\sigma,\tau)]^3 \\
& + \left[3f_1 p_1 [Z(\tau)]^2 - \frac{3}{2}(f_2 p_1 + f_1 p_2)Z(\tau) + \frac{1}{2}\left(f_2 p_2 + f_1 p_3 - \frac{\tau}{\sigma_*}p_1 \right) \right][Y(\sigma,\tau)]^2 \\
& + \left[-4f_1 p_1 [Z(\tau)]^3 + 3(f_2 p_1 + f_1 p_2)[Z(\tau)]^2 - 2\left(f_2 p_2 + f_1 p_3 - \frac{\tau}{\sigma_*}p_1 \right)Z(\tau) \right]Y(\sigma,\tau) \\
& + \left[f_1 p_1 [Z(\tau)]^4 - (f_2 p_1 + f_1 p_2)[Z(\tau)]^3 + \left(f_2 p_2 + f_1 p3 - \frac{\tau}{\sigma_*}p_1 \right)[Z(\tau)]^2 \right. \\
& \left. - \left(f_2 p_3 - \frac{\tau}{\sigma_*}p_2 \right)Z(\tau) - \frac{\tau}{\sigma_*}p_3 \right]\ln Y(\sigma,\tau) + \left(f_2 p_3 - \frac{\tau}{\sigma_*}p_2 \right)\frac{\sigma}{\sigma_*} \Bigg\}(G(\tau)+1) + g(\tau),
\end{aligned} \tag{5.2.53}
$$

with $Y(\sigma,\tau)$ defined by (4.4.2), $g(\tau)$ by (5.2.35), and

$$
G(\tau) := u_1 \frac{\tau}{\sigma_*} + u_2\left(\frac{\tau}{\sigma_*}\right)^2 + u_3\left(\frac{\tau}{\sigma_*}\right)^3 + u_4\left(\frac{\tau}{\sigma_*}\right)^8 ,
$$
$$
Z(\tau) = -r\frac{\tau}{\sigma_*} - s\left(\frac{\tau}{\sigma_*}\right)^6 + \tau_o = Y(\sigma,\tau) - \frac{\sigma}{\sigma_*} ,
\tag{5.2.54}
$$

with $u_1 = 0.036$, $u_2 = -0.00265$, $u_3 = 5.256\times10^{-5}$, $u_4 = 1.57\times10^{-12}$. Further in (5.2.53) $p_1 = -9.83\times10^{-7}s^{-1}$, $p_2 = -5.226\times10^{-5}s^{-1}$, $p_3 = 9.84\times10^{-5}s^{-1}$, and f_1, f_2 given in (4.3.5).

The dash-dot line is the C/D boundary as determined from tests while the full line is the short-term failure surface also as determined from tests. Again, for stress states close to the hydrostatic line and to the failure line the two surfaces H= constant and F=constant are distinct and a nonassociated constitutive law is to be used. That follows also from Figure 5.9, from where one can see that the line $\partial H/\partial\sigma = 0$ is considerably distinct from the C/D boundary as determined from tests. However, for high pressures both surfaces approach Mises kind of surfaces and as such they are not too distinct from one another. Thus Figure 5.16 shows that at **very high pressures rock salt behavior is very close to what can be labeled Mises viscoplasticity,** with significant ductility and no irreversible volumetric changes, a behavior similar to that exhibited by some metals. Thus for very high pressures one can try to apply an associated kind of constitutive equation. Most practical problems involve pressures which are not really so high (i.e. $\sigma<\sigma_o$).

Other variants of constitutive equation for rock salt describing transient creep are due to Nicolae [1996] for the Slănic-Prahova salt, and to Jin *et al.* [1996b], Jin and Cristescu [1997], for the Gorleben salt.

An elastic/viscoplastic constitutive equation for **granite** has been given by Cristescu [1987, 1989a], for **coal** by Cristescu [1989a] and Cristescu *et al.* [1989], and for **andesite** by Cristescu *et al.* [1985] and Cristescu [1989a].

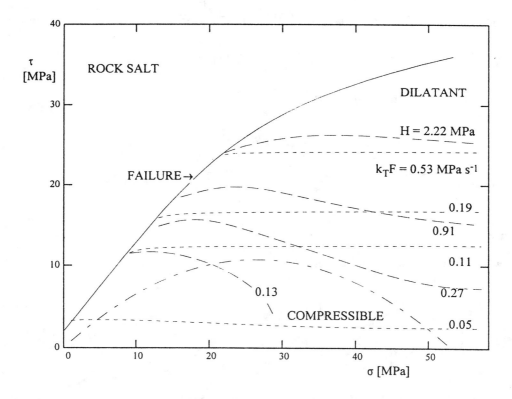

Figure 5.16 *Yield surfaces H = constant and surfaces F=constant for rock salt; the dash-dot line is the C/D boundary.*

5.2.5   Simplified variants of the constitutive equation.

From the above examples follows that the C/D boundaries $\partial F/\partial \sigma = 0$ are considerably distinct from the corresponding stability threshold boundaries $\partial H/\partial \sigma = 0$. We have concluded that for most geomaterials and particulate system materials **associated constitutive equations are inappropriate**: the model will not predict correctly the exact position of the C/D boundary if the associated constitutive equation is used. Generally, for such materials an associated constitutive equation can be used only for high pressures, and only seldom elsewhere. However, this is to be judged in conjunction not only with a specific geomaterial, but also with the specific problem which is to be solved with a particular variant of the constitutive equation.

For instance, quite often when dealing with rock salt one is assuming that the surfaces F = constant are very close to Mises kind of surfaces, and therefore that $\dot{\varepsilon}^I$ is proportional to σ', i.e.

$$F(\sigma) := II_{\sigma'} = \frac{3}{2}\, \tau^2 ,$$ (5.2.55)

with

$$\tau^2 = \frac{1}{3}\left(\sigma'^2_1 + \sigma'^2_2 + \sigma'^2_3\right) .$$ (5.2.56)

From here it follows that

$$\dot{\varepsilon}^I_T = k_T\left\langle 1 - \frac{W(t)}{H(\sigma)}\right\rangle \sigma' .$$ (5.2.57)

This constitutive equation implies incompressibility, and therefore this simplified model will be unable to describe any irreversible volumetric changes, nor will the short term failure be incorporated into the constitutive equation. In this case short term failure is to be considered as an additional condition attached to the constitutive equation. If dilatancy, i.e. microcracking and creep failure, are of great significance for the problem to be solved, such as in the design of radioactive waste repositories or the study of the stability of underground excavations, then a complete model involving a correct and general expression for F must be used. Recall that in the constitutive equation the function F (or N) is involved *via* the partial derivative $\partial F/\partial \sigma$ and $\partial F/\partial \tau$ (or N_1 and N_2) only. The meaning of various terms involved in $\partial F/\partial \sigma$ is obvious from (5.2.27). In $\partial F/\partial \tau$ we have two kinds of terms: these already involved in $\partial F/\partial \sigma$ and which describe the influence of σ on $\partial F/\partial \tau$, and the function $g'(\tau)$ which is the dominant term in $\partial F/\partial \tau$ (see formulae (5.2.31) and (5.2.50), (5.2.53), where the two groups of terms are evident). If the simplified constitutive equation (5.2.57) is accepted, then k_T is to be determined independently, using formulae (5.2.14) or (5.2.15) or similar ones.

One could try to simplify the function F by disregarding sometimes altogether the rock compressibility and considering it to be everywhere dilatant. Obviously that can be attempted only for some specific problems, while for the surfaces F = constant one is using either conical surfaces or Drucker-Prager surfaces, depending which one approximates better the surface

F = constant. For soils and particulate systems as well as for any other highly compressible material this approach is inappropriate.

One can try to simplify the expressions of the functions H as well. For this purpose, in order to give an example, one can write the yield function for rock salt in the form

$$H(\sigma,\tau) := H_1 \tau^{14} + H_2 \tau^3 + H_3 \tau + H_H \ , \tag{5.2.58}$$

with an obvious meaning of the coefficients. If for a deviatoric loading (σ = constant, τ increases) we estimate the contribution of these four terms, we obtain the results shown in Figure 5.17 (Cristescu [1994b]) for the confining pressure σ = 20 MPa, when τ is increased. The location of the C/D boundary is shown as a dashed line. The relative contribution of the term H_H, shown as diamonds, is dominant in the compressibility domain; let us recall that this term represents the energy of microcrack closing when the mean stress increases. For higher values of τ some other terms involved in H may become dominant, while close to failure the term H_1 is the dominant one. It follows from here that the term H_H, maybe combined with the last term from (5.2.8), which has a similar meaning, may be dominant in many mining and petroleum applications and not only in triaxial tests done in the laboratory. In Chapters 8-10 examples will be given to show that an underground excavation sometimes produces a significant variation of the mean stress in the surrounding rock besides a variation of the octahedral shear stress. Generally a significant irreversible volumetric change is to be expected as a result of an excavation. Finally let us mention that the term H_H plays a more important role in the overall balance of dissipated energies for materials with a more pronounced initial porosity (where significant compressibility is to be expected).

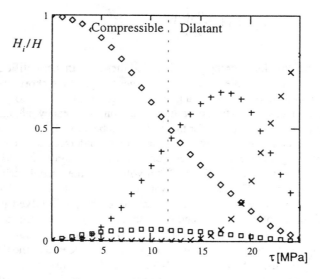

Figure 5.17 *Comparison of magnitude of various terms involved in function H during a deviatoric test on rock salt for σ: H_1/H shown by x x, H_2/H by ++, H_3/H by $\square\square$, H_H/H by $\diamond\diamond$.*

5.2.6   *Creep formula and comparison with the data*

Let us find out first how transient creep can be described with the model. We assume that at time t_o a loading is produced, i.e. the stresses are increased quite fast from σ^P up to $\sigma(t_o)$ and that for $t > t_o$ they are kept constant. By tensorial multiplication on both sides in (5.2.1) with σ and by integration, one obtains quite easily (see Cristescu [1989a]) for $t > t_o$

$$1 - \frac{W(t)}{H(\sigma,\tau)} = \left(1 - \frac{W(t_o)}{H(\sigma,\tau)}\right) \exp\left[\frac{k_T}{H}\frac{\partial F}{\partial \sigma}\cdot\sigma\,(t_o - t)\right], \qquad (5.2.59)$$

where $\sigma(t) = \sigma(t_o) = \textbf{Constant}$ and $W(t_o) = W^P$, the initial value of W at time t_o. The superscript P stands for "primary". If the initial state is on a yield surface then $W(t_o) = H(\sigma^P)$, but generally one can consider cases when $W(t_o) < H(\sigma^P)$. This latter case occurs if unloading has taken place at the location where the excavation will be performed, as for instance if a significant volume of the rock above that location has been removed either by excavation, or by natural causes (erosion, land slides, etc.). The relation (5.2.59) gives the variation in time of W under a constant stress state. If the bracket involved in (5.2.1) is replaced by the right-hand-side term from (5.2.59), by another integration one obtains the formula describing transient creep. Adding the elastic, instantaneous, components one obtains finally the **creep formula for transient creep** as

$$\varepsilon(t) = \varepsilon^E + \frac{\left(1 - \dfrac{W(t_o)}{H}\right)\dfrac{\partial F}{\partial \sigma}}{\dfrac{1}{H}\dfrac{\partial F}{\partial \sigma}\cdot\sigma}\left\{1 - \exp\left[\frac{k_T}{H}\frac{\partial F}{\partial \sigma}\cdot\sigma\,(t_o - t)\right]\right\}, \qquad (5.2.60)$$

with the initial conditions

$$t = t_o: \begin{cases} \varepsilon^I = 0, \quad W(t_o) = W^P \\[2mm] \varepsilon^E = \left(\dfrac{1}{3K} - \dfrac{1}{2G}\right)\sigma\mathbf{1} + \dfrac{1}{2G}\,\sigma. \end{cases} \qquad (5.2.61)$$

If the initial stress state is non-zero, i.e. $\sigma^P \neq 0$, then the stresses involved in (5.2.61) have the meaning of "relative" stresses, as discussed in Chapter 7 (i.e. the strains are the result of the stress variation $\sigma(t_o) - \sigma^P$ and in formula (5.2.61) stress components are to be replaced by this stress variation). Formula (5.2.60) can be used to describe creep starting from any initial existing stress and strain state, with an appropriate choice of initial data.

If we would like to describe creep that is the result of several successive loadings, or if we would like to describe a triaxial test in which the loading is the result of a great number of successive stress increments, we can use the same kind of formulae. For instance, let us assume that the stresses are increased in small successive steps, according to the same law as in the experiments done to establish the model and using the same global loading rate as in the experiments. For each small stress increment from $\sigma(t_o)$ to $\sigma(t) = \sigma(t_o) + \Delta\sigma$ which afterwards is held constant, the corresponding variation in time of the strain is obtained from

$$\Delta\varepsilon_1(t) = \Delta\varepsilon_1^E(t) + \frac{\left\langle 1 - \dfrac{W(t_o)}{H(\sigma(t))} \right\rangle \dfrac{\partial F}{\partial\sigma_1}}{\dfrac{1}{H}\left(\dfrac{\partial F}{\partial\sigma}\sigma + \dfrac{\partial F}{\partial\tau}\tau \right)} \left\{ 1 - \exp\left[\frac{k_T}{H}\frac{\partial F}{\partial\sigma} \cdot \sigma(t_o - t) \right] \right\} , \qquad (5.2.62)$$

and similar formulae for the other components. Here all the functions are computed for $\sigma(t)$ (assumed constant in the interval $t - t_o$), Δ is the variation of the function in the time interval $t - t_o$, $W(t_o)$ is the value of W just before the new reloading, and t_o is the moment of reloading (assumed to be sudden). Thus these formulae can be used to describe creep tests or any other loading histories which can be approximated by successive step-wise stress increments, as for instance tests done with constant global loading rate.

Let us give some examples of **model prediction compared with data** for various kinds of tests. For instance, for rock salt tested in a true triaxial test with $\sigma = 35$ MPa with an average loading rate $\dot{t} = 21.4$ MPa/min, the experimental data due to Hunsche (see Section 2.1) are shown in Figure 5.18. The full lines are the prediction of the model (in which the function in $\bar{G}(\tau)$ (5.2.27) was disregarded since it has little influence) obtained with successive small increases of the octahedral shear stress, each followed by a short term dimensionless time interval $k_T(t_o - t) = 0.56$, during which creep is taking place. As one can see, the model predicts the data reasonably well.

The validity of the model was also checked by describing with FEM the deformation of the specimen subjected to the stress loading history used in performing the tests (Jin and Cristescu [1997]) (Figure 5.19). Again the matching is reasonable. Similar FEM calculations

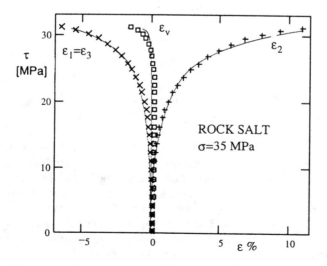

Figure 5.18 *Triaxial stress-strain curves for rock salt for σ=constant; model prediction (solid line) compared with data (symbols).*

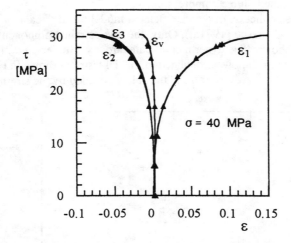

Figure 5.19 *Stress-strain curves for rock salt as obtained by a test (solid line) and by model incorporated in FEM (triangles).*

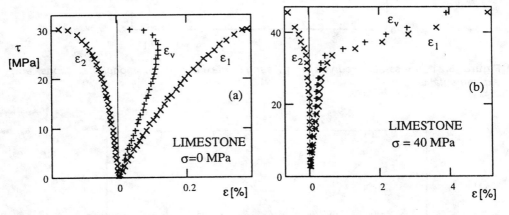

Figure 5.20 *Stress-strain curves for "Pietra Leccese": a) uniaxial test; b) test with confining pressure p =40 MPa*

have been performed by Glabisch [1997].

Another example is given for limestone ("Pietra Leccese") as obtained by Brignoli and Sartori [1993]. Their model is an associated one for a rock of high initial porosity of 34%, which in uniaxial tests is compressible and afterwards dilatant but at high confining pressure is compressible only (Figure 5.20). The results shown in Figure 5.20a correspond to a uniaxial test and could be considered "standard" rock behavior. However, the results shown in Figure 5.20b obtained in a triaxial test show compressibility only. The authors were able to formulate so well an associated model for transient creep, i.e. (5.2.1) with $H = F$, that the prediction of the model is simply superposed on the experimental data and cannot be distinguished from them. It is also interesting to mention that failure at high confining pressures occurs during the volume compressibility.

A comparison of the prediction of a nonassociated model for dry sand with the data is shown in Figure 5.21 (Cristescu [1991a]). Only the two strain components involved in the deviatoric test are shown. The experimental data are by Hettler *et al.* [1984]. Similar stress-strain curves for alumina powder obtained in triaxial tests are shown in Figure 5.22 (Jin and Cristescu [1996]). The curves start from the values reached by the strains at the end of

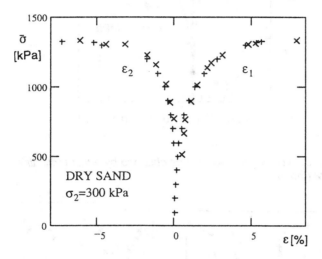

Figure 5.21 *Stress-strain curves for dry sand in triaxial test: ++: model predictions; xx: experimental data.*

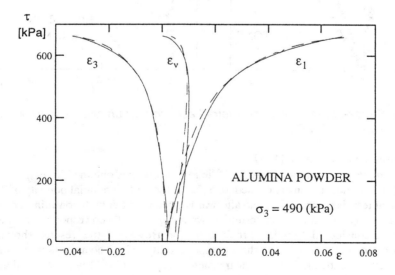

Figure 5.22 *Stress-strain curves for alumina powder as obtained by test (interrupted lines) or by model (solid line).*

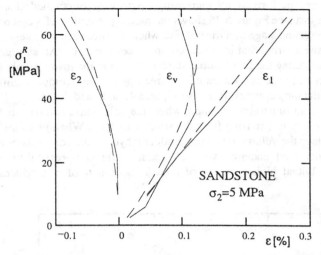

Figure 5.23 *Stress-strain curves for sandstone in triaxial test: solid lines: test; dashed lines: associated model.*

the hydrostatic first part of the triaxial test. The solid lines are the predictions of the nonassociated model given above, while the interrupted lines are experimental data.

The associated model can still predict compressibility and/or dilatancy, but the C/D boundary is never correctly located. That is why such models will overestimate or under-estimate either compressibility or dilatancy. For instance Figure 5.23 shows stress-stress curves for sandstone (Cristescu [1989b]) obtained in triaxial tests. The dashed lines are predictions of an associated model, while the solid lines are the experimental results. While the model describes reasonably well the overall behavior, the dilatancy threshold is located at much too high a value of τ. That means that with such a model we are unable to determine precisely up to where around an excavation the rock becomes dilatant, and that may be very important in designing a radioactive waste repository or storage for other hazardous wastes or even for petroleum products, etc. That is why, if enough data of good quality are available, an effort is to be done to determine a nonassociated constitutive equation.

The prediction of the model was compared also with some other kind of experimental data. For instance the nonassociated model for transient creep determined for porous chalk by Dahou [1994] was compared with his creep data obtained in triaxial and uniaxial tests. An example of model prediction as compared with triaxial creep data is shown in Figure 5.24 for the axial strain and volumetric strain. Note that this rock is compressible only. Its initial porosity is 38% and it is due to pores ranging in size mainly between 0.35μm and 0.6μm. Dahou has also done some triaxial relaxation tests. First the specimen was subjected to a hydrostatic pressure which was kept constant allowing for the volumetric creep to stabilize. Afterwards the confining pressure was kept constant and the axial stress was increased until the axial strain reached 0.7%. Then the axial piston was blocked (axial strain constant) and the confining pressure was held constant: the variation in time of the axial stress was recorded as described in Section 2.2.2. The corresponding stress relaxation is shown in Figure 5.25 and compared with the model prediction (solid line). The matching is good.

The model presented in this chapter can describe quite sophisticated volumetric behavior when triaxial tests are performed by increasing the pressure or octahedral shear stress in successive steps. For instance Figure 5.26 shows the model prediction of a creep test on rock salt performed in the second stage of a triaxial test, when the mean stress is kept constant and it is the octahedric shear stress that is increased in successive steps. An enlargement of the creep curves obtained during the first loading steps is shown in the lower graph. One can see that the model describes at lower values of τ a creep producing compressibility, while at higher values of τ dilatancy appears (see also Figures 4.9a, b, and 2.13).

Figure 5.27 shows an illustrative example when the volumetric behavior is followed all along a true triaxial creep test starting from the stress-free state. When pressure is increased in three successive steps the volumetric creep produces hydrostatic compressibility. The next steps, in which τ is increased under constant σ, produce little compressibility (most of the variation is elastic), but at higher values of τ each increment of τ produces an elastic

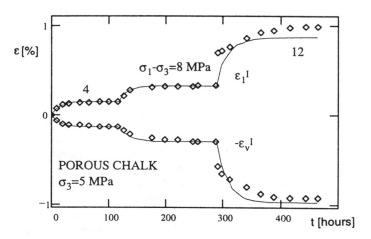

Figure 5.24 *Triaxial creep curves for chalk: solid lines : model prediction; diamonds: test results.*

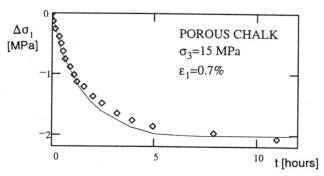

Figure 5.25 *Stress relaxation in triaxial test: diamonds: tests; solid line: nonassociated model.*

$$\left[\left(\dot{\varepsilon}_1^I - \dot{\varepsilon}_2^I\right)^2 + \left(\dot{\varepsilon}_2^I - \dot{\varepsilon}_3^I\right)^2 + \left(\dot{\varepsilon}_3^I - \dot{\varepsilon}_1^I\right)^3\right]^{\frac{1}{2}} = k_T \left\langle 1 - \frac{W(t)}{H(\sigma)} \right\rangle N_2 + k_S \frac{\partial S}{\partial \tau} \tag{5.3.8}$$

if formulae (5.3.3) are used. If classical triaxial tests are available formula (5.3.6) is written as

$$\sqrt{2}\left|\dot{\varepsilon}_1^I - \dot{\varepsilon}_2^I\right| = k_T \left\langle 1 - \frac{W(t)}{H(\sigma)} \right\rangle \frac{\partial F}{\partial \tau} + k_S \frac{\partial S}{\partial \tau} \ . \tag{5.3.9}$$

Let us consider a typical uniaxial or triaxial creep curve for a rock, showing the variation of one of the stress components under a constant loading. Figure 5.29 shows such a curve for rock salt: the first non-linear portion corresponds primarily to the transient creep, though a steady-state creep may also be present with a negligible effect. In the last, linear portion steady-state creep is dominant, and it is expected that the contribution of the transient creep is negligible. From a microphysical point of view the situation is certainly more involved (see Section 3.1)

Let us discuss now the restrictions to be imposed on the function S in order to correctly describe the irreversible volumetric behavior. We use Figure 5.30 for this purpose, where the constitutive domain of all possible stress states is shown. This domain is bounded by the short-term failure surface and can be divided into two subdomains: one is the compressibility domain where $X > 0$ and the other one is the dilatancy domain where $X < 0$. The boundary between these two domains is just the C/D boundary $X = 0$. In the compressibility domain, where $X > 0$, the irreversible volumetric rate of deformation for steady-state creep must be zero:

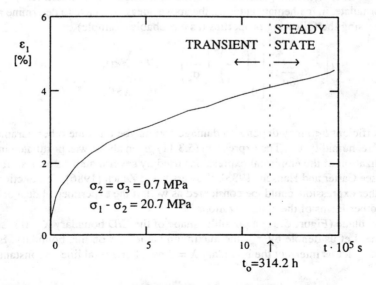

Figure 5.29 *Typical creep curves for rock salt (Mellegard et al. [1981]) showing a guessed boundary between transient and steady-state creep.*

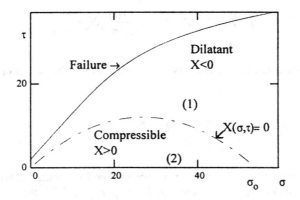

Figure 5.30 *Domains of dilatancy and compressibility in the σ-τ plane.*

$$\dot{\varepsilon}_v^I\big|_S = 0 \qquad\qquad (5.3.10)$$

since the volume cannot shrink to zero by steady-state creep. Thus in the compressibility domain the steady-state creep must result in a change in shape only and therefore $\dot{\varepsilon}_v^I = \dot{\varepsilon}_v^I\big|_T$, i.e., the irreversible volumetric component of the strain rate is due to the transient component only. In the dilatancy domain where $X < 0$, a steady-state volumetric creep producing dilatancy is possible, $\dot{\varepsilon}_v^I\big|_S < 0$, besides a creep producing a change in shape, since, as already mentioned, the microcracking generation and multiplication is here possible.

Let us formulate, in mathematical terms, the above ideas. We have to determine a function $S(\sigma,\tau)$ possessing the following properties (as a probable example)

$$\dot{\varepsilon}_v^I\big|_S = k_s \frac{\partial S}{\partial \sigma} := \begin{cases} b\left(\dfrac{\tau}{\sigma_*}\right)^m \left(\dfrac{\sigma}{\sigma_*}\right)^n & if \quad X<0, \\ 0 & if \quad X\geq 0. \end{cases} \qquad (5.3.11)$$

Here the coefficient $b<0$ may depend on damage and maybe on some other parameters such as temperature, humidity, etc. The expression (5.3.11) given above was postulated in this form as a generalization of the empirical expressions used by several authors to describe uniaxial creep data (see Carter and Hansen [1983], Wawersik and Zeuch [1986], and Sections 2.2 and 3.1). Any other expression could be considered as well, if the experimental data or theory are suggesting other forms of the generalization.

In the σ-τ plane (Figure 5.31) a possible shape of the C/D boundary $X = 0$ is shown as a dash-dot line. Let us denote by τ_m the maximum value of τ on this boundary. For a fixed value $0<\tau<\tau_m$ let us intersect the boundary $X = 0$ by a horizontal line τ=constant, to get

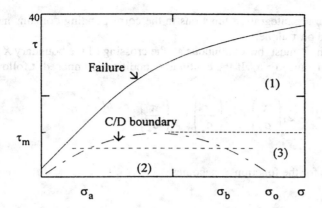

Figure 5.31 *Notations and domains used for the determination of the function S.*

$$\left.\begin{aligned}\frac{\sigma_a}{\sigma_*}\\[4pt]\frac{\sigma_b}{\sigma_*}\end{aligned}\right\}=\frac{-f_2\pm\left(f_2^2+4f_1\dfrac{\tau}{\sigma_*}\right)^{\frac{1}{2}}}{2f_1}\qquad(5.3.12)$$

for the points of intersection. Here for rock salt, the boundary $X = 0$ was written in the form (4.3.5) and with the maximum being $\tau_m = -(f_2^2/4f_1)$. The notation used in Figure 5.31 is self-explanatory.

If we integrate (5.3.11) with respect to σ (τ fixed) three cases have to be distinguished, corresponding to the three regions shown in Figure 5.31. One obtains quite easily

$$k_S S(\sigma,\tau)=b\left(\frac{\tau}{\sigma_*}\right)^m\left(\frac{\sigma}{\sigma_*}\right)^{n+1}\frac{\sigma_*}{n+1}+Q\left(\frac{\tau}{\sigma_*}\right)\quad if\ \begin{cases}\tau\ge\tau_m\ \ or\\ \sigma\le\sigma_a\ and\ \tau\le\tau_m\ ,\end{cases}\qquad(5.3.13a)$$

$$k_S S(\sigma,\tau)=b\left(\frac{\tau}{\sigma_*}\right)^m\left(\frac{\sigma_a}{\sigma_*}\right)^{n+1}\frac{\sigma_*}{n+1}+Q_1\left(\frac{\tau}{\sigma_*}\right)\quad if\ \begin{cases}\sigma_a\le\sigma\le\sigma_b\ and\\ \tau\le\tau_m\ ,\end{cases}\qquad(5.3.13b)$$

$$k_S(\sigma,\tau)=b\left(\frac{\tau}{\sigma_*}\right)^m\left[\left(\frac{\sigma_a}{\sigma_*}\right)^{n+1}-\left(\frac{\sigma_b}{\sigma_*}\right)^{n+1}+\left(\frac{\sigma}{\sigma_*}\right)^{n+1}\right]\frac{\sigma_*}{n+1}$$
$$+Q_2\left(\frac{\tau}{\sigma_*}\right)\quad if\ \begin{cases}\sigma\ge\sigma_b\ and\\ \tau\le\tau_m\ ,\end{cases}\qquad(5.3.13c)$$

where Q, Q_1 and Q_2 are integration functions in the corresponding domains mentioned in (5.3.13), and depend on τ alone.

But the function S must be continuous at the crossing of the boundary $X = 0$, i.e. for τ = constant at $\sigma = \sigma_a$ and $\sigma = \sigma_b$. If the continuity condition is imposed it follows straightforwardly that

$$Q_1\left(\frac{\tau}{\sigma_*}\right) = Q_2\left(\frac{\tau}{\sigma_*}\right) = Q\left(\frac{\tau}{\sigma_*}\right) . \tag{5.3.14}$$

Thus the expression for the function S becomes

$$k_s S(\sigma,\tau) = b\left(\frac{\tau}{\sigma_*}\right)^m \left(\frac{\sigma}{\sigma_*}\right)^{n+1} \frac{\sigma_*}{n+1} + Q\left(\frac{\tau}{\sigma_*}\right) \quad \text{if} \quad \begin{cases} \tau \geq \tau_m \quad or \\ 0 \leq \sigma_a \ and \ \tau \leq \tau_m \end{cases}, \tag{5.3.15a}$$

$$k_s S(\sigma,\tau) = b\left(\frac{\tau}{\sigma_*}\right)^m \left(\frac{\sigma_a}{\sigma_*}\right)^{n+1} \frac{\sigma_*}{n+1} + Q\left(\frac{\tau}{\sigma_*}\right) \quad \text{if} \quad \begin{cases} \sigma_a \leq \sigma \leq \sigma_b \ and \\ \tau \leq \tau_m \end{cases}, \tag{5.3.15b}$$

$$k_s S(\sigma,\tau) = b\left(\frac{\tau}{\sigma_*}\right)^m \left[\left(\frac{\sigma_a}{\sigma_*}\right)^{n+1} - \left(\frac{\sigma_b}{\sigma_*}\right)^{n+1} + \left(\frac{\sigma}{\sigma_*}\right)^{n+1}\right] \frac{\sigma_*}{n+1}$$
$$+ Q\left(\frac{\tau}{\sigma_*}\right) \quad \text{if} \quad \begin{cases} \sigma \geq \sigma_b \ and \\ \tau \leq \tau_m \end{cases}. \tag{5.3.15c}$$

The three expressions given above correspond to the three regions shown in Figure 5.31 as 1, 2 and 3.

If we now differentiate (5.3.15) with respect to τ, taking into account (5.3.12) as well, we get

$$k_s \frac{\partial S}{\partial \tau} = \frac{bm}{n+1}\left(\frac{\tau}{\sigma_*}\right)^{m-1}\left(\frac{\sigma}{\sigma_*}\right)^{n+1} + Q'\left(\frac{\tau}{\sigma_*}\right)\frac{1}{\sigma_*} \quad \text{if} \quad \begin{cases} \tau \geq \tau_m \quad or \\ 0 \leq \sigma_a \ and \ \tau \leq \tau_m \end{cases}, \tag{5.3.16a}$$

$$k_s \frac{\partial S}{\partial \tau} = \frac{bm}{n+1}\left(\frac{\tau}{\sigma_*}\right)^{m-1}\left(\frac{\sigma_a}{\sigma_*}\right)^{n+1} + \frac{b}{\sigma_*}\left(\frac{\tau}{\sigma_*}\right)^m \left(\frac{\sigma_a}{\sigma_*}\right)^n \left(f_2^2 + 4f_1\frac{\tau}{\sigma_*}\right)^{\frac{1}{2}}$$
$$+ Q'\left(\frac{\tau}{\sigma_*}\right)\frac{1}{\sigma_*} \quad \text{if} \quad \begin{cases} \sigma_a \leq \sigma \leq \sigma_b \\ and \ \tau \leq \tau_m \end{cases}. \tag{5.3.16b}$$

$$k_s \frac{\partial S}{\partial \tau} = \frac{bm}{n+1} \left(\frac{\tau}{\sigma_*} \right)^{m-1} \left[\left(\frac{\sigma_a}{\sigma_*} \right)^{n+1} - \left(\frac{\sigma_b}{\sigma_*} \right)^{n+1} + \left(\frac{\sigma}{\sigma_*} \right)^{n+1} \right] - \frac{b}{\sigma_*} \left(\frac{\tau}{\sigma_*} \right)^m$$

$$\times \left(f_2^2 + 4f_1 \frac{\tau}{\sigma_*} \right)^{-\frac{1}{2}} \left[\left(\frac{\sigma_a}{\sigma_*} \right)^n + \left(\frac{\sigma_b}{\sigma_*} \right)^n \right] + Q' \left(\frac{\tau}{\sigma_*} \right) \frac{1}{\sigma_*} \quad if \ \begin{cases} \sigma \geq \sigma_b \ ana \\ \tau \leq \tau_m \end{cases}. \qquad (5.3.16c)$$

Using the formulae (5.3.5) or (5.3.7) together with (5.3.11) one can determine b, m and n from volumetric creep data available in the dilatancy domain. Afterwards we use the formulae (5.3.6) (or (5.3.8), or (5.3.9)) in conjunction with (5.3.16) in order to determine the function Q from data in the whole constitutive domain. Uniaxial or triaxial deviatoric creep tests obtained with τ ranging from zero to failure will do it.

With this procedure the potential function can be obtained for a general constitutive law aimed at describing steady-state creep of geomaterial or of other similar materials:

$$\dot{\varepsilon}_s^I = k_s \frac{\partial S}{\partial \sigma} = k_s \left[\frac{\partial S}{\partial \sigma} \frac{\partial \sigma}{\partial \sigma} + \frac{\partial S}{\partial \tau} \frac{\partial \tau}{\partial \sigma} \right], \qquad (5.3.17)$$

which may exhibit irreversible dilatancy during creep besides a change in shape. This procedure to determine the function S is general and can be applied to any other compressible/dilatant material, with an appropriate choice of the expressions involved in (5.3.11), if necessary.

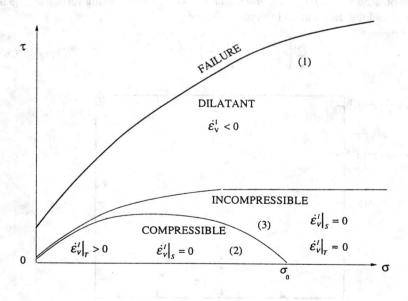

Figure 5.32 *Domains used in the determination of the function S.*

Concerning Figure 5.31 and the procedure to determine the potential $S(\sigma,\tau)$ as described by the formulae (5.3.11)-(5.3.17), it is useful to make the following remark. The domain labeled (3) is in fact a domain of practical incompressibility, i.e., Figure 5.31 should rather be plotted as Figure 5.32 (see also Figures 4.28 and 6.11). In other words domain (2) is a domain of compressibility for the transient creep but it is a domain of incompressibility for the steady-state creep. Domain (3) is anyway a domain of incompressibility for both phases of creep since at high confining pressures significant dilatancy is not to be expected except for very high values of τ. That is why in the above analysis both domains (2) and (3) can be considered to be domains of incompressibility for steady-state creep, and the formula (5.3.16) can be extended into the domain (3) as well. Thus the cumbersome domain (3) can be disregarded altogether when function S is determined. Also the formulae (5.3.15c) and (5.3.16c) can be disregarded. For various materials the boundary between the dilatancy domain and the incompressibility one may have various shapes, either as shown in Figure 5.32 or with the C/D boundary approaching very much the failure surface.

5.3.2 Example for rock salt

In order to give an example only, we use the data obtained by Herrmann et al. [1980] for New Mexico rock salt, by Hansen and Mellegard [1980], Mellegard et al. [1981], Herrmann and Lauson [1981], and Senseny [1986] for the Avery Island rock salt, for the determination of the constitutive parameters, since they have determined the volume change ε_v in their creep tests. From these data and various values of σ we get the irreversible volumetric rate of deformation shown in Figure 5.33. Using the data in conjunction with formulae (5.3.5) and (5.3.11) we obtain $b = -1 \times 10^{-14}$ s^{-1}, $m=5$, and $n = -0.1$. Further using formulae (5.3.9) and (5.3.16a,b,c) and the data shown in Figure 5.34 we get

$$Q'\left(\frac{\tau}{\sigma_*}\right) := \frac{p}{\sigma_*}\left(\frac{\tau}{\sigma_*}\right)^5 \ , \tag{5.3.18}$$

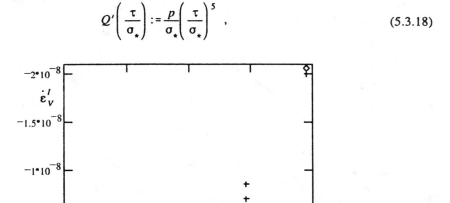

Figure 5.33 *Data used for determination of the constants involved in (5.3.11).*

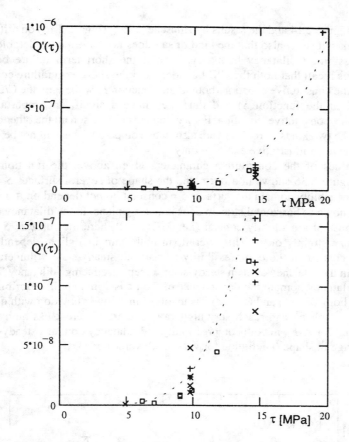

Figure 5.34 *Data used for the determination of the function Q and prediction of the model (dashed line, see (5.3.18)).*

with $p = 3 \times 10^{-13}$ MPa s^{-1}. The prediction of (5.3.18) is shown in Figure 5.33 as a dashed line. The lower portion of the Figure 5.34 shows an enlargement of the upper figure for smaller values of $Q'(\tau)$.

Let us remark that the steady-state term in the constitutive equation was determined for some other kind of rock salt, different from that for which the transient part was determined. In order to compare the mechanical behavior of the two kinds of salt, Figure 5.35 shows the compressibility/dilatancy boundary and the short term failure surface as obtained by Hunsche for Gorleben salt (see Cristescu and Hunsche [1992]) and that of the Avery Island rock salt as obtained by Mellegard et al. [1981], in triaxial tests obtained with several confining pressures. The mechanical behavior of the two types of rock salt is certainly not identical, but they are not too far apart either; we have to take into account the possible errors in the data and also that (mainly at high pressures) when τ increases, the passing from compressibility to dilatancy is very smooth and taking place in a quite wide interval of variation of τ. (As shown in Figures 2.11 and 4.25, it is not so easy to pinpoint this boundary.) Thus the two types of rock salt seem to exhibit comparable mechanical properties with respect to the C/D boundary. Moreover, as evidenced by Figure 5.35 the short-term failure surface for the two kinds of rock salt are similar. Figure 2.14 and the results by Hunsche [1996] (Figure 2.15) on different types of Gorleben rock salt show also that the kind of salt does not have a considerable influence on the compressibility/ dilatancy boundary, even if the short term failure boundary is influenced. Let us recall that both the C/D boundary and the short term failure condition are incorporated in the constitutive equation (both in the function F, while only the C/D boundary is incorporated in the function S) and thus they have a significant importance in the formulation of the constitutive equation. For a complete comparison all the other constitutive coefficients for the two kinds of rock salt are also to be compared. This will not be done, since our aim was to give an illustrative example only.

With the values of the constitutive parameters given above, the function S is fully determined. Figure 5.36 shows in a σ-τ plane the shape of several surfaces S = constant. Inside the compressibility domain the surfaces S = constant do not depend on σ, i.e. these are Mises surfaces shown as horizontal straight lines in the σ-τ plane. In the dilatancy domain the surfaces S = constant are slightly conical (i.e. $\partial S/\partial \sigma < 0$); here the surfaces S = constant depend only very slightly on σ. This overall (in both domains) slight dependence on σ ensures, however, that in the compressibility domain no steady-state volumetric creep is possible, while in the dilatancy domain steady-state creep producing dilatancy is possible. Also in the dilatancy domain the dependence of S on τ is dominant in describing the rock deformation; in both (5.3.11) and (5.3.18) τ is involved in a power function with quite a large exponent and as such $\dot{\varepsilon}^i_s$ can reach very high values there where τ has high values (see Chapters 8 - 10). Moreover, in both compressibility and dilatancy domains steady-state creep produces a change in shape (negligible for very small values of τ).

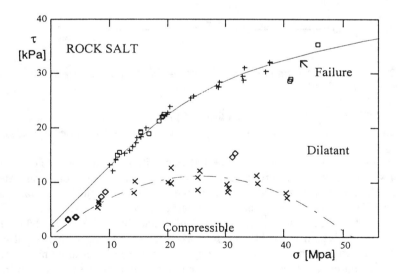

Figure 5.35 *Data for Gorleben (xxx and +++) and Avery Island ($\Diamond\Diamond\Diamond$ and $\Box\Box\Box$) types of rock salt, compared for failure surface and C/D boundary, with prediction of the model.*

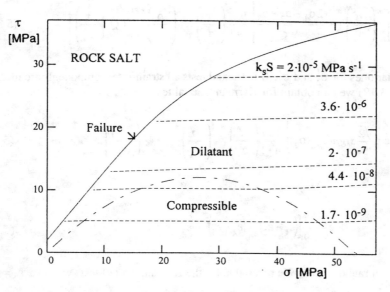

Fig.5.36   *Surfaces S = constant shown in a σ-τ plane*

5.3.3 *Triaxial generalization of uniaxial creep laws* for steady-state

One can find in the literature a great number of uniaxial creep laws which are either based on extensive experimental data or/and on physical arguments (see Tables 2.2 ans 2.3). Let us consider now the following problem (Cristescu [1994a]): if a certain uniaxial steady-state creep formula is already well formulated in the form $\dot{\varepsilon}_1 = f(\sigma_1)$, how can we generalize such a formula for a triaxial stress state in the general form (5.3.17) say, by taking into account the volumetric changes as well (obviously no volumetric changes are described by uniaxial creep laws). For this purpose we consider the steady-state creep only. The procedure we use is just the opposite of what one is usually doing: we do not try to give three-dimensional generalization of a uniaxial creep law, but use a *vice versa* method. We particularize the general triaxial creep law to a uniaxial one and compare the coefficients with those involved in the experimentally established uniaxial constitutive law. In this way we determine these coefficients.

The main effort is to determine the function $Q'(\tau)$. Since this function depends on τ alone, it can be determined either from triaxial data or from uniaxial data, whatever is available. This very much facilitates the determination of $Q'(\tau)$. Let us first present the formulae necessary to evaluate the constitutive function from the data.

Let us write the general constitutive equation (5.3.17) for triaxial tests and steady-state creepinthedilatancy region, i.e. the region 1 from Figure 5.30, for a single rate-of- deformation component as

$$\dot{\varepsilon}_1' = \frac{b}{3}\left(\frac{\tau}{\sigma_*}\right)^m\left(\frac{\sigma}{\sigma_*}\right)^n + \frac{2\sigma_1 - \sigma_2 - \sigma_3}{9\tau}\left[\frac{bm}{n+1}\left(\frac{\tau}{\sigma_*}\right)^{m-1}\left(\frac{\sigma}{\sigma_*}\right)^{n+1} + Q'\left(\frac{\tau}{\sigma_*}\right)\right] \qquad (5.3.19)$$

and two similar ones for $\dot{\varepsilon}_2'$ and $\dot{\varepsilon}_3'$. If in triaxial tests all strain-rate components are measured, then from (5.3.19) we can obtain for Kármán triaxial tests

$$\bar{\varepsilon} = \frac{\sqrt{2}}{3} sign(\sigma_1 - \sigma_2)\left[\frac{bm}{n+1}\left(\frac{\tau}{\sigma_*}\right)^{m-1}\left(\frac{\sigma}{\sigma_*}\right)^{n+1} + Q'\left(\frac{\tau}{\sigma_*}\right)\right], \qquad (5.3.20)$$

where

$$\bar{\varepsilon}^2 = \frac{2}{9}\left[\left(\dot{\varepsilon}_1 - \dot{\varepsilon}_2\right)^2 + \left(\dot{\varepsilon}_2 - \dot{\varepsilon}_3\right)^2 + \left(\dot{\varepsilon}_3 - \dot{\varepsilon}_1\right)^2\right] \qquad (5.3.21)$$

is the equivalent rate of strain. We recall that for the Kármán kind of tests we have $\varepsilon_2 = \varepsilon_3$ and $\sigma_2 = \sigma_3$, i.e.,

$$\bar{\varepsilon} = \frac{2}{3}\left|\dot{\varepsilon}_1 - \dot{\varepsilon}_2\right| \quad , \quad \tau = \frac{\sqrt{2}}{3}\left|\sigma_1 - \sigma_2\right| \quad . \qquad (5.3.22)$$

For true triaxial creep tests the formula (5.3.20) is written as

$$\bar{\varepsilon} = \frac{\sqrt{2}}{3}\left[\frac{bm}{n+1}\left(\frac{\tau}{\sigma_*}\right)^{m-1}\left(\frac{\sigma}{\sigma_*}\right)^{n+1} + Q'\left(\frac{\tau}{\sigma_*}\right)\right] \qquad (5.3.23)$$

and can be used if all strain-rate components are measured. Compared with (5.3.19) the two formulae (5.3.20) and (5.3.23) contain a single right-hand term and as such are more appropriate to use when volumetric data are less accurate.

The function $Q'(\tau/\sigma_*)$ can be determined using either (5.3.20) if both $\dot{\varepsilon}_1$ and $\dot{\varepsilon}_2$ are measured, or (5.3.23) if all rate-of-strain components are measured, or finally (5.3.19) if only $\dot{\varepsilon}_1$ is measured. All these approaches are possible, but again the first two are more accurate.

In order to approximate the data with the above formulae we assume for $Q'(\tau/\sigma_*)$ various particular expressions, for instance

$$P_0\left(\frac{\tau}{\sigma_*}\right)^{q_0} \quad , \quad P_1\left(\frac{\tau}{\sigma_*}\right)^{q_1} + P_2\left(\frac{\tau}{\sigma_*}\right)^{q_2} \quad , \quad p\left(\frac{\tau}{\sigma_*}\right)^r sinh\left(q\frac{\tau}{\sigma_*}\right) \quad . \qquad (5.3.24)$$

If used for rock salt these three expressions correspond to the case when the dislocation climb mechanism is dominant, to the case when an additional unidentified, but experimentally well defined, mechanism is also present, or finally to the case when the cross-slip mechanism is dominant at high stresses. All these mechanisms are thermally activated so that it is expected that some of the constitutive coefficients in (5.3.23) are temperature-dependent (see also

Jeremic [1994], Munson *et al.* [1996] and Chapter 3).

If only uniaxial data are available for Kármán tests when $\sigma_2 = \sigma_3$ and $\sigma_1 > 0$, we have also

$$\sigma = \frac{\tau}{\sqrt{3}} \quad , \quad \tau = \frac{\sqrt{2}}{3}\sigma_1 \quad , \quad \sigma = \frac{\sigma_1}{3} \quad . \tag{5.3.25}$$

One can still use formula (5.3.20) if both rate-of-deformation components are measured. However, if only $\dot{\varepsilon}_1$ is recorded one has to use (5.3.19) which for the uniaxial tests becomes

$$\dot{\varepsilon}_1 = \frac{b}{3}\left(\frac{\sqrt{2}}{3}\frac{\sigma_1}{\sigma_*}\right)^m \left(\frac{\sigma_1}{3\sigma_*}\right)^n + \frac{\sqrt{2}}{3}(sign\,\sigma_1)\left[\frac{bm}{n+1}\left(\frac{\sqrt{2}}{3}\frac{\sigma_1}{\sigma_*}\right)^{m-1}\left(\frac{\sigma_1}{3\sigma_*}\right)^{n+1}\right.$$
$$\left. + Q'\left(\frac{\sqrt{2}}{3}\frac{\sigma_1}{\sigma_*}\right)\right] \quad . \tag{5.3.26}$$

Using this formula one can determine Q' from uniaxial data. One can use, for instance, the expressions (5.3.24) for Q', or maybe some other one in which, however, one has to replace τ by $(\sqrt{2}/3)\sigma_1$.

If one wishes to disregard the irreversible volumetric changes during steady-state creep, then in all the above formulae we have to put $b = 0$. Generally it is to be expected that the last term from the right-hand side of (5.3.19), (5.3.20), (5.3.21) and (5.3.26), the one which involves Q', is the dominant one. Since most of the uniaxial experimentally determined creep formulae assume no irreversible volumetric changes, it is natural to compare these laws with various variants of (5.3.26) where the volumetric changes are disregarded, or to consider the cases when these changes are small anyway. Let us give some examples.

First let us consider the power law of the form

$$\dot{\varepsilon}_1 = A\left(\frac{\sigma_1}{\sigma_*}\right)^N \tag{5.3.27}$$

for the steady-state creep of rock salt. We assume that this law was well formulated from the data so that both coefficients A and N are well determined. It is expected that A depends mainly on temperature. We must compare (5.3.27) with (5.3.26) where the volumetric changes are disregarded ($b = 0$) and Q is replaced by the expression (5.3.24)$_1$. By identification of the coefficients it follows that

$$q_o = N \quad , \quad p_o = \left(\frac{3}{\sqrt{2}}\right)^{q_o+1}(sign\,\sigma_1)\,A \tag{5.3.28}$$

for the determination of the coefficients p_o and q_o involved in the triaxial law, in terms of the coefficients A and q involved in the uniaxial law. Adding the volumetric changes (quantitatively these terms are of relatively small magnitude), the triaxial model (5.3.19) becomes

$$\dot{\varepsilon}_1 = \frac{b}{3}\left(\frac{\tau}{\sigma_*}\right)^m\left(\frac{\sigma}{\sigma_*}\right)^n + \frac{2\sigma_1-\sigma_2-\sigma_3}{9\tau}\left[\frac{bm}{n+1}\left(\frac{\tau}{\sigma_*}\right)^{m-1}\left(\frac{\sigma}{\sigma_*}\right)^{n+1} + p_o\left(\frac{\tau}{\sigma_*}\right)^{q_o}\right]. \tag{5.3.29}$$

Thus for the general expression (5.3.17) we have obtained

$$k_S\frac{\partial S}{\partial \sigma} = b\left(\frac{\tau}{\sigma_*}\right)^m\left(\frac{\sigma}{\sigma_*}\right)^n \quad , \quad k_S\frac{\partial S}{\partial \tau} = \frac{bm}{n+1}\left(\frac{\tau}{\sigma_*}\right)^{m-1}\left(\frac{\sigma}{\sigma_*}\right)^{n+1} + p_o\left(\frac{\tau}{\sigma_*}\right)^{q_o} \tag{5.3.30}$$

so that the general constitutive equation (5.3.17) is entirely determined.

In a similar way let us assume that a uniaxial steady-state creep law of the form

$$\dot{\varepsilon}_1 = B\left(\frac{\sigma_1}{\sigma_*}\right)^2\sinh\left(C\frac{\sigma_1}{\sigma_*}\right) \tag{5.3.31}$$

was well established from uniaxial creep data (i.e. B and C are determined). We obtain with the same procedure, but using the expression $(5.3.24)_3$ for Q'

$$p = \left(\frac{3}{\sqrt{2}}\right)^3(\mathrm{sign}\sigma_1)B \quad , \quad q = \frac{3}{\sqrt{2}}C \quad , \quad r = 2 \quad , \tag{5.3.32}$$

which determine the coefficients p and q in terms of B and C as well as the constant value of r. The general form of the derivative

$$k_S\frac{\partial S}{\partial \tau} = \frac{bm}{n+1}\left(\frac{\tau}{\sigma_*}\right)^{m-1}\left(\frac{\sigma}{\sigma_*}\right)^{n+1} + p\left(\frac{\tau}{\sigma_*}\right)^2\sinh\left(q\frac{\tau}{\sigma_*}\right) \tag{5.3.33}$$

is thus determined and so is the general form of the constitutive equation ($k_S \, \partial S/\partial \sigma$ is taken from (5.3.30)).

For any existing uniaxial steady-state creep formula the procedure is similar: one assumes for Q' an appropriate general expression and one afterwards compares this general expression with the uniaxial existing creep formula. If for a specific geomaterial several creep mechanisms are involved, i.e. several expressions of the form (5.3.24) (or maybe some other ones) are acting simultaneously, the procedure is the same, the terms being additive if we assume additivity for the rate-of-deformation components due to several mechanisms. It is advisable to examine whether one of the creep mechanisms is dominant and to determine the corresponding coefficients from tests done for such cases. Thus step by step one can determine all coefficients involved.

Other variants of the viscoplastic constitutive equation taking into account temperature changes are due to Desai and Varadarajan [1987], Desai and Zhang [1987], Desai [1990], Desai and Ma [1992], Stormont et al. [1992], Chia and Desai [1994], Desai et al. [1995].

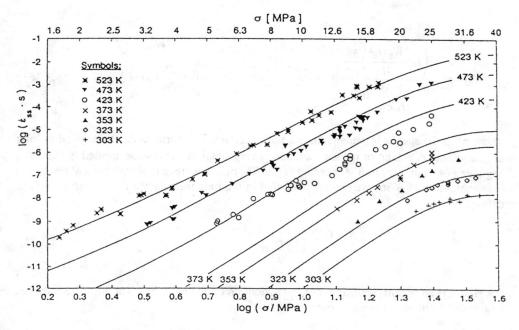

Figure 5.37 *Equivalent steady-state creep rate of axial strain stress differences σ ; model prediction compared with data (after Hampel et al. [1996] improved) (see Figure 3.5).*

5.3.4 The creep formula

Let us describe now how transient and steady-state creep are combined in practice. We can follow the procedure described in Section 5.2.5 to get a formula for the case when both transient and steady-state creep are present. We start from the idea that in the first time interval which follows a loading, the transient creep is dominant. Afterwards there follows a time interval in which neither the transient nor the steady-state creep can be disregarded. Finally in very long time intervals the transient creep becomes negligible compared with the steady-state one, though no complete stabilization or cessation of the transient creep occurs. In order to formulate these ideas in mathematical terms (Cristescu [1994a]) we decompose the irreversible stress work into two terms, corresponding to transient and steady-state creep respectively:

$$W(t) = \int_0^T \sigma \cdot \dot{\varepsilon}^I \, dt = \int_0^T \sigma \cdot \dot{\varepsilon}_T{}^I \, dt + \int_0^T \sigma \cdot \dot{\varepsilon}_S{}^I \, dt = W_T(t) + W_S(t) \quad . \tag{5.3.35}$$

We can write a formula similar to (5.2.52) but written for the transient part W_T of W only. Following a similar procedure as in Section 5.2.5 one obtains quite easily the **creep formula**

$$\varepsilon(t) = \varepsilon^E + \frac{\left\langle 1 - \dfrac{W_T(t_o)}{H} \right\rangle \dfrac{\partial F}{\partial \sigma}}{\dfrac{1}{H}\dfrac{\partial F}{\partial \sigma} \cdot \sigma} \left\{ 1 - \exp\left[\frac{1}{H}\frac{\partial F}{\partial \sigma} \cdot \sigma [k_T(t_o - t)]\right] \right\} + \frac{\partial S}{\partial \sigma} k_S(t - t_o) \ , \tag{5.3.36}$$

where ε^E is the elastic, instantaneous, part of the strain. This formula can be applied as long as stresses are constant, or if their variation is very small. It can also be applied to show the beginning of strain variation. According to (5.3.36) transient creep is always present up to $t \to \infty$. However, from (5.3.36) we can estimate that in the time interval $t > t_o$ and $t \ll t_T$, with t_T defined by

$$t_T := \frac{H}{k_T \dfrac{\partial F}{\partial \sigma} \cdot \sigma} \ , \tag{5.3.37}$$

transient creep is dominant. For $t \gg t_T$ transient creep is negligible. The ultimate value of strains due to transient creep is obtained for $t \to \infty$, i.e.,

$$\varepsilon_{T\infty}^I = \frac{\left\langle 1 - \dfrac{W_T(t_o)}{H} \right\rangle \dfrac{\partial F}{\partial \sigma}}{\dfrac{1}{H}\dfrac{\partial F}{\partial \sigma} \cdot \sigma} \ . \tag{5.3.38}$$

One can use this last formula to determine the time t_S when the two parts are equal $\varepsilon_S^I(t_S) = \varepsilon_{T\infty}^I$. Further for $t > t_S$ the contribution of the steady-state creep is non-negligible. Time t_S is obtained as

$$t_S = t_o + \frac{\left\langle 1 - \dfrac{W_T(t_o)}{H} \right\rangle \dfrac{\partial F}{\partial \sigma}}{k_S \dfrac{\partial S}{\partial \sigma} \dfrac{1}{H}\dfrac{\partial F}{\partial \sigma} \cdot \sigma} \ . \tag{5.3.39}$$

For experiments or underground excavations involving time intervals much longer than $t_S - t_o$, the steady-state creep is dominant.

From (5.3.36) follows also the formula for the volumetric creep as

$$\varepsilon_V(t) = \frac{\sigma}{K} + \frac{\left\langle 1 - \dfrac{W_T(t_o)}{H} \right\rangle \dfrac{\partial F}{\partial \sigma}}{\dfrac{1}{H}\dfrac{\partial F}{\partial \sigma} \cdot \sigma} \left\{ 1 - \exp\left[\frac{1}{H}\frac{\partial F}{\partial \sigma} \cdot \sigma [k_T(t_o - t)]\right] \right\} + \frac{\partial S}{\partial \sigma} [k_S(t - t_o)] \ . \tag{5.3.40}$$

which can describe the evolution of dilatancy during the creep of rocks.

Examples of the application of these formulae will be given in the next chapters.

5.4   *ROCK ANISOTROPY*

5.4.1 *Introduction*

Most rocks are anisotropic. We can distinguish between an intrinsic anisotropy and an anisotropy induced by the stress-strain history. To characterize the anisotropic response both quasistatic and dynamic tests can be performed. For instance, by measuring the travel times of seismic waves propagating in various directions in a rock sample, the type of anisotropy as well as the elastic parameters (see (8.6.3)) can be determined. However, to characterize the global behavior, including irreversible deformation, uniaxial and triaxial compression tests in the directions that are necessary for the formulation of the constitutive law are required.

Since the failure surface is the boundary of the constitutive domain, the anisotropy of this surface reflects the cumulative effects of the structural and induced anisotropy. Thus, the formulation of an anisotropic failure condition is a necessary step towards the modeling of the overall behavior of the rock. In order to solve rock mechanics or mining boundary value problems, rock is frequently considered to obey a linear elastic constitutive law. The directionality of the strength is accounted for by macroscopic anisotropic failure criteria. A brief review of the existing strength criteria is presented in the next subsection.

Most rocks, however, exhibit non-linear behavior prior to failure. A linear viscoelastic law for initially transversely isotropic rock, with application to the uniaxial creep (including volumetric compressibility) was used by Massier [1989]. For an initially anisotropic rock, the type of anisotropy may not change during the deformation process. Thus, to reduce the complexity of the constitutive law, the anisotropy may be considered as fixed, and can be described in terms of the strength characteristics (see, for example, Nova [1986]). However, a more realistic model of the material behavior should account for the combined effect of initial and induced anisotropy. Other geomaterials, such as sands or normally consolidated clays, or rock salt, may be essentially isotropic under zero effective stress and become anisotropic due to the deformation process. The induced anisotropy models for rock salt will be presented in Section 5.4.3.

5.4.2 *Macroscopic failure criteria for anisotropic rocks* *

Many rocks are characterized by a structural ineherent anisotropy.This ineherent anisotropy is the result of the process of rock formation and of various environmental factors (natural transformations such as diagenesis, methamorphosis or weathering). At the macroscopic level the directionality of the mechanical properties is related to the existence of well defined rock fabric elements such as bedding, layering, foliation and lamination planes, or the existence of linear stuctures. Different domains of the mechanical behavior can be influenced by this intrinsic anisotropy. The study of the influence of the intrinsic anisotropy on the strength characteristics is a topic which has attracted a great deal of interest in recent years.

It is not the purpose of this section to present an exhaustive summary of the studies devoted to anisotropic rock behavior. Comprehensive reviews of the state-of-the-art in the development

---

* Section written by O. Cazacu.

of anisotropic strength criteria can be found in Amadei [1983] and in the papers by Ramamurthy [1993] and Kwasniewski [1993]. In this section we limit attention to short-term quasi-static macroscopic failure of intact rocks.

For a given loading history and temperature, a failure surface is defined as the locus of points that separates the states of stress that can be reached in a given material from states of stress that cannot be attained. The term failure is used to describe either the stress state at which macrofracture occurs, or the peak stress attained during ductile deformation. To determine this surface, laboratory experiments of various types may be performed to obtain stress points that produce failure in the rock. For layered rocks, the tests are generally carried out on cylindrical specimens, cored in several directions, with respect to the anisotropy plane and subjected to axisymmetric compressive stresses, and on cubic specimens under true triaxial compression.

Experimental triaxial Kármán test results on cylindrical specimens have been reported by a great number of authors (Donath [1961, 1964, 1972]; Hoek [1964]; Deklotz et al. [1966]; McLamore and Gray [1967]; Akai et al. [1970]; Attewell and Sandford [1974]; Allirot and Boehler [1979], Niandou et al. [1997] etc.). From the strength data obtained several general conclusions can be drawn:

- The compressive strength behavior of anisotropic rocks is a function of both the confining pressure and the orientation of the plane of anisotropy with respect to the maximum compressive stress.
- The maximum compressive strength occurs at an orientation angle β of $0°$ or $90°$ while the minimum value of strength occurs at $\beta \in (30,45)$. β is the angle between the direction of the axial stress and the strata.
- Evidence exists to suggest that the anisotropic strength behavior tends to decrease with increasing confining pressure. However, Allirot and Boehler [1979] reported that for Montagne d'Andance diatomite the effect of anisotropy does not decrease with increasing pressure (see also Popp and Kern [1994] and Figure 6.9).
- Three types of failure were noted in anisotropic rocks (see McLamore and Gray [1967]; Donath [1972]): shear, both across and along the bedding or cleavage planes; slip along the bedding planes; failure and formation of kink bands.

Theories to explain the strength anisotropy of rocks have been proposed by several authors. Jaeger [1960] has formulated two anisotropic failure criteria. The first theory, known as the "single plane of weakness theory", describes an isotropic material that has one plane (or parallel planes) of weakness with different values of τ_0 and μ (the cohesive strength and the coefficient of friction, respectively) than the surrounding matrix material. While this theory reproduces reasonably well the strength characteristics of a material with very weak joints in a particular direction, it provides a less good description of a rock in which a continuous variation of the mechanical properties with respect to the oriented structure might be expected (Jaeger [1960]). To address this topic, Jaeger [1960] proposed a second theory, known as "continuously variable shear strength theory" . This theory is a generalization of the Coulomb-Navier criterion for isotropic materials. It is assumed that the shear strength τ of the material in a plane inclined at β to the direction of maximum compression has the value

$$\tau = \tau_0 - \tau_1 \cos 2(\alpha - \beta) ,$$

$$(5.4.1)$$

where α corresponds to the plane of minimum shear strength. However, as shown by Donath [1961], the coefficient of internal friction can vary with the inclination of the anisotropy. McLamore and Gray [1967] proposed a law of variation of μ of the same nature as that for τ. Also, these authors have noted that a better fit might be obtained if the cosine term is raised to a power and a different pair of constants are used for the range $0 < \beta < \alpha$ than for the remaining range of β (see equation (5.4.1)). Assuming that bedding planes, cleavage planes or schistosity planes represent oriented Griffith cracks, Walsh and Brace [1964] extended the McClintock and Walsh theory (McClintock and Walsh [1962]), to describe the fracture of anisotropic rocks. These authors supposed that the body is composed of long non-randomly oriented cracks that are superposed on an isotropic array of randomly distributed smaller cracks. Failure occurs due to tensile stress through the growth of either the long or the small cracks, depending upon the orientation of the long crack system with respect to the applied deviatoric stress. The theories discussed so far require a wide range of tests and a large amount of curve fitting.

A more general approach was proposed by Goldenblat and Kopnov [1965]. Based on the idea that an anisotropic failure criterion must be invariant with respect to the choice of the coordinate axes, these authors suggested the use of strength tensors. They proposed an anisotropic criterion in the following general form

$$(F_i \, \sigma_i)^\alpha + (F_{ij}\sigma_i\sigma_j)^\beta + (F_{ijk}\sigma_i\sigma_j\sigma_k)^\gamma + ... = 1 \ , \tag{5.4.2}$$

where the contracted notation is used and $i, j, k = 1,2,...6$ ($\sigma_1 = \sigma_{11}$, $\sigma_2 = \sigma_{22}$, $\sigma_3 = \sigma_{33}$, $\sigma_4 = \sigma_{23}$, $\sigma_5 = \sigma_{13}$, $\sigma_6 = \sigma_{12}$). They investigated the special case $\alpha = 1$, $\beta = 1/2$, $\gamma = -\infty$.

Though proposed and investigated in the context of fiber-reinforced composites, the Tsai and Wu [1971] criterion is widely used in engineering for all types of anisotropic materials. For geological materials such as rocks a widely used failure criterion is Pariseau's criterion (Pariseau [1972]). For anisotropic rocks having 3 mutually orthogonal planes of symmetry (orthotropic), Pariseau's criterion (Pariseau [1972]) is expressed as

$$\left[(F(\sigma_2 - \sigma_3)^2 + G(\sigma_3 - \sigma_1)^2 + H(\sigma_1 - \sigma_2)^2 + L\sigma_4^2 + M\sigma_5^2 + N\sigma_6^2\right]^{\frac{n}{2}} - (U\sigma_1 + V\sigma_2 + W\sigma_3) = 1 \ , \tag{5.4 3}$$

where the nine coefficients of the stresses and n ($n \geq 1$) are material constants. The stress components correspond to the rectangular coordinate system which defines the planes of symmetry. If the material to be described is transversely isotropic in the S_1 direction, then equation (5.4.3) is invariant with respect to arbitrary rotations about the S_1 axis and the number of independent coefficients is reduced from 9 to 5. This leads to the relations

$$H = G, \quad N = M, \quad W = V, \quad L = 2G + 4F \ . \tag{5.4.4}$$

If there is also rotational symmetry with respect to the S_2 axis belonging to the isotropy plane, then

$$F = G = H, \quad L = M = N = 6F, \quad U = V = W \tag{5.4.5}$$

and equation (5.4.3) reduces to the isotropic case

$$(J_2)^{\frac{n}{2}} = A_3 I_1 + B_3 , \tag{5.4.6}$$

where

$$A_3 = U(6F)^{\frac{n}{2}}, \quad B_3 = (6F)^{\frac{n}{2}} \tag{5.4.7}$$

and I_1 is the first invariant of the stress, and J_2 is the second invariant of the stress deviator. If the hydrostatic stress does not affect failure then $U = V = W = 0$, and for $n = 1$, equation (5.4.3) reduces to Hill's criterion (Hill [1948]) for metal plasticity. Since the Hill criterion is obtained by generalizing the von Mises criterion, the coefficients associated with the cross-products $\sigma_{11}\sigma_{22}, \sigma_{11}\sigma_{33}, \sigma_{22}\sigma_{33}$, are dependent on the coefficients corresponding to $\sigma_{11}^2, \sigma_{22}^2$, and σ_{33}^2, respectively. Accordingly, the quadratic term in stresses in equation (5.4.3) does not allow dependence on hydrostatic stress.

Kaar et al. [1989] have proposed a generalization to anisotropic media of the following isotropic strength criterion

$$\sqrt{J_2} + C_2 \exp(C_3 I_1) - C_1 = 0 \tag{5.4.8}$$

where C_1 through C_3 are constants. Assuming linear elastic deformation, equation (5.4.8) can be considered to establish a limiting value of the distortion energy which is exponentially dependent upon the hydrostatic pressure. A generalization of equation (5.4.8) for anisotropic materials is then sought by replacing the second invariant J_2 with the generalized distortion energy and, I_1 with a linear term in σ. The concept of generalized distortion energy was introduced by Olszack and Urbanowschi [1956]. These authors emphasized that, unlike isotropic materials, the strain energy of anisotropic materials cannot, in general, be decomposed into two terms : Φ_V associated with the energy of volume change and Φ_f associated with distortion. However, for elastic anisotropic materials certain properties of Φ_f and Φ_V can be preserved. Thus, they derived the following decomposition of the strain energy:

$$\Phi = \Phi_f + \Phi_V \tag{5.4.9}$$

where Φ_f does not change if the material is subjected to hydrostatic pressure. Φ_f, called the generalized distortion energy, is defined by :

$$\Phi_f = A_{ijkl}\, \sigma_{ij}\, \sigma_{kl} \; , \tag{5.4.10}$$

where A_{ijkl} is a fourth-order symmetric tensor. The anisotropic exponential criterion thus takes the form

$$\sqrt{\Phi_f} + K_1 \exp(K_2\sigma_1 + K_3\sigma_2 + K_4\sigma_3) - K_5 = 0 \; , \tag{5.4.11}$$

where K_1 through K_5 are constants. The number of independent components of A depends on the material symmetries. In attempting to establish a failure criterion, suitable defining the maximum stresses attained during post yield deformation, Kaar *et al.* [1989] considered the components of the tensor A as independent strength parameters, rather than elastic constants. Let us note that as a limiting case (first-term approximation of $\exp(C_3 I_1)$) equation (5.4.8) reduces to the Drucker-Prager criterion. Also, for $C_2 = 0$, equation (5.4.8) reduces to the von Mises criterion. Thus, as special limiting cases the criterion (5.4.11) reduces to Hill's and Tsai and Wu's criteria, respectively.

Nova [1980] proposed a theory to describe the failure of transversely isotropic rocks in compression. The theory is able to predict reasonably well the dependence of the peak strength on the inclination of the cleavage or bedding plane to the axis of greatest principal stress.

Let us note that one of the main assumptions of this theory implies a linear dependence of the axial stress at failure on the confining pressure. However, this is in agreement with data only for a moderate range of variations of the confining pressures. Nova and Zaninetti [1990] have developed a criterion that describes the strength anisotropy in tension. Theocaris [1989a, b, 1991] has proposed an elliptic paraboloid failure criterion that accounts also for the differential strength effect. This criterion was applied to a great number of transversely isotropic materials such as fiber-reinforced composites, cellular solids, and brittle foams.

The directional material properties impose definite restrictions on the form of failure surface. Accordingly, an appropriate framework for modeling the failure of anisotropic solids is furnished by the theory of invariance. A comprehensive review of the principles involved, toghether with the application of the theory of invariance in solid mechanics, can be found in Boehler [1987] and Zheng [1994]. For an orthotropic material the symmetry group is generated by the set of reflections in the 3 mutually orthogonal planes, characterized by two unit vectors S_1 and S_2, say, and the normal to them. For a transversely isotropic material, the symmetry group is the group of rotations about a single "privileged" direction, say S_1. The dyadic products of the preferred directions form symmetric structural tensors associated with these two types of material symmetries (responses) :

$$M = S_1 \otimes S_1, \quad M_2 = S_2 \otimes S_2, \quad M_3 = \frac{1}{2}(S_1 \otimes S_2 + S_2 \otimes S_1) S_1 \cdot S_2 \; . \tag{5.4.12}$$

We emphasize that the anisotropy characterized by the structural tensors is the initial material symmetry, if the undeformed state is taken as the reference configuration.

Any constitutive relation must be invariant with respect to the symmetry group of the material. It was proven (Boehler [1978], Liu [1982]) that any scalar, vector or second-order tensor valued anisotropic function of vectors and second-order tensors, can be expressible as an isotropic function of the original arguments, and of the structural tensors as additional arguments. Therefore, any scalar property, such as the failure function, say, $f(\sigma)$, can be represented relative to its symmetry group by an isotropic function f_1 of σ and the structural tensors. For a transversely isotropic material, it follows that $f_1 = f_1(\sigma, M)$. Thus, the requirement that $f_1(\sigma, M)$ is isotropic implies that f_1 is a function of the five independent invariants :

$$I_1 = tr\sigma, \; J_2 = \frac{1}{2} tr(\sigma')^2, \; J_3 = tr\sigma^3, \; J_4 = trM\sigma, \; J_5 = trM\sigma^2 .$$

(5.4.13)

Then, the most general form of a stress failure criterion for a transversely isotropic material is given by

$$f_1(I_1, J_2, J_3, J_4, J_5) = 1$$

(5.4.14)

with f_1 an arbitrary function of its arguments.

In order to obtain simple forms of f_1 as well as to be able to reduce the established anisotropic failure criteria to failure criteria for isotropic materials, Boehler [1975] proposed a simplified theory. The underlying idea of this theory is to replace in the expression of a given isotropic failure criterion, the Cauchy stress tensor σ by a transformed tensor Σ defined as

$$\Sigma_{ij} = A_{ijkl}\sigma_{kl}$$

(5.4.15)

where A is a fourth-order tensor which accounts for the structural anisotropy. The tensor A satisfies the usual symmetry conditions with respect to the pairs of indices (i,j) and (k,l), i.e.,

$$A_{ijkl} = A_{jikl} = A_{klij} = A_{lkij} .$$

(5.4.16)

From (5.4.15) and (5.4.16) it follows that Σ is a second-order symmetric tensor. Additional simplified assumptions are made regarding the form of A :

$$A_{ijkl} = 0 \qquad for \; i = j, \; k = l \; and \; i \neq k$$

(5.4.17)

For transversely isotropic materials, the restriction (5.4.17) implies that the number of independent components of A is reduced from five to three. Thus, the truncated matrix of A in the coordinate system (S_1, S_2, S_3) associated with the material symmetry is given by

$$A = diag\left[\alpha, \gamma, \gamma, \frac{\gamma}{2}, \frac{\beta}{2}, \frac{\beta}{2}\right]$$

(5.4.18)

where S_1 is the symmetry axis and α, β, and γ are independent material constants.

By substituting (5.4.18) in (5.4.15) we obtain the expression of Σ as a function of the structural tensor M (see equation (5.4.12)), and of σ:

$$\Sigma = [(\alpha + \gamma - 2\beta) tr M\sigma] M + \gamma\sigma + (\beta - \gamma)(M\sigma + \sigma M) . \tag{5.4.19}$$

Simple computations give the expressions of the invariants of Σ in terms of the stress-invariants and of the mixed invariants of the stress and the structural tensor M:

$$tr\Sigma = [(\alpha - \gamma) tr M\sigma] M + \gamma\sigma + (\beta - \gamma)(M\sigma + \sigma M]$$
$$tr(\Sigma')^2 = a(tr\sigma^2) + c(tr M\sigma)^2 + \gamma^2 tr(\sigma')^2) - 9a(tr M\sigma')^2 + b(tr\sigma) tr M\sigma , \tag{5.4.20}$$

where a, b, and c are quadratic combinations in α, β, and γ, while σ' and Σ' are the deviators of σ and Σ, respectively. The first invariant of Σ is linear in σ and thus can be sought as a generalization to transversely isotropic conditions of I_1. The second invariant of the deviator of the transformed stress tensor is a quadratic homogeneous function of σ, and reduces to J_2 for isotropic conditions ($\alpha = \beta = \gamma = 1$). Let us note that in the case of hydrostatic pressure ($\sigma = pI$), $tr\Sigma'^2 = p(9a + 3b + c) \neq 0$. Thus, the choice of this invariant as generalization of J_2 allows the modeling of the observed distortion of anisotropic materials under hydrostatic pressure. Generalizations of von Mises and Coulomb-Navier isotropic criteria, obtained by substituting σ with Σ, can be found in Boehler [1975].

Using an approach similar to Boehler [1975], a failure criterion for transversely isotropic solids was proposed by Cazacu [1995] and Cazacu and Cristescu [1995]. The proposed criterion is expressed in the following form:

$$\frac{3}{2} tr(\Sigma')^2 - \frac{m}{3} tr\Sigma - 1 = 0 , \tag{5.4.21}$$

where m is a material constant, and Σ' is the deviator of the second-order tensor Σ defined by

$$\Sigma_{ij} = B_{ijkl}\sigma_{kl} . \tag{5.4.22}$$

In contrast to Boehler [1975], all five components of the fourth-order tensor B are considered as independent strength parameters. Thus, in the structural coordinate system (S_1, S_2, S_3) the truncated matrix of B is

$$B = \begin{bmatrix} a & b & b & 0 & 0 & 0 \\ b & d & e & 0 & 0 & 0 \\ b & e & d & 0 & 0 & 0 \\ 0 & 0 & 0 & \dfrac{d-c}{2} & 0 & 0 \\ 0 & 0 & 0 & 0 & \dfrac{c}{2} & 0 \\ 0 & 0 & 0 & 0 & 0 & \dfrac{c}{2} \end{bmatrix} , \qquad (5.4.23)$$

where a, b, c, d, e are independent material parameters.

For isotropic conditions, the proposed criterion reduces to the Mises-Schleicher paraboloid surface (see, e.g., Lubliner [1990])

$$3 J_2 + (\sigma_C - \sigma_T) I_1 - \sigma_T \sigma_C = 0 , \qquad (5.4.24)$$

where σ_T and σ_C denote the tensile and compressive strength, respectively.

In order to ensure that the failure surface (5.4.21) has the same shape for any orientation θ of the principal stresses system (X_1, X_2, X_3) with respect to the structural system (S_1, S_2, S_3), the following relationship between the engineering strengths of the material must be fulfilled:

$$\frac{1}{Q^2} = \frac{4}{Y_T Y_C} - \frac{1}{X_T X_C} , \qquad (5.4.25)$$

where Q is the shear strength in the symmetry plane (S_2, S_3); X_C, X_T are the uniaxial compressive and tensile strengths along S_1, while Y_C, Y_T are the uniaxial compressive and tensile strengths along S_2.

If (5.4.25) holds, the failure surface is an elliptic paraboloid in the three-dimensional space of the principal stresses. The intersection of the failure surface with the triaxial plane (σ_3, $\sqrt{2}\,\sigma_1 = \sqrt{2}\,\sigma_2$) is a parabola, for any orientation θ, θ being the angle between the normal to the strata and the maximum principal stress. Thus, the failure locus is "open" on the compression side, showing that the axial stress may be increased without limit, if the confining pressure is increased proportionally. However, the failure curve is "closed" on the tensile side, the hydrostatic tensile strength being

$$p = \frac{1}{2(1/Y_C - 1/Y_T) + (1/X_C - 1/X_T)} . \qquad (5.4.26)$$

In a plane stress test, on the other hand, the failure curve is "closed" and the strength can only achieve a finite magnitude. As an example, Figure 5.38 shows the intersection of the failure surface with the plane $\sigma_1 = \sigma_2$, for $\theta = 0°$, and $\theta = 90°$, for a **diatomite** (data after Allirot and Boehler [1979]). The intersections of the parabola corresponding to $\theta = 0°$ with the σ_3 axis are at the uniaxial compressive strength (Y_C), and the uniaxial tensile strength ($-Y_T$), respectively. Point C represents the limiting loading condition for a hydrostatic tensile stress state. The intersections with the ($\sigma_1 = \sigma_2$, $\sigma_3 = 0$) axis represent the biaxial failure strengths in compression and tension, respectively. Similarly, for $\theta = 90°$, the intersections with the σ_3 axis are the uniaxial compressive strength (X_C) and the uniaxial tensile strength ($-X_T$), respectively. The parabola passes through the same point C, expressing that hydrostatic strength does not depend on the orientation of the applied loading with respect to the structural axis of the material. The biaxial compressive strength for $\theta = 90°$ is 51 MPa and it is not represented on Figure 5.38. From (5.4.26) it follows that if the condition $2X_T < Y_C$ is fulfilled, as is the case for most rocks, then $|p| < Y_T$. Similarly, if condition $2Y_T < Y_C$ is satisfied, then: $|p| < X_T$. Thus, the hydrostatic absolute value of the hydrostatic tensile strength $|p|$ is lower than Y_T and/or X_T. This is in contrast with most existing criteria that postulate that failure occurs when the major principal stress is equal to the uniaxial tensile strength. The use of tension cutoffs on the failure surface produces numerical instabilities when used in computer codes. The

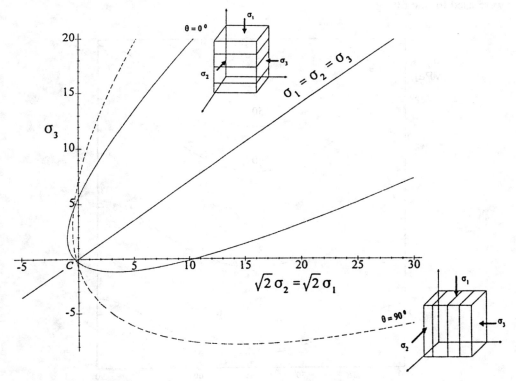

Figure 5.38 *Intersection of the criterion (5.4.21) with the triaxial plane (σ_3, $\sqrt{2}\,\sigma_1 = \sqrt{2}\,\sigma_2$) for $\theta = 0°$ and $\theta = 90°$ (data after Allirot and Boehler [1979]).*

application to boundary-value problems is rather difficult, since these criteria are defined by several expressions in the three-dimensional stress space. For a multiaxial tensile stress however, it can be shown that crack propagation that ultimately leads to failure may occur if none of the principal stresses reaches the uniaxial strength (Aubertin and Simon [1997]). Also it is reasonable to believe that a smooth failure curve in the neighborhood of C (as in the case of the proposed criterion) is physically sound.

For the determination of the 5 independent parameters of the criterion, only two types of tests need to be performed:

a) uniaxial compression and uniaxial tensile tests in the S_1 and S_2 direction, respectively:

b) shear test in the (S_1, S_2) plane.

Since for rocks shear tests are difficult to perform and to interpret, the parameter C (where $C = 1/(R\sqrt{3})$, and R is the shear strength in the (S_1, S_2) plane) can be estimated by a least squares fit, using the compression strengths at a given confining pressure, for the orientations $\theta = 0°$, $\theta = 90°$, and at least another intermediate orientation. As an example, Figure 5.39 shows the variation of the peak axial stress σ_a with the orientation θ, for several confining pressures, for Tournemire shale (data from Niandou [1994]). The solid lines correspond to model predictions, while symbols correspond to the data. The influence of the confining pressure on the strength characteristics is well described although only the test results for $p_c = 50$ MPa were used for the fitting.

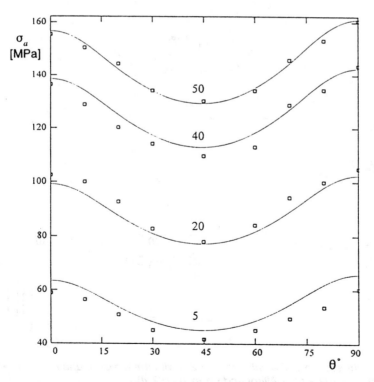

Figure 5.39 *Comparison between the experimental results on Tournemire shale and the theoretical predictions of the criterion (5.4.21) (data after Niandou [1994]).*

The criterion was further used as a short-term failure criterion in an elastic/viscoplastic constitutive model for initially transversely isotropic rocks.

To reduce the complexity of the constitutive law, Cazacu and Cristescu (1995) supposed that the degree and type of anisotropy of the material does not change during the deformation process. Thus, the anisotropy can be described by a unique fourth-order tensor. In a first approximation this tensor is considered constant. It is involved in the expression of the yield function, of the flow rule and of the failure criterion. The determination of its independent components, namely the anisotropic coefficients, was done from the strength characteristics of the material, in conjunction with the failure criterion (5.4.21). The hypothesis of fixed anisotropy reduces considerably the complexity of the constitutive equation. However, a more accurate model of the material behavior should account for the evolution of the anisotropy as a function of the irreversible deformation.

5.4.3 Induced anisotropy

A model which accounts for induced anisotropy was proposed by Aubertin and co-workers (Aubertin *et al.* [1991, 1994, 1996a], Aubertin [1996]). It was developed for rock salt with the following assumptions (In the presentation of the model we follow the paper by Aubertin *et al.* [1996b]). In many situations the inelastic behavior of salt corresponds to a ductile process that induces purely plastic (isovolumetric) flow, as for most metals. For temperatures T below 200°C, strain rates above 10^{-10} s^{-1} and deviatoric stresses smaller than about 40 to 50 MPa, it is considered that inelastic flow of salt in the ductile regime is controlled by dislocation motions that include glide, cross-slip and climb. Both kinematic and isotropic hardening are considered (see also Zaman *et al.* [1992], Senseny *et al.* [1993], Pudewills and Hornberger [1996]). In other circumstances the rock salt behaves as semi-brittle or purely brittle. That is

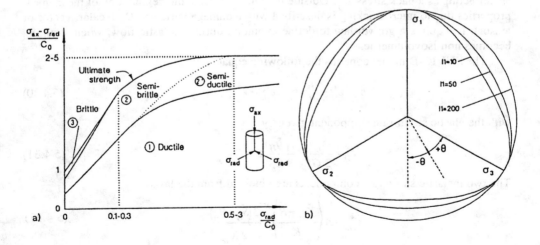

Figure 5.40   *a) Schematic representation of the various inelastic flow regimes; b) Shape of the damage initiation threshold in the π plane for b=0.75.*

shown in a $\sigma - \tau$ plane in Figure 5.40. The latter is expected to occur at fairly high strain rate and low mean stress and temperature, especially under a tensile stress state. Crack initiation and propagation is in this case the main deformation mechanism. A semi-brittle regime occurs when the deviatoric stress state goes beyond the crack initiation threshold (see Figure 5.40). In this regime, dislocation motion and microcrack propagation happen simultaneously. Here, hardening due to isovolumetric deformation processes is accompanied by some reduction of the mechanical strength associated with microcracking (producing dilatancy). The inelastic behavior is attained beyond the yield strength associated to the elastic range. When the material behaves in a ductile manner (isovolumetric, fully plastic), the influence of the loading history tends to fade as time goes by, and it can eventually disappear completely when a steady-state is reached and when the stress-strain relationship is unique.

History dependence is represented by an internal state variable Y_i of scalar or tensor values. For a theory of internal state variables see Cristescu and Suliciu [1982], Lemaitre and Chaboche [1990], and Lubliner [1990].

It is assumed that the deformation is infinitesimal so that the strain-rate components are additive:

$$\dot{\varepsilon}_{ij} = \dot{\varepsilon}_{ij}^e + \dot{\varepsilon}_{ij}^i , \tag{5.4.27}$$

with

$$\dot{\varepsilon}_{ij}^e = C_{ijkl}^{-1} \dot{\sigma}_{kl} , \tag{5.4.28}$$

$$\dot{\varepsilon}_{ij}^i = \dot{\varepsilon}_{ij}^i \left[\sigma_{ij}, T, Y_i \right] \tag{5.4.29}$$

where C_{ijkl}^{-1} is the temperature-dependent elastic compliance tensor of rank 4. The internal state variables requires a kinematic (flow) law complemented by one evolution equation for each such variable. The isotropic hardening is handled with a yield stress R and/or a normalizing drag stress K, while kinematic hardening is represented through a second-rank tensor acting as a back stress B_{ij}. Outside the ductile regime the degradation of mechanical properties due to microcracking is described with a damage variable D_V (scalar, vector or tensor). It requires a growth law to define evolution during inelastic flow, when straining becomes non-isovolumetric.

The SUVIC-D model contains the following equations:

$$\dot{\varepsilon}_{ij} = \dot{\varepsilon}_{ij}^e + \dot{\varepsilon}_{ij}^i = \dot{\varepsilon}_{ij}^e + \dot{\varepsilon}_{ij,v}^i + \dot{\varepsilon}_{ij,d}^i , \tag{5.4.30}$$

with the elastic strain-rate component given by

$$\dot{\varepsilon}_{ij}^e = \frac{\hat{S}_{ij}}{2G} + \frac{\hat{I}_1 \delta_{ij}}{9 K_b} , \tag{5.4.31}$$

The two inelastic strain rate components are obtained from the laws

$$\dot{\varepsilon}_{ij,v}^i = A \left\langle \frac{\hat{X}_{ac} - R}{K} \right\rangle^N \frac{3 \left(\hat{S}_{ij} - B_{ij} \right)}{2 \hat{X}_{ac}} , \tag{5.4.32}$$

$$\dot{\varepsilon}_{ij,d}^{i} = g_{1} \left\langle 1 - \frac{\varepsilon_{c}^{i}}{\varepsilon_{L}} \right\rangle^{-g_{2}} \left\langle \frac{\sqrt{\hat{J}_{2}} - F_{o} F_{\pi}}{F_{r}} \right\rangle^{M} \frac{\delta Q}{\delta \hat{\sigma}_{ij}} \ , \tag{5.4.33}$$

where

$$\hat{X}_{ac} = \left[\frac{3}{2} (\hat{S}_{ij} - B_{ij})(\hat{S}_{ij} - B_{ij}) \right]^{1/2} \ , \tag{5.4.34}$$

$$B_{ij} = B_{ij,s} + B_{ij,l} \ , \tag{5.4.35}$$

$$F_{o} = a_{1}\left(1 - \exp(-a_{2} I_{1})\right) + a_{2} \ , \tag{5.4.36}$$

$$F_{\pi} = \left\{ b\left[b^{2} + (1 - b^{2})\sin^{2}(45^{o} - 1.5\,\theta) \right]^{-1/2} \right\}^{\exp(-v_{1} I_{1})} \ , \tag{5.4.37}$$

$$Q = \sqrt{\hat{J}_{2}} - \left[a_{4}(1 - \exp(-a_{5}\hat{I}_{1})) \right] \hat{F}_{\pi q} \ . \tag{5.4.38}$$

In these equations the following parameters are used

$$\hat{\sigma}_{ij} = \frac{1}{2}\left[\sigma_{ik}(I - D_{V})_{kj}^{-1} + (I - D_{V})_{ik}^{-1}\,\sigma_{kj} \right] \tag{5.4.39}$$

$$S_{ij} = \sigma_{ij} - \frac{1}{3}\delta_{ij}\,tr(\sigma_{ij}), \qquad \hat{S}_{ij} = \hat{\sigma}_{ij} - \frac{1}{3}\delta_{ij}\,tr(\hat{\sigma}_{ij}) \ , \tag{5.4.40}$$

$$I_{1} = tr(\sigma_{ij}) \ , \qquad \hat{I}_{1} = tr(\hat{\sigma}_{ij}) \ , \tag{5.4.41}$$

$$J_{2} = \frac{1}{2}S_{ij}S_{ij} \ , \qquad \hat{J}_{2} = \frac{1}{2}\hat{S}_{ij}\hat{S}_{ij} \ , \tag{5.4.42}$$

$$J_{3} = \frac{1}{3}S_{ij}S_{jk}S_{ki} \ , \qquad \hat{J}_{3} = \frac{1}{3}\hat{S}_{ij}\hat{S}_{jk}\hat{S}_{ki} \ , \tag{5.4.43}$$

$$\theta = \frac{1}{3}\sin^{-1}\left[\frac{3\sqrt{3}}{2} \frac{J_{3}}{(J_{2})^{3/2}} \right] \quad (-\pi/6 \le \theta \le \pi/6) \ . \tag{5.4.44}$$

In these previous equations, σ_{ij} is the Cauchy stress tensor, and $\hat{\sigma}_{ij}$ is the symmetric net stress tensor defined by Murakami [1990]. S_{ij} is the deviatoric stress tensor and $\hat{S}_{ij}$ the net deviatoric stress tensor. I_{1} (and $\hat{I}_{1}$), J_{2} (and $\hat{J}_{2}$), and J_{3} (and $\hat{J}_{3}$) are the corresponding stress invariants. The function F_{o} defines the location of the damage initiation threshold in the $\sqrt{J_{2}} - I_{1}$ plane, while F_{π} does the same in the octahedral (π) plane as a function of the Lode angle θ. Q is the damage induced in the elastic potential; it uses $\hat{F}_{\pi q}$ (with $\hat{\sigma}_{ij}$) instead of F_{π}. Also in equation (5.4.31), G and K_{b} are elastic constants, while in equation (5.4.32), A and N are material constants for the ductile flow. In equation (5.4.33), ε_{L} is a parameter associated with localization of deformation that limits the application fields in this continuum model ($\varepsilon_{e}^{i} < \varepsilon_{L}$). In equation (5.4.39), I is a unit tensor.

For each of the internal state variables the following evolution laws are proposed:

$$\dot{B}_{ij,s} = A_{1s} \left[\frac{2}{3} \dot{\varepsilon}^i_{ij,v} - \frac{B_{ij,s}}{B'_{c,s}} \dot{\varepsilon}^i_{c,v} \right] - A_{2s} \left\langle \frac{B_{c,s} - B''_{c,s}}{C} \right\rangle^q \left[\frac{B_{ij,s}}{B_{c,s}} \right],$$

(5.4.45)

$$\dot{B}_{ij,l} = A_{1l} \left[\frac{2}{3} \dot{\varepsilon}^i_{ij,v} - \frac{B_{ij,l}}{B'_{c,l}} \dot{\varepsilon}^i_{c,v} \right] - A_{2l} \left\langle \frac{B_{c,l} - B''_{c,l}}{C} \right\rangle^q \left[\frac{B_{ij,l}}{B_{c,l}} \right],$$

(5.4.46)

$$\dot{R} = A_3 \left[\dot{\varepsilon}^i_{c,v} - \frac{R}{R'} \dot{\varepsilon}^i_{c,v} \right] - A_4 \left\langle \frac{R - R''}{C} \right\rangle^p,$$

(5.4.47)

$$\dot{K} = A_5 \left[\dot{\varepsilon}^i_{c,v} - \frac{K}{K'} \dot{\varepsilon}^i_{c,v} \right] - A_6 \left\langle \frac{K - K''}{C} \right\rangle^u,$$

(5.4.48)

where

$$B'_{c,s} = B_{os} \left(\ln(\dot{\varepsilon}^i_{c,v}/\dot{\varepsilon}_o) \right)^m, \qquad B'_{c,l} = B_{ol} \left(\ln(\dot{\varepsilon}^i_{c,v}/\dot{\varepsilon}_o) \right)^m$$

(5.4.49)

$$R' = R_o \left(\ln(\dot{\varepsilon}^i_{c,v}/\dot{\varepsilon}_o) \right)^m$$

(5.4.50)

$$K' = \left\langle (\sigma'_c - B'_c - R')/(\dot{\varepsilon}^i_{c,v}/A)^{1/N} \right\rangle$$

(5.4.51)

with:

$$\sigma'_c = \sigma_o \ln(\dot{\varepsilon}^i_{c,v}/\dot{\varepsilon}_o) \quad \text{(equations valid for } \dot{\varepsilon}^i_{c,v} \geq \dot{\varepsilon}^i_o)$$

(5.4.52)

$$B''_{cs} = a_s B'_{c,s}, \qquad B''_{c,l} = a_l B'_{c,l}, \qquad R'' = a_r R', \qquad K'' = a_k K'.$$

(5.4.53)

For the damage variable, the following is used

$$\dot{D}_{ij} = \frac{D_o \dot{\varepsilon}^i_c}{(1 - D_{kk}/D_c)^k} \left\langle 1 - \frac{\varepsilon^i_c}{\varepsilon_L} \right\rangle^{-g_2} \left\langle \frac{\sqrt{J_2} - F_o F_\pi}{F_d} \right\rangle^r d_{ij}$$

(5.4.54)

with

$$D_{kk} = tr(D_{ij}), \qquad 0 \leq D_{kk} \leq D_c \leq 1$$

(5.4.55)

where d_{ij} is a directional tensor that makes the damage variable evolve in the direction of the major principal stress increments. When $D_{ij} = \dot{D}_{ij} = 0$, the above equations reduce to those of SUVIC, developed for ductile regime (Aubertin et al. [1993b]). All the other quantities above are material constants, some of them temperature-dependent.

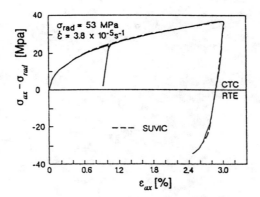

Figure 5.41 *Measured and simulated behavior of a salt sample submitted to conventional compression (CTC) and extrusion (RTE).*

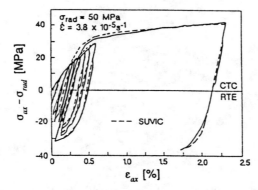

Figure 5.42   *Measured and simulated behavior of a salt sample during cyclic loading.*

The prediction of the SUVIC model has been compared with experimental data. Figure 5.41 shows such a comparison for polycrystalline (artificial) rock salt subjected to a conventional triaxial test. Figure 5.42 shows a comparison for a cyclic test; the mixed hardening of salt being reproduced properly.

6 Damage and Creep Failure

6.1 EXPERIMENTAL BACKGROUND

It has been shown by several examples given in Section 4.2 that **failure is a strongly time-dependent phenomenon,** while obviously ultimate failure depends also on confining pressure, temperature, humidity, and geological factors (see Chapters 1 and 2 and Plumb [1994]). Let us give additional examples. In Figure 6.1 are given several uniaxial stress-strain curves for limestone (Cristescu [1989c]) obtained with four distinct loading rates shown. Failure is marked by a star. Thus failure depends on the loading rate (see also Figure 4.3). If we found out by tests performed in the laboratory a failure condition formulated in terms of stresses only, since tests done in the laboratory are performed in quite short time intervals, we may give misleading information to the mining engineers. In the very long time intervals involved in mining or petroleum applications failure may occur at a much lower stress state, but after a certain long time interval. That is shown in Figure 6.1 by a full line: the test is a creep test in which stress was increased in steps, and after each increase it was held constant 3 days, then 4, 9, and finally 15 days. Thus failure occurring after 31 days is taking place at a much smaller stress state than in conventional tests performed with constant loading rate. The strain at failure is also **loading-history dependent.** Figure 6.1 also shows the elastic slope, but that has been obtained not from the initial slope of the stress-strain curve but using the method described in Section 4.2

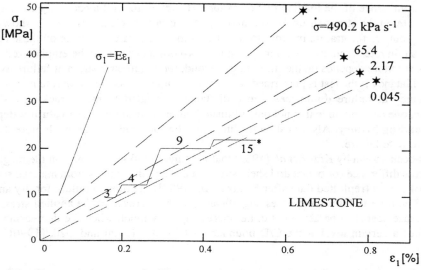

Figure 6.1 *Stress-strain curves for limestone showing a strong influence of loading history on failure (stars).*

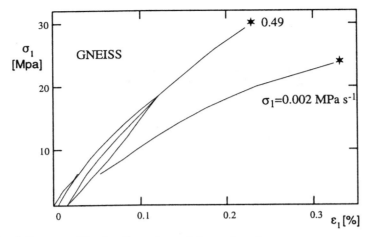

Figure 6.2   *Influence of the loading rate on failure of gneiss.*

Similar results for gneiss are shown in Figure 6.2, for two loading rates. The conclusions are similar. Figure 6.2 shows also an unloading process. A part of the strain component is irreversible but the hysteresis loop reminds us that (see Section 4.2) in this test time effects are combined with unloading.

When checking if the response of a rock is, or is not, loading-history-dependent one should also record the diameter strain, either in uniaxial tests or triaxial tests. The reason is that the behavior of the diameter strain combined with that of axial strain may reveal the volumetric behavior. Sometimes the axial stress-strain $\sigma_1 - \varepsilon_1$ curves show apparently little loading rate influence, but the $\sigma_1 - \varepsilon_2$ or $\sigma_1 - \varepsilon_V$ curves show an obvious loading-rate influence. That is shown in Figure 4.6a for granite (using the data by Sano *et al.* [1981]). While the overall influence of the loading rate on the $\sigma_1 - \varepsilon_1$ curves is rather small, even these curves show a significant influence of the loading rate on failure. However, the influence of the loading rate or strain rate on the stress-strain $\sigma_1 - \varepsilon_2$ and $\sigma_1 - \varepsilon_V$ curves is clear. Figure 6.3 shows only the $\sigma_1 - \varepsilon_V$ curves for granite in order to see clearly the strain-rate influence on these curves. The four strain rates used are shown in the figure. Two observations can be emphasized. First, the stress at failure depends on the strain rate. Second, the volumetric strain at failure exhibits a higher dilatancy if the test is performed with a slow strain rate. As the strain rate is increased, the dilatancy at failure diminishes. Since dilatancy is related to microcracking and pore formation, one can conclude that not only ultimate failure but also **damage evolution depends on the loading history.** Also, in slow loading rates the rock can sustain much more damage before ultimate failure.

It has been shown by Kranz *et al.* [1982] that the time to failure depends on the magnitude of the stress difference (or octahedral shear stress), a concept that comes from material science. Figure 6.4 shows (replotted data after Kranz *et al.* [1982]) the time to failure for dry and wet granite. The time to failure increases significantly with decrease of the applied stress. This time increase seems to be asymptotic, i.e. increasing very much when the decreasing stress approaches a certain level —the C/D boundary— (see also Lajtai and Dzik [1996], Tharp [1996]).

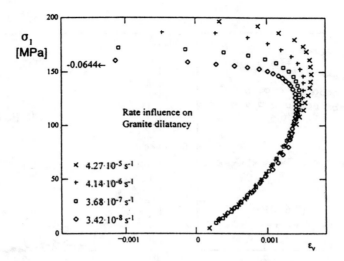

Figure 6.3   *Influence of the strain rate on dilatancy and failure (last point shown) of granite*

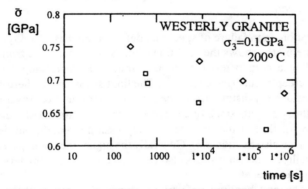

Figure 6.4   *Time to failure for various levels of applied stress for wet (squares) and dry (diamonds) granite.*

Failure after a long-term loading depends obviously on the stress level. Figure 6.5 shows several creep curves (Cristescu [1975]) obtained in uniaxial creep tests on limestone from Palazu Mare (initial porosity 4 to 6%). All curves have been obtained with a single specimen loaded successively with increasing stress level. The short term uniaxial compression strength of this rock is $\sigma_c = 72.4$ MPa. For smaller loading stresses the transient creep ends by stabilization after about 20 days, as shown by Figure 6.5. If the loading stress is less than a certain limit, which depends on the short term compression strength of the rock, practically only transient creep is apparently exhibited by the rock, i.e. stabilization is considered to be realized if after 5 - 10 additional days under the same constant stress no strain increase is exhibited. For this particular limestone axial loading stresses $\sigma_1 < 0.6\,\sigma_c$ will result more or less in transient creep only. For $\sigma_1 = 0.672\,\sigma_c$ the creep becomes steady-state and failure is obtained 14 days after the last reloading. No tertiary creep was exhibited, but for other specimens of the same rock tertiary creep was also observed several days before failure. For the test shown in

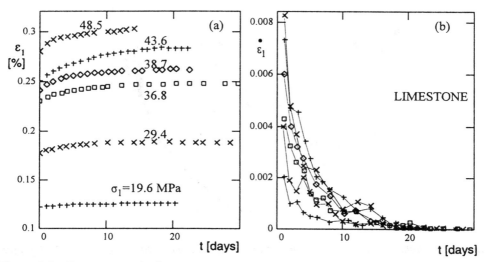

Figure 6.5 *Creep curves for limestone in uniaxial compression test showing that creep failure is possible for $\sigma_1 > 0.6\,\sigma_c$ only.*

Figure 6.5 the total duration up to failure of specimen deformation was 119 days (upper curve in Figure 6.5(a)). Figure 6.5(b) shows the variation of the strain rate, since from such figures one can distinguish if in a certain test the creep is transient (decreasing $\dot{\varepsilon}$), steady-state (constant $\dot{\varepsilon}$) or tertiary creep occurs (increasing $\dot{\varepsilon}$). The limit stress up to where only transient creep is observed is considerably distinct from one rock to other one. For several rocks (such as rock salt) this limit stress is quite small or even not existent, but for most other rocks this limit stress is a certain fraction of the short-term compression strength, but decreases with increasing temperature (see also Section 2.3). Generally the higher the applied stress difference the shorter the time to failure. This limit stress also depends on other parameters, for instance temperature, confining pressure, etc.

Creep tests in tension and bending on various stones (Carrara marble, two types of sandstone, and one type of limestone) have been carried on by Sorace [1996]. All these stones exhibit significant deformation by creep. Failure by creep takes place after a long period of time (more than one year) at stress levels well below the short term failure stress. The author had in mind to explain why ancient constructions and monuments in Italy have shown considerably low tensile and bending resistance capacities of load-carrying members, compared with instantaneous strength limits.

Kranz [1980] has shown that "higher pressure require more volumetric strain to accumulate prior to onset of instability" (see also Kranz [1979], Carter *et al.* [1981], Kranz *et al.* [1982], Schmidtke and Lajtai [1985], Lajtai and Schmidtke [1986]). By testing small circular openings in samples of jointed coal, Kaiser *et al.* [1985] have found that the rupture process is time-dependent and more readily detectable by observing creep deformation than from the instantaneous response to loading.

The influence of the loading rate on the failure mechanism has been studied also by Sano *et al.* [1981, 1982], who have also used the AE technique (see next paragraph). It was found that for **granite** the failure is **loading-rate-dependent** in the sense that the ratio $\Delta\varepsilon_v/\Delta\sigma_1$ is dependent on the loading rate (or rate of deformation). Several authors (see Schock *et al.* [1973], Schock and Heard [1974], Schock [1977]) have shown that an increase of the strain rate raises the failure surface, i.e. the strength increases. Also, in ductile rocks exhibiting work-hardening, the failure surface is not unique (see Section 2.1 and Schock [1977]). However, the strength can occasionally also decrease with increasing strain rate.

6.2   ACOUSTIC EMISSION

The evolutive damage which is developing in a rock specimen or in the rock surrounding an underground excavation can be revealed by the methods of acoustic emission (*AE*). These are transient elastic waves emitted by the rock due to microcrack and crack formation or microcrack closure. For the literature on *AE* see Hardy and Leighton [1977, 1980, 1984, 1989], Cristescu [1989a] and Hardy [1995].

The *AE* methods are important for mining, petroleum and civil engineering. One expects that by in-field source location (Rindorf [1984], Niitsuma and Chubachi [1986], Niitsuma *et al.* [1987], Yanagidani *et al.* [1985], Kranz *et al.* [1994], Kranz and Estey [1996]) it will be possible to predict future failure location. For this purpose research carried out in the laboratory must lead to the formulation of a general constitutive equation able to describe both rock compressibility and dilatancy, i.e. both closure of microcracks and pores as well as microcrack opening. A correlation between the mechanical behavior of the rock and its *AE* must also be established, if possible. Afterwards, by monitoring *AE* around an excavation one would be able to predict its mechanical state, and the location where dilatancy is progressing very fast and thus an impending failure is to be expected. In the laboratory one can study during triaxial creep tests the *AE*, i.e. one has to monitor stresses, strains and *AE*. With *AE*, one is usually recording the total number N of *AE*, or their rate $\dot{N}$ (number of events per unit time). For more advanced evaluations one can also measure the location of the events, the amplitude of *AE*, the energy of the *AE* (square of the amplitude) or the total energy of *AE* (sum of the squares of all amplitudes). A relationship between *AE* and mechanism of fracture propagation is due to Rao *et al.* [1989], Holcomb [1993].

Some revealing data have been obtained by Fota [1983] (unpublished results from 1983) during short term uniaxial compressive creep tests. The stress was increased in successive steps and after each such increase it was kept constant for several minutes. Only events between 500 and 2000 Hz were monitored. It was found that at each reloading the *AE* rate $\dot{N}$ suddenly increased and afterwards decreased in time, during the time interval in which stress was held constant. The *AE* characteristics are quite similar with those of the irreversible volumetric strain rate, which has also been shown in Figure 2.10 for salt. This is shown in Figure 6.6 for an **andesitic** rock (for which $K=13.4$ MPa, $G=18.2$ MPa, $E=37.5$ MPa, $v=0.031$). Axial stress σ is increased in successive steps and after each increase is held constant for several minutes (Fig.6.6c). The corresponding variation of $\dot{N}$ and $\dot{\varepsilon}_v$ are shown in Figures 6.6a and b.

The $\sigma_1 - \varepsilon_v$ curve can be divided into three intervals. In the first one when compressibility takes place, the characteristics of $\dot{\varepsilon}_v$ and those of $\dot{N}$ are alike. Figure 6.6 shows just this case. On the second portion of the $\sigma_1 - \varepsilon_v$ curve, corresponding to the smooth transition from compressibility to dilatancy, N has only a very small variation, while ε_v is nearly constant. Finally in the portion of the $\sigma_1 - \varepsilon_v$ curve where dilatancy occurs, the characteristic variation of $-\dot{\varepsilon}_v$ and $\dot{N}$ are alike. That is why a relationship of the form

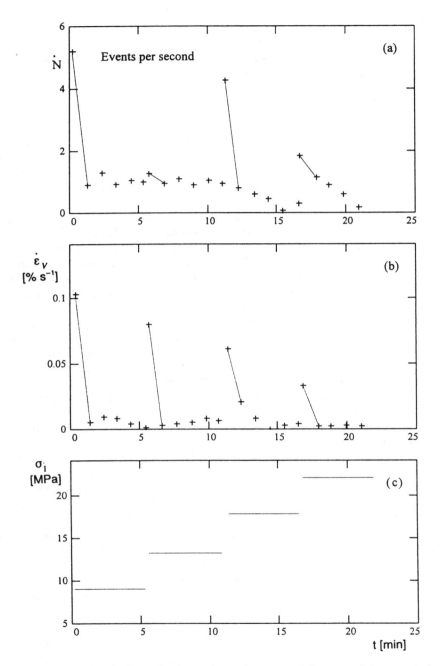

Figure 6.6   *Variation in time of volumetric strain rate and the rate of the events in uniaxial creep tests with incremental stress increase for andesite.*

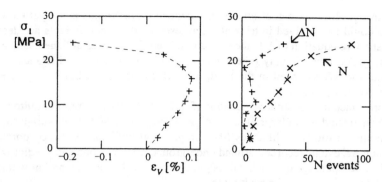

Figure 6.7 *Variation of volumetric strain, of total events N and of the total events per each loading level with axial stress, for andesite.*

$$|\dot{\varepsilon}_V| = A(\sigma)\dot{N} + B(\sigma) \qquad (6.2.1)$$

was proposed (Cristescu [1983, 1989c]) for the mechanical response-*AE* correlation (*MR-AE* correlation, for short). It was found that the coefficients *A* and *B* are stress-dependent, but dependence on other parameters may be possible as well.

Let us consider now the sequence of final values of ε_V and N as obtained successively at the end of various time intervals during which the stress level is held constant in an incremental form of loading. Figure 6.7 shows for andesite the successive final values of ε_V and those of $\dot{N}$ at the end of each time interval in which stress was held constant; the successive loading stress increment is $\Delta\sigma_1 = 24.5$ MPa and after each increment the stress is held constant for 15 minutes. During this time interval creep takes place and the number of events is recorded. The initial porosity of the rock is $n = 1.84$ %. So long as the rock is in the compressibility state, the total number of events ΔN as recorded at the end of each of the successive loading steps (i.e. ΔN is the total number of events recorded during a single loading stress) is somewhat higher than that recorded during those loading steps which correspond to the stress interval during which the rock is passing from compressibility to dilatancy. Finally, during dilatancy the number of events per loading step increases quite fast from one step to the next one. Figure 6.7 shows also the total number of events *N* recorded during the test. The slowest increase of *N* takes place during the passage from compressibility to dilatancy. Similar results to those shown in this figure were obtained for dried sandy soils by Tanimoto and Nakamura [1984]. Terada *et al.* [1984], by testing granite arrived at the conclusion that "the inelastic volumetric strain rate is proportional to A.E. rate". A similar result was reported by Kranz and Estey [1996].

If the specimen has some imperfections, the variation of *N* is rather irregular, but generally the smallest values of $dN/d\sigma_1$ always correspond to the passage from compressibility to dilatancy and the largest ones to the advanced stage of dilatancy just preceding failure. A relative maximum of $dN/d\sigma_1$ is obtained during non-linear compressibility. From the tests performed by Fota it was not possible to make any qualitative distinction between the events produced by the closing of pores and cracks (when $\dot{\varepsilon}_V^I > 0$) and those due to crack multiplication (when $\dot{\varepsilon}_V^I < 0$).

The coefficient functions *A* and *B* involved in (6.2.1) are determined from experimental data. From the uniaxial creep tests it was found that both coefficients are second-degree polynomials in σ_1. How these coefficients may depend on invariants is not yet known. Such a relationship could be obtained in triaxial creep tests. If these coefficients are determined, using the constitutive equation (5.1.1) the full MR-AE correlation is established. Since damage and failure of rocks, and of many other materials, are also related to dilatancy, using this correlation and the AE recorded in the field, one can expect to predict damage and failure around underground openings.

Furthermore, since the constitutive equation (5.1.1) can be used to predict quite accurately where and when around an underground opening, during stress and strain variation, the rock passes from compressibility to dilatancy (the compressibility/dilatancy is incorporated in the model), by using AE events recorded around underground openings one can predict where and when around these openings the rock will begin to be in a dilatant state and how this state is migrating and developing into the rock mass.

Another correlation between the variation in time of N and that of ε_1 during creep tests has been obtained by Hardy et al. (see Hardy [1972]): it was observed that the overall characteristics of $\varepsilon_1 - t$ curves are quite close to those of the corresponding $N - t$ curves. Assuming linear viscoelastic behavior for the rock, a linear relationship between N and ε_1 is established.

A correlation between the inelastic volumetric strain rate during dilatancy and the AE rate was also pointed out by Sano [1981], Sano et al. [1981, 1982], Terada et al. [1984].

From the above follows that it should be possible in the future to establish the correlation between the irreversible volumetric deformation of the rock and AE as a correlation between ε_v^I and N. A similar relationship could be established between $\dot{W}_v^I$ and $\dot{N}$. Since $\dot{W}_v^I$ is related to energy of microcrack closing or opening, a correlation between $\dot{W}_v^I$ and $\dot{N}$ may be revealing.

6.3   DAMAGE ESTIMATION BY DYNAMIC PROCEDURES

The amount of damage existing in a rock specimen, as well as the evolution of porosity in a rock, can be determined by measurements of travel times of various waves crossing the specimen. Generally such tests are performed during triaxial tests, during which stresses and strains are also recorded. Besides various kinds of surface waves disregarded here, in extended solids with instantaneous elastic response, longitudinal and shear waves propagate, while in thin bars longitudinal waves are of relevance in what follows. Their corresponding velocities of propagation are

$$v_P = \left[\frac{E}{\varrho} \frac{1-\upsilon}{(1+\upsilon)(1-2\upsilon)} \right]^{\frac{1}{2}} = \left(\frac{\lambda + 2G}{\varrho} \right)^{\frac{1}{2}} ,$$

$$v_S = \left(\frac{G}{\varrho} \right)^{\frac{1}{2}} = \left[\frac{E}{\varrho} \frac{1}{2(1+\upsilon)} \right]^{\frac{1}{2}} , \qquad (6.3.1)$$

$$v_B = \left(\frac{E}{\varrho} \right)^{\frac{1}{2}} .$$

The first two velocities are known as **seismic velocities**: v_P for the longitudinal waves, v_S for shearing waves, while v_B is the **bar velocity**. Here E, G and υ are the elastic parameters, which are generally not constant. It is just by measuring the travel time of various waves that one determines the values of these parameters for various stress states using the formulae where K is the bulk modulus and ϱ the current density.

$$E = \varrho \frac{v_S^2 \left(3 v_P^2 - 4 v_S^2 \right)}{v_P^2 - v_S^2} = \varrho v_B^2 \quad , \qquad K = \varrho \left(v_P^2 - \frac{4}{3} v_S^2 \right) \quad ,$$

$$G = \varrho v_S^2 \qquad , \qquad \upsilon = \frac{v_P^2 - 2 v_S^2}{2 \left(v_P^2 - v_S^2 \right)} = \frac{v_B^2}{2 v_S^2} - 1 \quad , \tag{6.3.2}$$

The experimental procedures used for the recording of the travel times of various waves are described by several authors: Kolsky [1963], Obert and Duvall [1967], King [1970], Volarovich *et al.* [1974], Schreiber *et al.* [1973], Lama and Vutukuri [1978, Ch.7]. Generally is it a question of ultrasonic longitudinal and shear waves transmitted through the specimen.

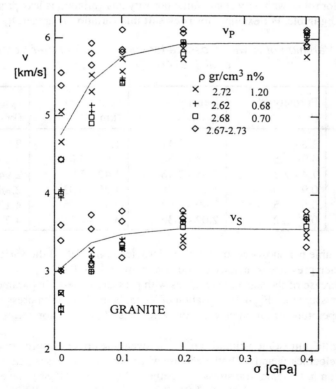

Figure 6.8 *Variation of longitudinal velocity of propagation v_P and of the shear velocity of propagation v_S with pressure, for granite.*

For the history of the method see Bell [1973, Section 3.39].

It is well known (see Kolsky [1963]) that if the ratio between the radius of the specimen and the wavelength is small, i.e. $R/\lambda < 0.3$, for instance, then the bar is considered "thin" and the longitudinal waves propagate with the bar velocity v_B. If, however, $R/\lambda > 1.3$, then the bar is considered thick and the longitudinal waves propagate with velocity v_p. If the frequency of the longitudinal waves can be changed, then both v_B and v_p can be measured on a single specimen. Since the velocities of propagation are quite large (several kilometers per second), and since the size of the specimen is generally small, the accuracy of the measurements must be high.

As a general trend obtained from tests, **all the velocities of propagation generally increase with increasing pressure.** To illustrate this statement, Figure 6.8 shows the variation of the velocities of propagation v_P and v_S with pressure for several granite specimens according to the experimental data by Bayuk [1966]. The densities and initial porosities are also given. The solid lines represent the mean of all cases shown. The increase in the velocities with pressure is more significant for relatively small pressures, while at high pressure this increase is very slow and tending towards a limit value, when all microcracks and pores are closed. On the other hand, the increase of v_P with pressure is more significant than that of v_S. The variation of velocities with pressure is also more important for rocks with high initial porosities, while for rocks with very small initial porosity this variation is less pronounced and sometimes even negligible. For example for **rock salt** this variation is generally quite small.

Table 6.1 *Variation of v_P with pressure for various rocks (after Volarovich et al. [1974])*

Rock	Porosity (%)	Density (g cm⁻³)	v_P at 1 atm (km s⁻¹)	v_P at 0.1 GPa (km s⁻¹)
Clay	3.8 - 3.0	1.77 - 2.41	1.37 - 2.7	2.0 - 4.2
Diorite	4.9 - 3.5	1.78 - 2.38	1.95 - 5.0	3.0 - 5.6
Siltstone	7.4 - 2.7	2.05 - 2.46	1.42 - 5.0	2.94- 5.4
Sandstone	0.9 - 3.0	1.9 - 2.78	1.25 - 6.03	2.88- 6.52
Limestone	0.6 - 1.5	2.40 - 2.82	3.4 - 6.35	4.17- 6.68
Rock salt	0.2 - 2.2	2.02 - 2.36	3.7 - 4.5	4.2 - 4.8

For illustration, Table 6.1 shows according to Volarovich *et al.* [1974] the variation of v_P in various rocks, when pressure is increased from 1 atm up to 0.1 GPa.

Due to the increase of the seismic velocities with pressure the elastic parameters are also increasing with pressure (see Fig.4.19, variation of Young's modulus with pressure for shale). Generally, a temperature variation changes the velocities of propagation (King and Paulson [1981]).

By measuring the ultrasonic velocities one can reveal the effect of porosity on the elastic parameters: the velocities v_P and v_S both decrease when porosity increases (Zimmerman *et al.* [1986]). Using an ultrasonic transmission method, Lo *et al.* [1986] have shown that Chelmsford **granite**, Chicopee **shale**, and Berea **sandstone** are elastically anisotropic up to 0.1 GPa confining pressure; the elastic anisotropy decreases with increasing confining pressure. That is generally true for rocks which at high pressures are isotropic. For wet and dry

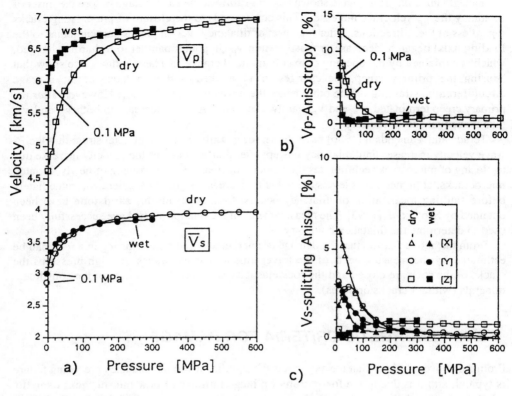

Figure 6.9 *Effect of pressure on averaged P and S-wave velocities a), v_P-anisotropy; b) and shear wave splitting (parallel to the main axes of the reference system); c) in an amphibolite/gneiss under dry and wet (saturated) conditions (Popp and Kern [1994]).*

amphibolite and **amphibolite/gneiss** Popp and Kern [1994] have shown that both v_P and v_S increase with increasing pressure. However, their values at high pressure may not be the same, for wet and dry, in the sense that the ultimate value of v_S at high pressure, for instance, depends on the preferred orientation of the major minerals. Figure 6.9 shows the results obtained by Popp and Kern [1994]. Figure 6.9a shows that the P-wave velocities are about 20% higher at atmospheric conditions in saturated rocks, whereas the v_P-anisotropy is significantly lower than in dry rocks. The corresponding S-wave data reveal that S-wave velocities and S-wave splitting are only weakly affected by intergranular fluids. The authors have found that the effect is greater for greater rock porosity. If velocities of propagation are measured in various directions, and if Z is the direction perpendicular to the foliation, the results are as shown in Figures 6.9b and c. As pressure is increased, differences in the v_P of dry and wet rocks become progressively smaller, approaching a constant value above 100 - 200 MPa. The pressure-independent part of velocity anisotropy is a result of preferred orientation of the major minerals (e.g. hornblende). Figure 6.9c shows a significant intrinsic shear wave splitting (above 200 MPa) as a result of crystal orientation. For some other results see Nur and Simmons [1969], Johnston and Christensen [1993], Scott *et al.* [1993], Schutjens *et al.* [1994], Martin and Haupt [1994], Hesler *et al.* [1996].

Yanagidani *et al.* [1985] have shown that in **granite** at the earlier stage before the onset of dilatancy the v_p velocities increase: "This corresponds to the closure of pre-existing cracks regardless of their directions. After the onset of dilatancy, $v_{P_\perp}$ (in planes perpendicular to the loading axis) began to decrease gradually while $v_{P_\parallel}$ (in planes parallel to it) hardly changed, which is explained by the opening of axially induced cracks". These authors also show that "during the primary creep, the changes in $v_{P_\perp}$ tracked well with those of the average circumferential straining on the other hand, there was no change in $v_{P_\parallel}$. However, after the primary creep terminated, $v_{P_\perp}$ and $v_{P_\parallel}$ hardly changed". A rapid decrease in both $v_{P_\perp}$ and $v_{P_\parallel}$ take place just before faulting.

Terada and Yanagidani [1986] have shown for **granite** tested in uniaxial stress that the *P*-wave velocity increases until dilatancy sets up; after dilatancy begins the velocity increase due to closing of microcracks echilibrates the velocity decrease due to growing of newly initiated microcracks; at higher stress levels the velocity decrease is greatly accelerated, immediately before faulting localization of dilatancy occurs. Similar results for **sandstone** have been obtained by Scott *et al.* [1993]. The measurements of the relation v_P/v_S have successfully been used to determine the dilatancy boundary.

From the above follows that the damage or microcracks and pores existing in a rock can be estimated by the measurements of the travel time of seismic waves. At high pressures the cracks of the rock are closed and the rock elasticity is in fact the elasticity of the constituent minerals (without cracks or pores).

6.4 ENERGETIC CRITERIA FOR DAMAGE

From the above it follows that for most rocks the significant volume dilatancy preceding failure is typical, and it is due to the **formation of a large number of new microcracks over the whole volume of the specimen.** One can conclude that damage and failure of rock are progressive and **related to the same mechanisms which produce dilatancy.** Damage of the rock specimen (at least for rocks of small porosity) begins when dilatancy of the rock starts to increase. Thus it is very important to determine quite accurately, for general triaxial stress states, the point at which during deformation, the rock passes from compressibility into dilatancy, i.e. the location in a stress space of the compressibility/dilatancy boundary must be determined as accurately as possible (see Section 2.1).

Schock [1976] has found for Climax Stock **granodiorite** that "failure is always preceded by dilatant behavior" and that dilatant strain would be a better indication of failure than a $\tau - \sigma$ envelope, because accumulated shear strain is assumed to be directly related to failure. Surface deformation studies of Westerly granite have shown that the detection of the location of the future rupture cone is as late as the tertiary creep stage (Kurita *et al.* [1983]).

The observation that loading produces damage of the rock specimen over its entire volume is also confirmed by *AE* studies. Thus the distribution of microseismic hypocenters in the specimen has been found to be random over the entire volume of the specimen during most of the loading period (see Scholz [1968], Sondergeld and Estey [1981]). For Ahshima granite Yanagidani *et al.* [1985] have found that localization of dilatancy is delayed until the beginning of tertiary creep. Lockner [1993] has found that "in the absence of preexisting heterogeneity in crystalline rock, microcrack localization leading to fault nucleation occurs late in the loading history; near peak stress in a constant strain rate experiment or at the onset of accelerating tertiary creep in a constant stress experiment". The damage mechanism attributed to the microcracking in the rock medium, subjected to dynamic loading is due to Taylor *et al.* [1986].

In order to describe the process of damage leading to failure, we shall use the constitutive equation developed in the previous chapters. We recall that work-hardening by dilatancy and/or compressibility is described by the irreversible stress work per unit volume

$$W(T) = \int_0^T \sigma(t) \cdot \dot{\varepsilon}^I(t) \, dt \, , \tag{6.4.1}$$

and the compressibility/dilatancy boundary is defined by (see (4.2.18) and (5.2.39) and Section 4.3)

$$\frac{\partial F}{\partial \sigma} \cdot 1 = 0 \quad or \quad N_1 = 0 \, . \tag{6.4.2}$$

When during loading the boundary (6.4.2) is crossed, this relationship marks the **onset of dilatancy and damage of rock.** We would like to describe **damage and further creep failure** (failure occurring after some time), using elements already existing in the constitutive equation, **but only those related to dilatancy.** The classical failure criteria formulated in stress invariants only can describe short term, time-independent, failure only, and none of the failure peculiarities shown in Figures 6.1 - 6.3. The same is true for criteria only formulated in strain.

Let us mention some other time-dependent criteria for failure which have been formulated. For instance Kranz and Scholz [1977] have proposed as a criterion the transition from secondary to tertiary creep, i.e., for the instability in the final stage of deformation by creep, a certain limit value of the irreversible volumetric strain ε_V^I. Since tertiary creep appears in creep tests during the very last period of deformation (just before failure), the irreversible volumetric strain might possibly be used as a criterion to predict failure, or better the onset of tertiary creep. It should be pointed out, however, that in performing creep experiments with rocks, it is sometimes difficult to record the tertiary creep region at all. In some other cases, however, the tertiary creep period can be observed for several days before failure occurs. Thus a long time interval elapses from the beginning of the tertiary creep to failure, and therefore a criterion involving ε_V^I alone cannot be used. Also from Figure 6.3 it follows that for various loading rates, failure is to be expected at quite distinct values for ε_V^I: with increasing loading rate the dilatancy at failure decreases significantly. The same is true for different values of σ or p (see Figure 2.20). Both facts are proven by Figure 2.36.

By testing granite at room temperature, Kranz [1980] has shown that the dilatant volumetric strain at the onset of instability leading to failure increases with pressure. This increase is more significant in creep tests than in constant loading rate (of 10 MPa s⁻¹) fracture tests. Another approach to time-dependent damage and failure is due to Costin [1983, 1985, 1987].

We introduce an **energetic damage parameter**, which can describe the evolution in time of the damage of a geomaterial following Cristescu [1986, 1989a]. Let us recall that the irreversible stress work per unit volume (6.4.1) can be decomposed into two parts:

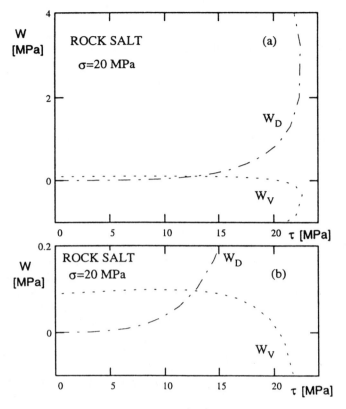

Figure 6.10 *Variation of the two terms W_V and W_D of the irreversible work per unit volume, during a triaxial test.*

$$W(T) = \int_0^T \sigma(t)\dot{\varepsilon}_v^I(t)\,dt + \int_0^T \sigma'(t)\cdot\dot{\varepsilon}^{I'}(t)\,dt = W_V(T) + W_D(T) \qquad (6.4.3)$$

corresponding to the volumetric deformation and to the change in shape (deviatoric).

During a true triaxial test as described in Sections 1.3 and 2.1, the variation of the two terms involved in (6.4.3) is shown in Figure 6.10 (see also Figure 2.19). The lower graph shows an enlargement of the initial variation shown in the upper graph. Let us consider now a typical low porosity compressible/dilatant isotropic rock and its mechanical behavior during true triaxial tests. Figure 6.11a shows two possible stress trajectories followed in such tests. In this figure the C/D boundary is shown as a "transition" zone of incompressibility, between the compressibility domain and the dilatant one (see Section 2.1 and Figure 2.12). During the hydrostatic portion of the test it is only W_V which increases while $W_D = 0$. (See also Figures 1.13, 1.14, 1.15, 2.10 and 2.11.) A schematic representation of the W_V increase during hydrostatic test is shown in Figure 6.11b (compare Figure 2.22b)). If this test is carried out up to very high pressures, W_V reaches a constant maximum value $W_{V(max)}$, at point E

d_f in triaxial tests need the exact recording of the time at failure as well. From this figure follows that d_f seems not to depend on σ. However, it should decrease for small values of σ for obvious reasons. (Failure is possible at $\sigma = 0$.) Therefore, a dependence according to $d_f \sim \sigma^{2/3}$ was proposed by Cristescu [1992], Hunsche [1996]. This is not a final conclusion and additional tests are necessary to check the influence of other parameters on d_f, such as temperature, humidity, etc. Theoretical considerations are very important at this point. From the results shown in Fig.6.13 one has determined (Cristescu [1994a]) the average value

$$d_f = 0.71 \; MPa \,. \tag{6.4.9}$$

This value has been used in all examples given in Cristescu [1994d, 1996c,d] and in the next chapters. Hunsche [1992,1996] also reports values for d_f ranging between 0.5 and 1 MPa. For granite the value obtained by Cristescu [1986] by analyzing the data by Sano *et al.* [1981] is $d_f = 0.167 \, MPa$. The granite was tested in uniaxial tests with four loading rates $\dot{\sigma}_1 = 1.96 \, MPa \, s^{-1}$, $1.92 \times 10^{-1} MPa \, s^{-1}$, $1.68 \times 10^{-2} MPa \, s^{-1}$, and $1.76 \times 10^{-3} Mpa \, s^{-1}$—see Figure 6.14 (also Figure 6.3). The last specimen was unloaded just before failure, but the first three specimens failed at the following stresses and times: failure at 199 MPa after 98 s; failure at 186 MPa after 928 s, and failure at 172 MPa after 9936 s. Thus, if the loading rate is decreased, the stress at failure also decreases and the loading duration increases, as mentioned in Section 6.1. Using the data, W_V was computed using the formula

$$W_V(T) = \frac{1}{3} \int_0^T \sigma_1(t) \dot{\varepsilon}_V(t) \, dt - \frac{\sigma_1^2(T)}{18 \, K} \tag{6.4.10}$$

where σ_1 is the only non-zero component. The results are shown in Figure 6.14. The arrows are pointing out the stresses at failure. Thus the above mentioned value for d_f for granite was obtained as an average of three tests performed with quite distinct loading rates. Also for granite the energy produced by microcracking has been estimated by Wong [1982], using a

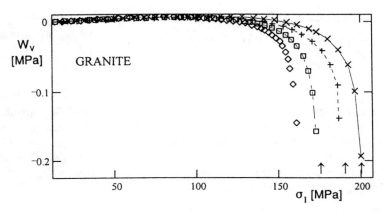

Figure 6.14 *Variation of W_v with σ_1 for granite in uniaxial compression tests. The arrows are showing stresses at failure.*

simplified formula and the experimental data by Taponnier and Brace [1976] obtained with confined pressure of 50 MPa. This energy was estimated to range between $6.67 \times 10^3 \, \text{J m}^{-3}$ and $6.67 \times 10^4 \, \text{J m}^{-3}$ ($= 6.67 \times 10^{-3} \text{b}$, and $6.67 \times 10^{-2} \text{MPa}$) and thus is much smaller than the one mentioned above. For some coals the value $d_f = 4 \times 10^{-3} \text{MPa}$ was found (Cristescu [1987] [1989d]). Alheid and Rummel (after Hunsche [1992]) obtained about $d_f = 0.5 \text{MPa}$ in triaxial tests on sandstone. Unfortunately the data basis for the description of creep fracture of geomaterials is not yet sufficient.

For classical type of triaxial tests the formula to compute W_V is

$$W_V(T) = \int_0^{T_H} \sigma(t)\dot{\varepsilon}_V(t)\,dt - \frac{\sigma^2(T_H)}{2K} + \int_{T_H}^{T} \sigma_1^R(t)\left[\dot{\varepsilon}_1^R(t) + 2\dot{\varepsilon}_2^R(t)\right]dt - \frac{1}{9K}\frac{(\sigma_1^R(T))^2}{2}, \qquad (6.4.11)$$

where the meaning of the superscript R is given in Section 5.2.

6.5 CREEP FAILURE

In order to show how the above concepts can be used to describe creep failure or long term failure of geomaterials, let us consider failure during creep tests. If during tests the stress is constant then

$$W_V(T) = \int_0^T \sigma(t)\,\dot{\varepsilon}_V^I(t)\,dt \qquad (6.5.1)$$

can be integrated using the constitutive equation to get

$$W_V(t) = \sigma\,\varepsilon_V^I(t) = \frac{\left\langle 1 - \dfrac{W_T(t)}{H(\sigma)}\right\rangle \dfrac{\partial F}{\partial \sigma}\,\sigma}{\dfrac{1}{H}\dfrac{\partial F}{\partial \sigma}\cdot\sigma}\left\{1 - \exp\left[\frac{k_T}{H}\frac{\partial F}{\partial \sigma}\cdot\sigma(t_o - t)\right]\right\} + k_S\frac{\partial S}{\partial \sigma}\sigma(t - t_o) + W_V^P. \qquad (6.5.2)$$

where t_o is the time of the beginning of the test (just after loading) and W_V^P stands for the initial "primary" value of W_V at time t_o. For instance, if we would like to describe a creep test performed during the deviatoric part of a triaxial test and we start the test from the end of a hydrostatic compression creep test which ended at time t_H, for example, then $W_V^P = H(\sigma^P, 0)$ with σ^P the "primary" value of σ. If we disregard the difference between $W_V(t_H)$ and $W_{V(\text{max})}$, then the damage parameter (6.4.4) becomes

$$d(t) = W_V^P - W_V(t), \qquad (6.5.3)$$

with the right-hand term obtained from (6.5.2). If $d(t)$ is replaced by its ultimate value d_f, then

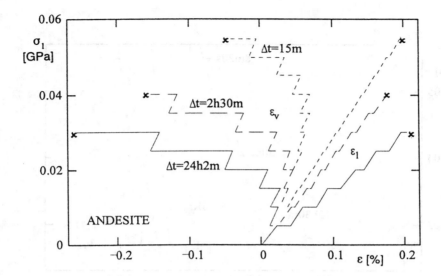

Figure 6.15 *Incremental creep tests for andesite: after each stress increase, the stress is kept constant for the time interval shown.*

the time when creep failure is to be expected can be estimated from (6.5.3).

As an example Figure 6.15 shows several stress-strain curves obtained numerically in uniaxial creep tests for andesite (Cristescu [1986]), where an associated model was used. The average loading rates are $\dot{\sigma} = 6.06 \times 10^{-8}$ GPa s^{-1}, 6.17×10^{-7} GPa s^{-1} and 5.7×10^{-6} GPa s^{-1}, respectively. After each stress increase, stress was held constant for the time interval shown. Following the last loading, failure occurred after the time intervals 12h 30m (bottom curve), 30m and 8m 20s (top curve) respectively. All the features described in the tests mentioned in Section 6.1 can be found in Figure 6.14 as well. Thus the influence of the loading rate on the stress-strain curve, on dilatancy, and on failure, are well described by the model.

The **dependence of creep failure on loading history** is also shown in Figure 6.16 for uniaxial numerical creep tests for the same andesite. Two loading histories have been considered. Short-dashed lines show where creep failure is taking place for loading paths described above, i.e., loading in successive steps of $\Delta\sigma_1 = 0.005$ GPa. The crosses correspond to the failure already shown in Figure 6.15. The times shown are the time intervals elapsed from the last loading up to failure. Failure at 50 hours after loading corresponds to the overall mean loading rate of $\dot{\sigma}_1 = 6.61 \times 10^{-9}$ GPa s^{-1}. If the loading history is changed so that a single loading step is done up to the desired ultimate stress level, and, further the stress is kept constant up to failure, then the loci of the stress and strain state at failure are shown by the long-dashed lines in Figure 6.16. For stresses below $\sigma_1 < 0.024$ GPa, the deformation by creep which follows after a single loading finally stabilizes so that no failure is possible. For higher stresses failure occurs (for the same stress levels) at a much smaller strains than in the previously described path (loading in several steps). Generally, if a certain stress level is reached in fewer steps, then failure occurs in a very short time interval and at very small strains (see Figure 6.15).

It is interesting to find out how long is the time to failure as a function of the magnitude of

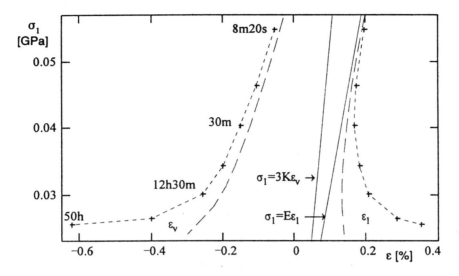

Figure 6.16 *Dependence of creep failure on the loading history.*

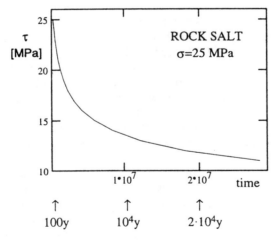

Figure 6.17 *Time to failure in true triaxial creep tests on rock salt (Cristescu [1993b, 1996d]).*

the octahedral shear stress τ. If the mean stress σ remains the same in all cases and we change only τ we get revealing results. Figure 6.17 shows the time to failure for rock salt in true triaxial creep tests using formula (6.5.2) for $\sigma = 25$ MPa and various values of τ. If the value of τ is close to the short term failure surface (which is at 26 MPa) then the time to failure is close to zero. As the value of τ is smaller, the time to failure increases and tends towards infinity if τ is approaching the C/D boundary. The time shown is dimensionless, due to the incertainty in finding out from test an accurate value for k_S (k_T is not important in long term creep tests). However, if we estimate k_S to range in the interval $10^{-5}\,\text{s}^{-1} < k_S < 10^{-4}\,\text{s}^{-1}$, then failure at approximately $\tau = 21$ MPa is expected after 100 years, for example, for $\tau = 13$ MPa after 10^4 years, and when τ approaches 11 MPa the time to failure tends towards infinity. Therefore, for each σ, the time to failure depends on the octahedral shearing overstresses $\tau - \tau_{C/D}$ above the C/D boundary. Thus failure is possible anywhere in the dilatancy domain if stationary creep produces microcracking leading ultimately to a loss of cohesion (see also Rokahr and Staudtmeister [1983]). Since in stationary creep laws power functions τ^m are generally involved, stress states involving a high value of τ will produce a fast creep failure. Thus for any practical purposes one has to check when and where (i.e., at what location around an excavation, for example) creep failure will take place (and that is expected there where τ reaches high values). Also progressive failure may slowly spread into the rock mass. Examples are given in the next chapters.

6.6   HISTORICAL NOTE

The idea of associating the damage concept to the history of microcracking as related to dilatancy, and thus that the octahedral shearing overstress above the compressibility/dilatancy boundary is responsible for the damage, was proposed by Cristescu [1986] (see also Cristescu [1989a, 1996b,d]). It was first applied to andesite, granite and several types of coal, where the damage parameter was defined as the past energy release due to microcracking during the dilatancy process.

This criterion was further used for rock salt by Hunsche [1991], and by Thorel and Ghoreychi [1996], for potash salt, and afterwards by very many authors (see also Allemandou and Dusseault [1996]).

It should be mentioned that this criterion can be used for compressible/dilatant rocks only. For rocks that are compressible only, this criterion cannot be used, or has to be adjusted accordingly.

Other approaches to damage in granular materials, rocks or concrete, or sometimes, variants of the above, are due to Bodner [1981], Mróz and Angelillo [1982], Spetzler et al. [1982], Levy [1985], Lemaitre [1986], Desai [1990], Holcomb [1991], Maier et al. [1991], Ofoegbu and Curran [1991], Margolin and Trent [1992], Aubertin et al. [1993a], Cheng and Dusseault [1993], Stumvoll and Swoboda [1993], Voyiadjis and Abu-Lebdeh [1993], Martin and Chandler [1994], Shao et al. [1994].

7 Mining and Petroleum Engineering Problems

In order to illustrate how the constitutive equations given in the previous chapters can be used, we intend to give some examples. In the present chapter we give the **mathematical formulation of the problem**. For more details see Cristescu [1989a Chap.11, and 1994a].

7.1 STRESSES IN SITU

The stresses existing *in situ* before excavation are called "primary" stresses and the notation $\sigma^P(X)$ is used for the primary stress at location X. These stresses are generally different in various locations, but at each location they are practically constant in time. Generally the vertical component σ_v of the primary stress is due to the overburden pressure and at depth h it is obtained from the overburden pressure as

$$\sigma_v(h) = \int_0^h \rho\, g\, dz = \gamma\, h \qquad (7.1.1)$$

with h in meters and σ_v in MPa, ρ is expressed in Mg m^{-3}, $g = 9.81$ m s^{-2}, and γ is expressed in MPa m^{-1}. The horizontal components of the primary stresses depend strongly on location and on depth. Sometimes we will make the simplifying assumption that at the depth considered the principal stresses in all horizontal directions are equal, and in this case we use the notation $\sigma_h(h)$. Measurements of σ_h have shown that generally at shallow depth $\sigma_h > \sigma_v$, while at greater depth $\sigma_h \approx \sigma_v$ (for literature see Martinetti and Ribacchi [1980], Rummel and Baumgärtner [1985], Rummel *et al.* [1986], Herget [1987], Cooling *et al.* [1988], Ribacchi [1988], Cristescu [1989a, Chap.10], Hyett and Hudson [1989], Cunha [1990], Bock [1993]). In the examples given in the following chapter various values for the ratio of the far field stresses will be considered, in order to discuss various possible cases. That is why we will use the formula $\sigma_h = n\sigma_v$, with the constant n taking various values in the interval $0.3 \leq n \leq 3$. Occasionally n can be either smaller than 0.3 or bigger than 3.

However, very many authors have stressed that in the horizontal planes there are directions of maximum value of primary stress, denoted by σ_H and orthogonal directions to the first ones where the horizontal primary stresses take a minimum value σ_h; pore pressure, lithology, geologic structure and tectonic setting have a great influence (see Kanagawa *et al.* [1986], Rummel [1986], Arjang [1989], Ljunggren and Amadei [1989], Zoback *et al.* [1989], Pahl and Heusermann [1991,1993], Burlet and Cornet [1993], Cornet [1993], Jeffery and North [1993], Haimson *et al.* [1993a, b], Moos and Zoback [1993], Klee and Rummel [1993], Stephanson [1993], Addis *et al.* [1994], Misbahi *et al.* [1994], Sheorey [1994], Yassir and Bell [1994], Haimson *et al.* [1996], Heusermann [1996], Sugawara and Jang [1996]). It was also found that in various locations the three principal components of the primary stress can be distinct up to very large depths (Herget [1987], Kern and Schmidt [1990], Kern *et al.* [1991], Baumgärtner *et al.* [1993], Herget [1993], Zoback *et al.* [1993], Plumb [1994]) (see also Hardy [1996] for the use of the Kaiser effect for the determination of the magnitude of the in situ stress).

Figure 7.1 shows the variation of σ_H, σ_h and σ_v with depth according to several authors. For excavations in the Canadian Shield, Herget [1987] has proposed the following empirical formulae to describe the variation of the primary stress components with depth:

$$\frac{\sigma_H}{\sigma_v} = \frac{357}{h} + 1.46 \quad , \quad \frac{\sigma_h}{\sigma_v} = \frac{167}{h} + 1.10 \ . \qquad (7.1.2)$$

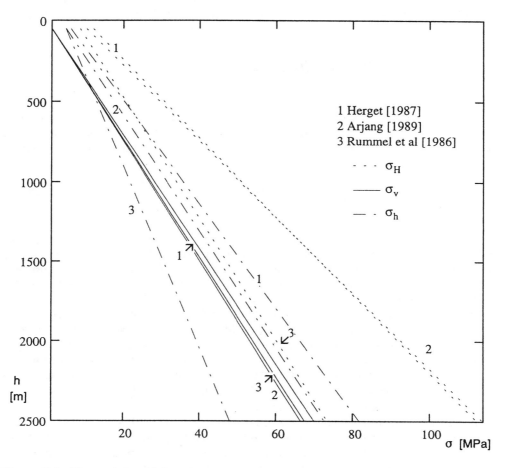

Figure 7.1 *The variation of the primary stress components with depth according to several authors.*

Similar formulae are found to hold for the "strike-slip type" stress regime ($\sigma_h < \sigma_v < \sigma_H$) down to a depth of 3 km by Baumgärtner *et al.* [1993]:

$$\sigma_h = 16.0\,MPa + (0.011\,MPa/m)h \quad , \quad \sigma_H = 30.4\,MPa + (0.023\,MPa/m)h \ ,$$
$$\sigma_v = [(0.0275 \div 0.0284)MPa/m]h \ . \tag{7.1.3}$$

For the Canadian Shield, Arjang [1989] has found that $1.2 \le \sigma_H/\sigma_v \le 4.0$ and $0.7 \le \sigma_h/\sigma_v \le 2.0$. Figures 7.1 and 7.2 have been plotted using the formulae or data given by various authors. As a general trend up to the depth shown, $\sigma_H > \sigma_v$. For the component σ_h in some locations it was found that $\sigma_h > \sigma_v$, while in some others $\sigma_h < \sigma_v$. Figure 7.2 shows the variation with depth of the ratios σ_H/σ_v and σ_h/σ_v. At shallow depth generally $\sigma_H > \sigma_v$ and $\sigma_h > \sigma_v$, but at greater depths σ_h can be close to σ_v.

Kern *et al.* [1994] have studied the effect of both hydrostatic and deviatoric stresses on the velocities of propagation of seismic waves and thus on the *in-situ* stress field at the superdeep borehole from Germany. The data obtained down to 3000 m were extrapolated down to 6000 m. It was found that at shallow depths $\sigma_H > \sigma_v$ and $\sigma_v \approx \sigma_h$. At greater depths ($> 4000$ M) σ_v approaches σ_H more closely and $\sigma_h < \sigma_v \approx \sigma_H$. A more general stress field $\sigma_h < \sigma_v < \sigma_H$ is characteristic for intermediate depth levels. The rock is foliated and σ_h is directed normal to the foliation plane and σ_v and σ_H are parallel to the foliation.

Since generally at shallow depths the ratios of the far field stresses can be any order of magnitude, in the examples given below various possible ratios σ_H/σ_v and σ_h/σ_v of the far

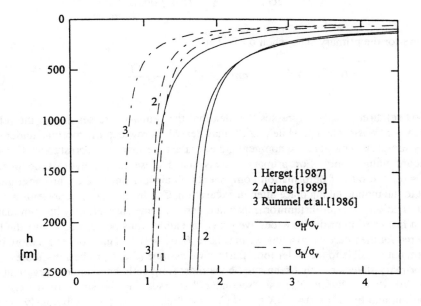

Figure 7.2 *The variation of the ratios σ_H/σ_v and σ_h/σ_v with depth according to various authors.*

field stresses will be considered. We will use the notation

$$\sigma_h = n\,\sigma_V \quad , \quad \sigma_H = N\,\sigma_V \tag{7.1.4}$$

and we will assign various constant values to the coefficients n and N. From the point of view of mining applications, i.e. of excavations possessing dimensions much smaller than the depths, we consider that the primary stress components depend on depth alone and that their variation with depth is quite smooth. For deep boreholes the primary stresses are also assumed to vary smoothly with depth.

In a simplified approach one assumes usually that σ_V is given by (7.1.1) and σ_h, equal in all horizontal directions and sometimes having the meaning of average horizontal stresses, depends on the depth alone and is related to σ_V by $\sigma_h = n\sigma_V$; various values will be assigned to the constant n to accommodate different possible cases. With these assumptions the primary stress states satisfy the Cauchy equilibrium equations.

7.2   INITIAL CONDITIONS

If a constitutive equation was formulated based on tests performed in the laboratory, this constitutive equation is to be satisfied by the primary stresses as well. Therefore from (see (5.1.1))

$$\dot{\varepsilon} = \frac{\dot{\sigma}}{2G} + \left(\frac{1}{3K} - \frac{1}{2G}\right)\dot{\sigma}\mathbf{1} + k_T\left\langle 1 - \frac{W(t)}{H(\sigma)}\right\rangle\frac{\partial F}{\partial\sigma} + k_S\frac{\partial S}{\partial\sigma} \tag{7.2.1}$$

it follows for the primary state when $\dot{\sigma} = 0$ that

$$H(\sigma^P) = W^P = constant \quad ; \quad \dot{\varepsilon}^P = k_S\frac{\partial S(\sigma^P)}{\partial\sigma}. \tag{7.2.2}$$

The first relation (7.2.2)$_1$ expresses the idea that the primary stresses are on the relaxation boundary for transient creep, while (7.2.2)$_2$ may describe slow tectonic motions under existing primary stress; such motions are non-negligible for the time spans of interest to us if τ^P reaches significant values only. For primary stress states close to the hydrostatic one, when $\sigma_V = \sigma_h = \sigma_H$ the rate of deformation components obtained from (7.2.2)$_2$ are negligible with respect to the motion taking place after an excavation. Condition (7.2.2)$_1$ represents a condition of "stabilization" or equilibrium for transient creep. In some cases, it may happen that, either due to a fast erosion or due to excessive previous excavations performed above the location considered for the future excavation, we may have $H(\sigma^P) < W^P$. This would be a more involved problem, but possible to consider too. Further we assume (7.2.2)$_1$ to be satisfied everywhere in the primary state, since that seems to be the most frequently encountered practical case.

We will denote by $\sigma^S(X,t)$ the "secondary" stresses, i.e. the stress field existing after excavation, and by $\sigma^R(X,t) = \sigma^S(X,t) - \sigma^P(X)$ the "relative" stresses. Obviously σ^S and σ^R vary in time. Sometimes, in order to obtain a very simple solution, we assume that the excavation is done quite fast with respect to the time-span during which the excavation (tunnel, shaft, cavern, etc.) is in use. Thus the initial stress variation due to the excavation will be assumed to be elastic, i.e. "instantaneous", and quite often an "exact" solution will be obtained for this initial stress state. Afterwards, in the long time interval which follows, the rock that is surrounding the excavation will deform by creep, possibly associated with a stress relaxation. The strains produced by the excavation will be estimated with respect to the state existing *in situ* before excavation, considered to be the reference configuration. Thus the "relative" stress components are responsible for the strains induced by the excavation.

The procedure we are using to get a simple solution is the following. We assume that $\sigma^P(X)$ is measured and therefore known at the location of excavation. Assuming a fast excavation we determine the secondary stress $\sigma^S(X,t)$ as an "elastic" solution, and therefore also $\sigma^R(X,t)$. Introducing the stress variation in the constitutive equation we determine the relative strains $\varepsilon^R(X,t)$ and afterwards the relative displacements $u^R(X,t)$. Let us observe that what it is possible to measure after excavation are just the **secondary stresses, relative strains** and **relative displacements**. These are just the components which are involved in the constitutive equation (7.1.1): secondary stresses and relative strains (involved through the derivatives only, in W as well). In order to be accurate, a solution must satisfy the following obvious requirements. If the distance from the wall of the excavation increases very much, then

$$\lim_{dist\to\infty} u^R = 0 \quad , \quad \lim_{dist\to\infty} \varepsilon^R = 0 \quad , \quad \lim_{dist\to\infty} \sigma^S = \sigma^P \quad , \tag{7.2.3}$$

i.e., far enough from the wall of the excavation the displacements and strains must become zero while the stresses must approach their primary values (the so called "far field stresses"). In the so called simplified solutions used in the following chapters the initial secondary stresses, i.e. stresses just after excavation, are obtained as elastic solutions. Moreover, the initial secondary state, i.e. the state at time t_o just after excavation is the "initial data" to be used in order to describe the time effects, i.e. (creep and/or relaxation) taking place in the long time intervals $t > t_o$ which follow.

7.3   BOUNDARY CONDITIONS

The boundary conditions must be formulated at infinity and on the surfaces of the excavation. The boundary conditions at infinity are just (7.2.3), to which we have to add the condition that for any t at far enough distances from the excavation, W_T must tend towards the primary value W^P defined by (7.2.2)$_1$. We can consider W^P to describe the state of hardening before excavation, due to the preliminary loadings to which the rock was subjected. W^P cannot be disregarded if, for instance, we would like to match the solution with the boundary conditions at infinity.

In order to understand this aspect let us assume that Hooke's law holds for a certain rock:

$$\varepsilon_{xx} = \frac{1}{E}\left[\sigma_{xx} - \nu\left(\sigma_{yy} + \sigma_{zz}\right)\right] \quad , \quad \varepsilon_{xy} = \frac{\sigma_{xy}}{2G} \quad , \ldots \tag{7.3.1}$$

If here stresses are replaced by primary stresses, we get "primary" strains which do not have any physical meaning. A physical meaning obtained by "unloading" would not be revealing. Only relative strains obtained from (7.3.1) (with reference state the one *in situ* before excavation), where stresses are replaced by relative components, have a physical meaning. The secondary strains and displacements do not satisfy the conditions (7.2.3).

For the boundary conditions at the surface of the excavation, several cases can be considered. If the **walls of the excavation are not supported,** we must express the condition that the stress vector acting on the wall has zero components (stress-free surface), i.e.

$$\left(\sigma^s\right)^T n = 0 \ , \tag{7.3.2}$$

where the superscript T stands for "transpose" and n is the normal to the surface and oriented towards the interior of the opening. If the **opening is filled with a fluid or gas** under pressure, then as a boundary condition we have to prescribe the pressure exerted on the walls of the cavity

$$\left(\sigma^s\right)^T n = p n \ , \tag{7.3.3}$$

where p is the pressure. The pressure depends on the depth in the case of a fluid (weight of the fluid column). If the cavern is filled with a gas one can consider either the case of a gas under a constant controlled pressure p_o, for instance, or the case when this pressure is varying between two well established limits. The case of a "closed" cavern, where the pressure of the gas deposited in the cavern is increasing due to the creep closure of the walls, can also be considered. In this case a constitutive equation expressing the compressibility law of the gas or the liquid is to be prescribed. Thus, for the pressure involved in (7.3.3) four variants can be considered:

$$p = \gamma_1 h \ , \quad p = p_o = constant \ , \quad p_1 \le p \le p_2 \ , \quad p = f(V) \ , \tag{7.3.4}$$

where V is the volume of the cavern.

The case when on the walls of the underground excavation there act some forces due to artificial support will not be discussed here. For a theory of rigid, elastic, and yieldable supports, i.e. "intelligent" supports which are designed to yield to the pressure of the creeping rock, see Cristescu *et al.* [1987,1988], Cristescu [1989a, Chap.13, 1996a], Cristescu and Duda [1989].

8 Closure and Failure of Vertical Caverns and Boreholes

From the various mining engineering problems, this is maybe one of the simplest, since quite often due to the symmetry of the opening and sometimes also due to the symmetry of the assumed boundary conditions, the problem formulation is relatively simple. In the present chapter we study the creep, dilatancy and/or compressibility, failure, damage and creep failure around a cylindrical vertical cavity, either a vertical shaft or a cavern used for the storage of petroleum products, or a borehole, etc. Some of these wellbores reach great depths (of several kilometers) and thus high stresses may be involved. Also about 75% of all stuck pipe events observed in drilling oil or gas wells are caused by wellbore stability problems (Steiger and Leung [1993]). Convergence of large boreholes has been measured in the Asse salt mine in Germany (Staupendahl et al. [1979], Prij et al. [1996]).

From the various approaches for the convergence of boreholes or stress-strain state around a borehole, we would like to mention Thallak and Gray [1993], who treated the medium as an assemblage of particles (discs) while particle motions are coupled with Darcy flow through the assembly. A statically admissible solution of the elastic interface, for the plane strain problem of a pressurized circular hole in an infinite medium, characterized by a Mohr-Coulomb yield strength and subjected to a non-hydrostatic stress at infinity is due to Detournay [1986]. Starting from the Mohr-Coulomb criterion a rupture model for deep boreholes, which is closer to the classical approach, is due to Charlez et al. [1990]. Elastic brittle plastic rock with Mohr-Coulomb dilatation flow rule was considered by Reed [1986, 1988] for analytical and numerical solutions. For the study of stress development around circular openings, Wang and Dusseault [1994] have considered rocks as Mohr-Coulomb materials, with strain-weakening as a sudden strength loss followed by a perfect plastic behavior after weakening has occurred. Solutions involving stress, displacement, and pore pressure field induced by the drilling and/or pressurization of a vertical borehole in a poroelastic rock are due to Detournay and Cheng [1988] and Coussy et al. [1991]. A semi-analytical solution for the radial dissipation of pore water pressure around a freshly created, vertical hole in an elastic soil is due to Carter [1988]. Thermoplastic and thermo-viscoplastic solutions are due to Wong [1995] and Wong and Simionescu [1996]. Thermal stresses related to borehole stability were studied by Wang et al. [1992, 1996]. Borehole instabilities as bifurcation phenomena were studied by Vardoulakis et al. [1988] for the study of wellbore stability. Modeling of damaged zones around cylindrical openings using radius-dependent Young's modulus is due to Nawrocki and Dusseault [1994].

The exposure below follows the papers by Cristescu [1985a,c, 1989a, Chap.11, 1994a]. See also Fischer et al. [1992], Cristescu [1992], Roegiers [1993], Brignoli and Sartori [1993], and Cristescu [1994c].

8.1 FORMULATION OF THE PROBLEM

It is assumed that the cavity is quite deep so that the influence of the free surface of the Earth will be disregarded. Thus the shaft or borehole, say, is considered a long vertical circular cylindrical orifice drilled in a semi-infinite space. Cylindrical coordinates r, θ and z will be used, with the z-axis coinciding with the symmetry axis of the cavity and directed vertically downwards. The initial radius is $r = a$. The problem is assumed to be a **"plane strain"** problem and solutions will be given for various fixed depths h. The solution in the neighborhood of the roof (if it exists, as in the case of caverns) and of the floor will not be analyzed here.

In the first case considered we assume isotropic boundary conditions at infinity, i.e. the horizontal far field stresses are equal in all directions and denoted by σ_h. The vertical component is σ_v and is obtained from (7.1.1). The boundary conditions (7.3.2) on the surface $r = a$ of the borehole and for a fixed depth h are

$$r = a \,,\, z = h: \quad \sigma_S^R = p - \sigma_h \,,\, \sigma_{rz}^R = \sigma_{r\theta}^R = 0 \,. \tag{8.1.1}$$

The depth h is involved in the particular constant values taken by σ_h and σ_v at that specific depth, and maybe in p too. The superscript R comes from "relative" (see previous chapter).

Due to the symmetry of the problem, the strain components in cylindrical coordinates are

$$\varepsilon_r^R = \frac{\partial u^R}{\partial r} \,,\quad \varepsilon_\theta^R = \frac{u^R}{r} \,,\quad \varepsilon_z^R = 0 \tag{8.1.2}$$

while the equilibrium equation and the compatibility equation are

$$\frac{\partial \sigma_r^R}{\partial r} + \frac{\sigma_r^R - \sigma_\theta^R}{r} = 0 \,,\quad \frac{\partial \varepsilon_\theta^R}{\partial r} + \frac{\varepsilon_\theta^R - \varepsilon_r^R}{r} = 0 \,. \tag{8.1.3}$$

If the excavation is performed very fast or if the rock is assumed to be elastic, the elastic stress, strain and displacement distributions are obtained quite easily (see for instance Cristescu [1985a,b, 1989a, Chap.11]). For convenience we give a short presentation of the analysis. Since the rock is assumed to be elastic

$$\varepsilon'^R = \frac{1}{2G} \sigma'^R \,,\quad \sigma^R = 3K\varepsilon^R \,, \tag{8.1.4}$$

where the values of the elastic parameters are determined for the particular depth considered, and all the components are relative. A prime indicates the deviator, i.e., $\sigma' = \sigma - \sigma\mathbf{1}$, etc. Introducing (8.1.4) in (8.1.3)$_2$ we get

$$\sigma_r^R + \sigma_\theta^R = 2A \,, \tag{8.1.5}$$

where A is an integration constant. From (8.1.5) and (8.1.3)$_1$ we further obtain

$$\sigma_r^R = A + \frac{B}{r^2} \quad , \quad \sigma_\theta^R = A - \frac{B}{r^2} \qquad (8.1.6)$$

with B another integration constant. Since these components must be zero at infinity (see (7.2.3)) we have $A=0$ and (8.1.1) determines B. Also from (8.1.6) and (8.1.4) follows that

$$\sigma^R = 0 \quad and \quad \varepsilon^R = 0 \quad , \qquad (8.1.7)$$

i.e., due to the drilling, in each plane $z =$ constant, the point representing the stress state remains in the same octahedral plane (σ^P and σ^S belong to the same octahedral plane). Afterwards the strains are obtained straightforwardly from (8.1.4) so that the complete elastic solution is

$$\sigma_r^S = (p - \sigma_h)\frac{a^2}{r^2} + \sigma_h \quad , \quad \varepsilon_r^R = \frac{p - \sigma_h}{2G}\frac{a^2}{r^2} \quad ,$$

$$\sigma_\theta^S = -(p - \sigma_h)\frac{a^2}{r^2} + \sigma_h \quad , \quad \varepsilon_\theta^R = -\frac{p - \sigma_h}{2G}\frac{a^2}{r^2} \quad ,$$

$$\sigma_z^S = \sigma_v \qquad\qquad , \quad \varepsilon_z^R = 0 \quad , \qquad (8.1.8)$$

$$\sigma^S = \sigma^P \qquad\qquad , \quad u^R = -\frac{p - \sigma_h}{2G}\frac{a^2}{r} \;, \; w^R = 0 \quad .$$

We recall that after an excavation the stress in the surrounding rock is the "secondary" stress, and the strain and displacement are the "relative" ones, i.e. with respect to the state before excavation. For mean stress and equivalent stress we have

$$\sigma^S = \sigma^P = \frac{2\sigma_h + \sigma_v}{3} \quad , \quad (\bar{\sigma}^S)^2 = 3(p - \sigma_h)^2\frac{a^4}{r^4} + (\sigma_v - \sigma_h)^2 \quad . \qquad (8.1.9)$$

$\bar{\sigma}$ is maximum for $r = a$ while for $r \to \infty$ we have $\bar{\sigma}^S \to \bar{\sigma}^P$.

When a viscoplastic model is used for the rock, if the excavation is performed quite fast with respect to the long time during which the borehole is in use, we will assume that the response of the rock is instantaneous, and therefore the state just after excavation is obtained from (8.1.8). Afterwards a deformation by creep will take place; in some domains around the cavity the rock becomes dilatant while in some others compressible. This will be described in the next sections.

The stress distribution along the walls of a deep borehole (or well) is shown in Figure 8.1. The example is given for rock-salt (Cristescu and Hunsche [1996]) and $\sigma_h = \sigma_v = 0.025\, h$. This figure is obtained by superposing the stress distribution (8.1.8) over the "map" of the constitutive model for that particular rock, which shows in the τ-σ plane the failure line (solid line) and the compressibility/dilatancy boundary (dash-dot line). This superposition of the "instantaneous" solution over the map of the rock allows us to reveal the main characteristics of the deformation of the rock which follows excavation. For an empty well the

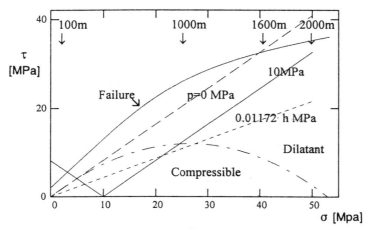

Figure 8.1 *Stress distribution in the walls of a deep well: empty well (interrupted line); well filled with brine (dotted line); well filled with gas under pressure p=10 MPa (solid line).*

stress distribution is shown as interrupted line ($p = 0$); at depths greater than 1600 m instantaneous failure will occur while down to this depth the stress states at the well wall are in the dilatancy domain. For depths approaching 1600 m, one can expect a creep failure to take place after a not too long time interval. If the well is filled with brine, the pressure along the walls is due to the fluid column if it is tight, and the corresponding stress state along the walls at various depths is shown by a dotted line. In this case it is only for depths greater than 1000 m, for example, that the rock surrounding the well becomes dilatant, while above this depth it is compressible. If the well (or cavern) is filled with a gas under constant pressure (10 MPa, say) the stress at various depths is shown as a full line. In this case, at greater depths the rock in the immediate neighborhood of the excavation is dilatant, at intermediate depths it is compressible, while at smaller depths again a dilatant region is possible, followed at shallow depths by failure (due to tensile stresses).

8.2   CONVERGENCE, DILATANCY, DAMAGE AND STABILITY

For the study of convergence of the walls of the cavern by creep, we use the formulae employed in the previous chapters with the assumption $\sigma_h = \sigma_v$. The main part of the convergence is due to the stationary creep (see also Frayne *et al.* [1996]). For instance, a simplified illustrative solution can be obtained if we assume that the stress state remains constant during creep. In this case we obtain from (5.1.1) for the radial displacement

$$\frac{u}{a} = -\frac{p-\sigma_h}{2G}\frac{a}{r} + \frac{r}{a}\left\{ \frac{\left\langle 1 - \frac{W_T(t_o)}{H}\right\rangle \frac{\partial F}{\partial \sigma_\theta}}{\frac{1}{H}\frac{\partial F}{\partial \sigma}\cdot\sigma}\left\{1 - \exp\left[\frac{k_T}{H}\frac{\partial F}{\partial \sigma}\cdot\sigma(t_o - t)\right]\right\} + k_S\frac{\partial S}{\partial \sigma_\theta}(t - t_o)\right\} \quad , \quad (8.2.1)$$

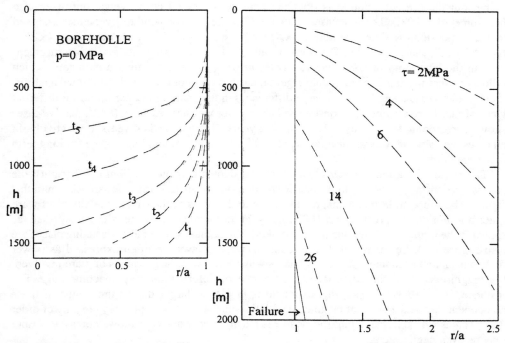

Figure 8.2 *Convergence of the walls of a deep cavern without pressure, at five successive times (left) and distribution of octahedral shear stress in the rock mass explaining the fast convergence at large depths (right); the full line markes the instantaneous failure domain (see Figure 8.1).*

where t_0 is the moment of excavation. The depth is involved in (8.2.1) via σ_h and the stress state. The time t_c of complete closure of the cavern can be obtained from (8.2.1) when for $r = a$ the displacement u becomes equal to a:

$$t_c \approx t_0 + \frac{1}{k_S \dfrac{\partial S}{\partial \sigma_\theta}}\left\{1 + \frac{p - \sigma_h}{2G} - \frac{\left\langle 1 - \dfrac{W_T(t_0)}{H}\right\rangle \dfrac{\partial F}{\partial \sigma_\theta}}{\dfrac{1}{H}\dfrac{\partial F}{\partial \sigma}\cdot\sigma}\right\} \tag{8.2.2}$$

Here the last right-hand term is the ultimate value (for $t \to \infty$) of the transient component of the displacement. Thus in the time interval $t_S < t < t_c$ the closure is entirely due to the stationary creep (see (5.3.39)). The stress variation during creep, which is significant mainly at great depths, will be considered in Section 8.3.

Figure 8.2 (left) shows the successive positions of the walls of an empty ($p=0$) vertical cavern excavated in rock salt, according to (8.2.1). At great depths the closure is very fast, but it slows down exponentially as the depth diminishes, i.e. at shallow depths very long time intervals are necessary for even a partially closure of the cavern. Apparently the bottom is rising, but the present solution is obtained considering radial convergence only (see also Munson *et al.* [1992]). The five cases shown in Figure 8.2 correspond to five values assigned to the dimensionless time variable $t_i = k_S(t - t_o)$ (i.e., $t_1 = 4 \times 10^4$, $t_2 = 1.6 \times 10^5$, $t_3 = 4 \times 10^5$, $t_4 = 1.52 \times 10^6$, and $t_5 = 8 \times 10^6$). The correct value of k_S needs further studies (long term creep tests in the laboratory or convergence studies *in situ*), but it is certainly a parameter which influences very much the speed of convergence, besides the depth. k_S needs additional analysis, if one wants to use creep formulae; large differences in the value of k_S are found for different types of salt. If we assume $k_S =$ constant and choose, for instance, $k_S = 3 \times 10^{-3} s^{-1}$, the five cases shown correspond to 154 days, 1.7 year, 4.2 years, 16 years and 84 years, respectively. In all cases the value of k_T was taken as $k_T = 5 \times 10^{-6} s^{-1}$, but it is of less importance for long term creep prediction.

The reason for a much faster closure at great depths is shown in Figure 8.2 (right): the curves $\tau =$ constant according to the elastic solution (instantaneous response) are shown in this figure. Therefore, at large depths, mainly close to the walls of the cavern the values of τ are quite big. For depths greater than 1600 m, say, an instantaneous failure is to be expected, as shown in the figure by a solid line that is bounding the domain where failure is taking place (see also Figure 8.1). Let us recall that all stationary creep laws are often expressed as power functions of τ (see Chapters 2 and 5). That is why there where large values of τ are involved, the magnitude of all the components of the rate of deformation may become huge. For instance, if for rock-salt, say, this exponent is $m = 5$, at larger depths the value of these components are of the order of magnitude of 10^8, while at shallow depth they are only of order 10^1. That is why at larger depths the closure is much faster, while at shallow depths the closure may be even negligible.

We will give now an example how dilatancy and creep failure spread into the rock mass

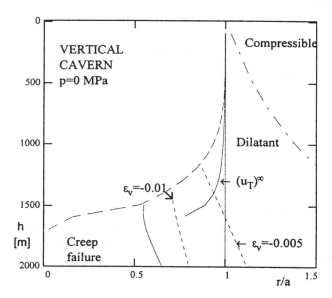

Figure 8.3 *Spreading of dilatancy and creep failure into the creeping rock during cavern closure (see also Figures 8.1 and 8.2).*

during wall closure. In Figure 8.3 the position of the cavern wall is shown by an long-dashed line at an arbitrary chosen time. We have chosen the case of rock salt and an empty cavern $(p = 0)$. At the initial moment the walls of the cavern are at $r = a$, while the initial compressibility/dilatancy boundary is shown as a dash-dot line. However, at far distances the dilatancy is quite small, the rock being practically incompressible, i.e., the very small rock deformation is essentially elastic. An instantaneous failure takes place just after excavation at depths surpassing 1600 m. The initial shape of the failed domain is not shown in Figure 8.3, but it is shown in Figure 8.2. Its ultimate location at the moment of full closure is evident by the slope discontinuity of the long-dashed line in Figure 8.3. The boundary of the domain up to where creep failure has already taken place at the time considered is shown as a solid line. Two short-dashed lines show how dilatancy progresses into the creeping rock. A full line marked $(u_T)^\infty$ shows where the ultimate position of the walls due to transient creep alone would be, according to formula (8.2.1) without the last steady-state creep term. It follows that transient creep alone is not able to describe a full closure of a cavern, and in long time intervals the main contribution to the creep deformation of rocks is that of steady-state creep.

Another example which will be presented shortly is that of a closed cavern filled with a gas under pressure, and the creeping rock is increasing this pressure according to $pV = $ constant, for $T = $ constant, with V the volume of the cavern. Any other law could equally be used in a similar way. In Figure 8.4 an example is shown. The lower dash-dot line was taken as "initial" shape of the cavern when the internal pressure was $p = 5$ MPa. At that moment we assume that

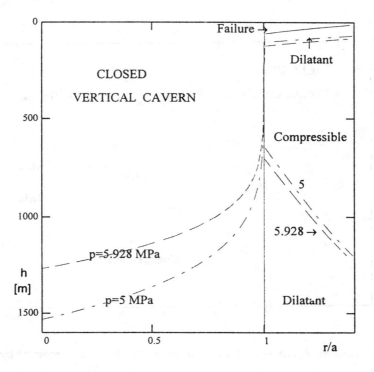

Figure 8.4 *Convergence of the walls of a closed cavern, dilatancy, compressibility, and failure around the cavern.*

the cavern is closed and afterwards the pressure is building up due to the creep convergence of the walls, which produces a decrease of the volume of the cavern. In turn, the increasing internal pressure is expected to reduce the creep rate. The upper long-dashed line shows the subsequent position of the walls when the internal pressure has increased up to 5.928 MPa. Due to the pressure built up, the compressibility/dilatancy boundary also changes its position: as the pressure increases both the upper and lower compressibility/dilatancy boundaries go down and the upper dilatant region spreadd. At shallow depths a domain of short term failure is also possible. This failure has no significance anyway since no cavern under pressure is excavated to shallow depths. Also it is interesting to mention that at small depths, due to internal pressure, there is a slight tendency of the cavern to expand by creep.

The stress distribution along the walls of the cavern, as a function of depth, is shown in Figure 8.5 for the initial moment when the cavern is closed (i.e. when $p = 5$ MPa) and for the actual time ($p = 5.928$ MPa). Thus due to the pressure built up the danger of failure at shallow depths is increased, while at great depths the danger of creep failure is diminished. This was already evident from Figure 8.4 since the compressibility domain is coming down as the pressure is building up. Also, Figure 8.5 shows that for each pressure there is a certain depth where this pressure is equal to the far field stress, so that at that location the cavern walls do not move by creep. Let us mention that Charlez and Heugas (1991) have presented a model predicting mud density which would prevent mechanical instability of wellbore; a Cam-Clay constitutive equation is used.

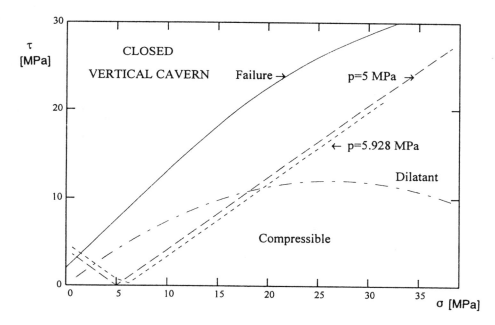

Figure 8.5 *Variation of pressure in a closed cavern due to the convergence by creep of the walls.*

8.3 A MORE GENERAL PRIMARY STRESS STATE

If the horizontal far field stresses are not equal in all directions, the problem is slightly more involved (Cristescu [1985a,c, 1989a, Chap 11]). Let us denote by σ_{h1} and σ_{h2} the maximal values of the orthogonal horizontal principal far field stresses. The primary stresses can be expressed in cylindrical coordinates as

$$\sigma_{rr}^{P} = \frac{1}{2}(\sigma_{h1} + \sigma_{h2}) + \frac{1}{2}(\sigma_{h1} - \sigma_{h2})\cos 2\theta \ ,$$

$$\sigma_{\theta\theta}^{P} = \frac{1}{2}(\sigma_{h1} + \sigma_{h2}) - \frac{1}{2}(\sigma_{h1} - \sigma_{h2})\cos 2\theta \ ,$$

$$\sigma_{r\theta}^{P} = -\frac{1}{2}(\sigma_{h1} - \sigma_{h2})\sin 2\theta \ ,$$

$$\sigma_{zz}^{P} = \sigma_{v} \ .$$

$$(8.3.1)$$

After the excavation of the vertical cavern or borehole, the secondary stresses are (see Cristescu [1989a])

$$\sigma_{rr}^{S} = p\frac{a^2}{r^2} + \frac{1}{2}(\sigma_{h1} + \sigma_{h2})\left(1 - \frac{a^2}{r^2}\right) + \frac{1}{2}(\sigma_{h1} - \sigma_{h2})\left(1 - \frac{4a^2}{r^2} + \frac{3a^4}{r^4}\right)\cos 2\theta \ ,$$

$$\sigma_{\theta\theta}^{S} = -p\frac{a^2}{r^2} + \frac{1}{2}(\sigma_{h1} + \sigma_{h2})\left(1 + \frac{a^2}{r^2}\right) - \frac{1}{2}(\sigma_{h1} - \sigma_{h2})\left(1 + \frac{3a^4}{r^4}\right)\cos 2\theta \ ,$$

$$\sigma_{r\theta}^{S} = \frac{1}{2}(\sigma_{h1} - \sigma_{h2})\left(-1 - \frac{2a^2}{r^2} + \frac{3a^4}{r^4}\right)\sin 2\theta \ ,$$

$$\sigma_{zz}^{S} = \sigma_{v} - \nu(\sigma_{h1} - \sigma_{h2})\frac{2a^2}{r^2}\cos 2\theta \ .$$

$$(8.3.2)$$

Now the stress distribution around the circumference of the cavern depends on orientation (on θ). The elastic relative displacements are

$$u_r^{R} = \frac{1 + \nu}{E}\left\{-p\frac{a^2}{r} + \frac{1}{2}(\sigma_{h1} + \sigma_{h2})\frac{a^2}{r} + \frac{1}{2}(\sigma_{h1} - \sigma_{h2})\left[4(1 - \nu)\frac{a^2}{r} - \frac{a^4}{r^3}\right]\cos 2\theta\right\} \ ,$$

$$u_\theta^{R} = -\frac{1 + \nu}{E}\frac{1}{2}(\sigma_{h1} - \sigma_{h2})\left[2(1 - 2\nu)\frac{a^2}{r} + \frac{a^4}{r^3}\right]\sin 2\theta \ .$$

$$(8.3.3)$$

The total displacement due to elastic deformation and creep as well is obtained with formulae similar to (8.2.1) but where the first right-hand terms are replaced with (8.3.3) and stresses are computed using (8.3.2). Along a deep borehole one can find depths where $\sigma_{h1} > \sigma_{h2}$ and some other ones where $\sigma_{h2} > \sigma_{h1}$.

Let us give an example of a borehole or cavern located between the depths $h = 500$ m and $h = 1500$ m and excavated in rock salt. For the sake of giving an example let us assume that the vertical far field stress component is obtained from $\sigma_v = \gamma h$ (with γ =0.025 Mpa m^{-1} and σ_v in MPa), and that at $h = 500$ m : $(2/3) \sigma_{h1} = \sigma_v = \sigma_{h2}$; at $h = 1000$ m : $\sigma_{h1} = \sigma_v = \sigma_{h2}$;and finally at $h = 1500$ m : $(5/4) \sigma_{h1} = \sigma_v = \sigma_{h2}$. The stress distribution at these three depths is shown in Figure 8.6. The isolated symbols are marking the corresponding primary stress states. The secondary stress state along the contour is given starting from $\theta = 0°$ corresponding to the direction of σ_{h1} up to $\theta = 90°$ corresponding to the direction of σ_{h2}. For these three depths

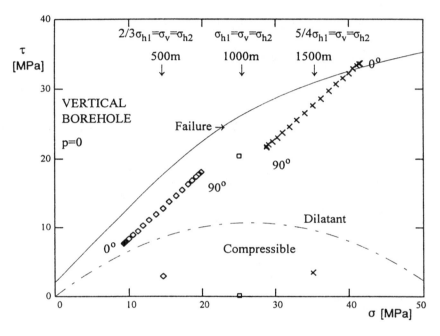

Figure 8.6 *Stress distribution along the contour of a deep borehole at three depths shown for three different depths and values for σ_{h1}/σ_{h2} .*

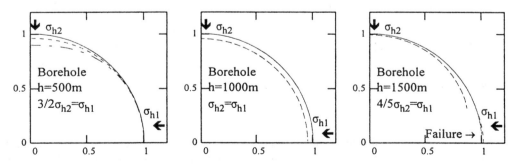

Figure 8.7 *Convergence of the wall of a borehole at three depths shown for three values of σ_{h1}/σ_{h2}.*

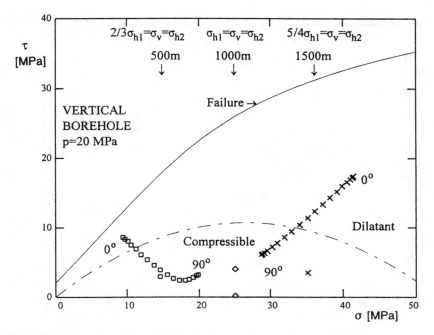

Figure 8.8 *Same as for Figure 8.6 but for internal pressure p = 20 MPa.*

the convergence of the walls is shown in Figure 8.7. The convergence is fastest in the direction of the smallest σ_h component. At the depth $h = 1500$ m an instantaneous failure is taking place again in the direction of the smallest σ_h component; the fastest creep is also there. In the case $\sigma_{h1} = \sigma_{h2}$ the convergence by creep is symmetric. Any other relationship between σ_v, σ_{h1} and σ_{h2} can be handled in a similar way. Very interesting laboratory results relating the failure around a borehole to the magnitude of the far field stresses are due to Bandis and Barton [1986], Möhring-Erdmann and Rummel [1987], Guenot [1989], Morales *et al.* [1989a], Choi *et al.* [1991], Haimson *et al.* [1993a], Haimson and Song [1993]. Field investigations of the same problem are due to Morales *et al.* [1989b]. For the theoretical explanation why sometimes the break outs are in the direction of the maximum far field component and why in some other times they are in the direction of the smallest far field stress component see Cristescu [1986, 1989a] and the next chapter.

The presence of an internal pressure changes the stress distribution. For instance Figure 8.8 shows the stress distribution along the contour for the same depth as in Figure 8.6 but for an internal pressure $p = 20$ MPa. Comparing Figures 8.6 and Figure 8.8, one can conclude that an internal pressure of 20 MPa makes a borehole or cavern excavated between $h = 500$ m and $h = 1500$ m, rather stable, at least for the relationship between far field stresses shown.

If the internal pressure we would like to use at the same depth is higher, the cavern may become unstable, as shown in Figure 8.9. This time the cavern is stable at great depths and unstable at depth 500 m. Here we can pose the problem where (i.e., at what depth) a cavern should be excavated, which would be stable if the internal pressure is varied between two limits. From the above figures it is already obvious and well known that too high (i.e. $p = 40$

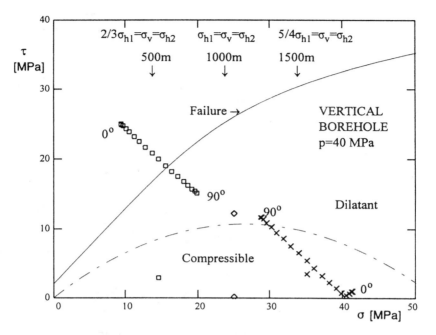

Figure 8.9 *Same as for Figure 8.6 but for internal pressure p = 40 MPa.*

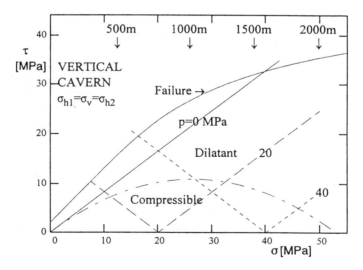

Figure 8.10 *Stress distribution on the wall of a vertical cavern for various internal pressures.*

MPa) or too low ($p = 0$) pressure are not suitable for a cavern excavated in the range 500 m
< h < 1500 m. Either the depth of the cavern is imposed by some arguments related to the
geometry and state of the rock mass (and in this case we have to choose the appropriate interval

pressure variation to keep the cavern stable and safe and not to collect too much damage and loosening), or we have the liberty to choose the depth of the cavern (if the conditions existing *in situ* allow us this). In this way we can find out the optimal location of the cavern for a prescribed interval of variation of the internal pressure (the one imposed by economic arguments, etc.; see also Rokahr and Staudtmeister [1996]).

Let us give an example: we would like to design a storage cavern in the rock salt somewhere between $h = 500$ m and $h = 2000$ m. To keep the description simple let us start with the case $\sigma_{h1} = \sigma_v = \sigma_{h2}$. We start by studying the stress distribution on the cavern wall if the cavern is very deep. This is shown in Figure 8.10. For instance, if the cavern is empty ($p = 0$) the stress state at the walls for various depths is shown in Figure 8.10 as a solid line. Again at depths greater than 1600 m an instantaneous failure is taking place. If the pressure on the wall is $p = 20$ MPa the stress distribution along the walls is as shown as the long-dashed line; this time it is at shallow depths (above approximately 200 m) that failure is taking place and no short term failure is possible except well below 2500 m, say. Similarly for an internal pressure of 40 MPa, the stress distribution as is shown as the short-dashed line; failure is expected above 600 m, say.

Drawings of the type shown in Figure 8.10 can suggest the way to choose the optimal interval of pressure variation inside a cavern located between two selected levels for the top and bottom of the cavern. Of course, there are additional arguments besides the purely mechanical ones. Here we consider only the cavern stability due to stress state and interval of variation of internal pressure. Let us assume now that the cavern lies between the levels $h_t \le h \le h_b$, i.e., h_t is the depth of the top of the cavern and h_b the depth of the bottom (Figure 8.11). Thus the primary stress state at the top is $\sigma(h_t) = \sigma_v(h_t) = \sigma_{h1}(h_t) = \sigma_{h2}(h_t)$ and similar for

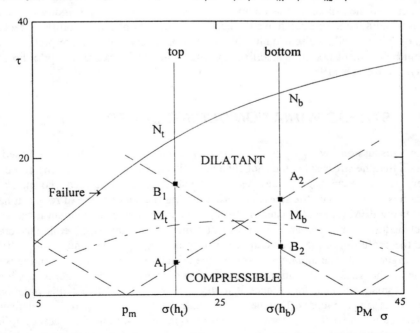

Figure 8.11 *How to relate the depth and height of a vertical cavern to the interval of internal pressure variation.*

the bottom. The corresponding secondary stresses are located on the vertical lines passing passing through $\sigma(h_t)$ and $\sigma(h_b)$. Their location on these lines depend on the internal pressures. If the minimum value of the internal pressure is p_m, then A_1 will represent the corresponding stress state at the top of the cavern and A_2 that at the bottom. We have to choose A_2 close enough to the compressibility/dilatancy boundary to ensure long term stability by creep. From the estimations presented at Chapter 10 one could suggest to choose A_2 in the first quarter of the segment $M_b N_b$. That will ensure long term stability. After choosing the position of A_2, the magnitude of p_m follows. In a similar way we choose the position of point B_1 which would determine the value of p_M. By increasing the interval Δh one has to decrease the interval $\Delta p = p_M - p_m$. Also by increasing the interval Δp we have to sacrifice convergence on the wall (which will increase).

If the cavern is filled with a liquid, the pressure on the walls is that of the column of the liquid at the corresponding depth. In this case the above considerations apply with the distinction that the lines $p = \gamma h$ have a much smaller slope than the lines $p = \text{constant}$ (see Figure 8.1).

Obviously the considerations related to Figure 8.11 are only a preliminary estimation of how the depth and height of the cavern are to be related to the interval of variation of internal pressure. Taking into account that the formulae used are for a semi-infinite cavern, the case of a cavern of finite height is more conservative, i.e. any relationship between Δh and Δp obtained from Figure 8.11 will certainly ensure the stability of a cavern of finite height, if some other parameters as temperature, humidity, etc., are not involved.

In the case $\sigma_{h1} \neq \sigma_v$ and $\sigma_v \neq \sigma_{h2}$ the matching of the cavern location and size, with an optimum interval of variation of internal pressure is a much more difficult task. However, if the relation between the primary stress components is known, one can plot figures of the type of Figures 8.8 and 8.9 and one can make the necessary estimation. The stability of rectangular galleries or caverns will be discussed in Chapter 10. For observation and interpretation of time-dependent behavior of borehole stability in the Continental Deep Drilling pilot borehole in Germany, see Kessels [1989].

8.4   STRESS VARIATION DURING CREEP

The above discussion involving creep of the rock around a wellbore has been based on the assumption that the stress state does not change or changes only slightly during creep. In this case very simple formulae can be used for the analysis as described. To check if this assumption is a correct one, **the stress variation during transient creep** of rock salt has been obtained using FEM (Jin and Cristescu [1997]). The most important element is the variation of the octahedral shear stress since this invariant is the main parameter governing the creep law and also the evolution of the creep damage. The result for $h=1200$ m and $\sigma_h = \sigma_v = 30\,\text{MPa}$ is shown in Figure 8.12. The thin lines represents the initial (elastic) stress distribution while the thick one is the stress distribution at stabilization of transient creep (after a very long time interval). At smaller depths, for instance at $h=400$ m with $\sigma_h = \sigma_v = 10\,\text{MPa}$, this variation is less pronounced, as seen in Figure 8.13. One can conclude that the stress variation in relatively short time intervals is not really too significant; close to the excavation, τ is decreasing in time, while at farther distances it is increasing, and the two effects may compensate somehow in the computation of wall convergence. Thus, the above assumption concerning the constancy of stress state during creep seems reasonable if applied to short time intervals and at not too great depths. The variation of individual stress components is shown in Figure 8.14 for h=1200 m. The thin line corresponds to the stress distribution at the moment when creep starts (elastic solution) while the thick line corresponds to the stress distribution at the stabilization of transient creep (after a very long time interval). The variation of the component σ_θ is the most remarkable, but again a very long time interval is necessary to get a significant stress variation.

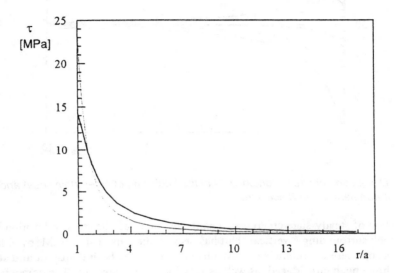

Figure 8.12 *Variation of octahedral shear stress during transient creep for h=1200 m.*

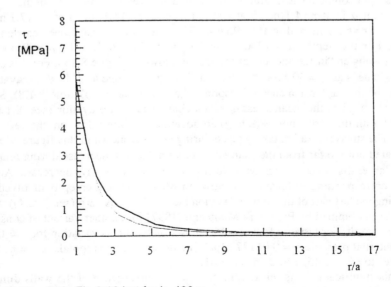

Figure 8.13 *As in Fig.8.12 but for h=400 m.*

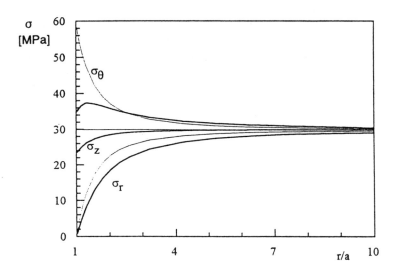

Figure 8.14 *Distribution of stress components at the beginning of creep (thin lines) and at the stabilization of transient creep (thick lines).*

Consideration of steady-state creep may change the picture. The stress relaxation during creep of the rock surrounding a vertical shaft has been studied also using FEM by Glabisch [1997]. He has considered the complete constitutive equation, i.e., **both transient and steady-state creep have been considered**, as well as very long time intervals. The computations have been carried out for a shaft of radius $a = 4$ m, excavated in rock salt. The stress relaxation has been computed at several points located at several distances from the shaft axis as folows: point a is at $r = 4.1$ m, b at $r = 5.1$ m, c at $r = 11.1$ m, d at $r = 17.1$ m, e at $r = 29.1$ m, f at $r = 61.6$ m, and g at $r = 96.6$ m. The computations have been carried out up to $r = 100$ m. For the depth $h = 1200$ m and $\sigma_v = 30$ MPa , $\sigma_h/\sigma_v = 1$, $p = 0$, the stress relaxation following an "instantaneous" excavation is shown in Figure 8.15a. For this case the primary stress state is at $\sigma = 30$ MPa, $\tau = 0$, and since $\sigma_h = \sigma_v$, due to a "fast" excavation the stress state is obtained as "instantaneous" response, i.e., as an elastic solution (8.1.8). Since in this solution $\sigma^S = \sigma^P$, i.e., the mean stress does not change due to excavation (see (8.1.8)), the stress states for all the above points (a to g) are located in Figure 8.15a on the vertical line $\sigma = 30$ MPa. The stress relaxation taking place during creep is shown in this figure. Generally at points located not too far from the shaft surface, both τ and σ decrease during relaxation. However, at larger distances, τ increases during creep while σ always decreases. An arrow at the curve corresponding to location a shows to where transient creep is involved in the process. During the last part of the stress relaxation the bracket $\langle \rangle$ is zero (see (4.2.4)). Similar results have been obtained by Paraschiv-Munteanu [1997] for another variant of constitutive equation for rock salt. For instance Figure 8.15b shows such a stress variation for $h = 1200$ m, $\sigma_h = \sigma_v$ and internal pressure $p = 0.01177$ Mpa. The upper line corresponds to $r = a$, and the following ones to $r/a = 1.45$, 1.9, 2.36 and 2.81.

Due to this significant stress relaxation, the rate of convergence of the walls diminishes continuously. Figure 8.16 shows the radial convergence of the walls at three depths. The

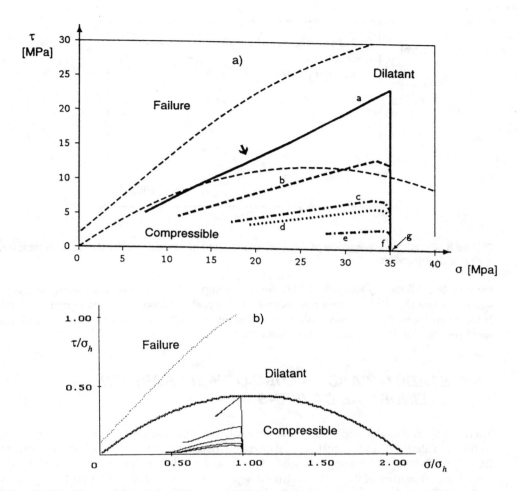

Figure 8.15 *Stress relaxation during creep of the rock surrounding a deep shaft: a} without internal pressure (Glabisch [1997]), and b) with the pressure of brine column (Paraschiv-Munteanu [1997]).*

solid lines have been obtained with FEM (Glabisch [1997]) with the following values of the viscosity parameters: $k_T = 1 \times 10^{-6} \, \mathrm{s}^{-1}$ and $k_S = 3 \times 10^{-3} \, \mathrm{s}^{-1}$. The dash-dot lines have been obtained with the simplified formula (8.2.1) which assumes that stress is constant during creep, but for a smaller value for $k_S = 1 \times 10^{-4} \, \mathrm{s}^{-1}$. It follows that for shallow and medium depths the formula (8.2.1) can give a reasonable description of the convergence for not too long time intervals. At greater depths, however, this formula can be used to show the beginning of the convergence process, only. Let us recall that according to the FEM solution, at smaller radial distances from the shaft wall, both τ and σ are decreasing, while at greater distances τ is increasing. The formula (8.2.1) assumes that everywhere the stresses remain constant during creep. In both solutions the steady-state creep plays a significant role in the radial convergence process.

A creep analysis of tunnels using FEM and considering both creep and stress relaxation,

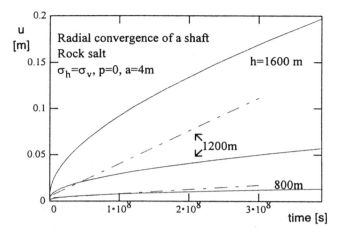

Figure 8.16 *Well convergence of the walls of the shaft at various depths due to both transient and stationary creep.*

has also been done by Roatesi [1997]. She has compared the numerical solution with the simplified one (8.2.1). Examples have been given for coal and sand,also using tunnel supports. It has been found that the two methods to compute the wall convergence, do not furnish the same results, but they are not too far apart either.

8.5 STABILITY AND FAILURE OF WELLBORE OR CYLINDRICAL CAVERNS

There are several modes of possible failure or evolutive failure in the rock surrounding a deep wellbore, a shaft or a cylindrical vertical cavern (Cristescu [1986, 1989a, Chap.11, 1993b]). Important factors influencing stability, failure and evolutive failure, are certainly the mechanical properties of the rock but also the depth and the magnitude of the far field stresses.

Let us first mention that **the most stable case is the case of equal far field stresses**, i.e., $\sigma_v = \sigma_{h1} = \sigma_{h2}$. In order to give an example Figure 8.17 shows several cases of far field stresses in a deep wellbore excavated at $h = 1000$ m in rock salt. If $\sigma_v = \sigma_{h1} = \sigma_{h2}$, the primary stress state is shown as a square on the hydrostatic axis. The secondary stress state is also shown as a square having the same mean stress but $\tau \neq 0$. Unless the depth is not too great, no instantaneous failure is to be expected. Since the secondary stress state is relatively far from the instantaneous failure surface, the wellbore is relatively stable in the sense that no significant convergence by creep is to be expected in relatively short time intervals. The stability depends mainly on the depth. For greater depths the secondary stress state is closer to the failure surface and the stability by creep may last for short time intervals only; at the end of this time interval creep damage steadily progressing in the rock is to be expected. Let us recall that the time to creep failure depends on the octahedral shear overstress above the compressibility/dilatancy boundary (see Figure 4.17). The closer is the stress state to the instantaneous failure, the shorter is the time interval up to failure. If this stress state is

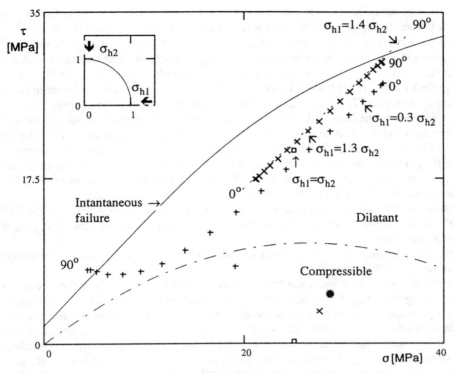

Figure 8.17 *Various possible failure modes around a wellbore (see text).*

approaching the C/D boundary, the time to failure increases to infinity. This type of evolutive damage (a time effect) is steadily spreading into the rock mass (see next chapter).

The **borehole is much more unstable if the in-plane far field stresses are unequal**, i.e., if for instance σ_{hl} is much bigger than σ_{h2}, $\sigma_{hl} \gg \sigma_{h2}$ or *vice versa* (see also Hoffers *et al.* [1994]). For instance, the stress distribution along the contour for $\sigma_{hl} = 1.4\sigma_{h2}$ is shown in Figure 8.17 as a short-dashed line; the corresponding primary stress is shown as a filled star. Thus, if one of the horizontal components is distinct from the other one, the stability is in jeopardy. The loosing of stability occurs in the direction of σ_{hmin} which coincides with σ_{h2} and with the direction of τ_{max}. An instantaneous failure is followed by evolutive damage in the same direction. In this direction occurs the fastest convergence (due to the maximal value of τ) and the evolutive damage by dilatancy spread into the rock mass in the same directions by losing the cohesion or by spalling.The volume and shape of the rock involved in instantaneous failure can be estimated quite easily. Also the way in which progressing damage afterwards spreads into the rock mass can also be described very accurately (see Figure 8.3 and many examples given in the next two chapters). Interesting field observations were presented by Zoback *et al.* [1985] and Martin [1990] besides others, and laboratory tests by Lee and Haimson [1993] and Haimson and Song [1993, 1995], Rokahr and Staudtmeister [1996].

If the ratio σ_{hl}/σ_{h2} is only slightly smaller than the previous one, for instance $\sigma_{hl} = 1.3\sigma_{h2}$, instantaneous failure is no longer possible, but instabilities in the direction of $\sigma_{hmin} = \sigma_{h2}$ coinciding with the direction of τ_{max} are still possible since in these directions the stress state is very close to the instantaneous failure surface: fast convergence by creep is followed by evolutive failure. In Figure 8.17 the stress state around the circumference is shown as x.

If the ratio σ_{h1}/σ_{h2} is much smaller than unity, for instance $\sigma_{h1} = 0.3\sigma_{h2}$, a short term failure (tensile stress) takes place in the direction of σ_{hmax}, which this time coincides with σ_{h2}, and the direction of τ_{min}. In the direction of σ_{hmin} coinciding with τ_{max} and σ_{h1}, a fast convergence by creep is expected followed after not too long a time interval by creep failure produced by dilatancy. The time elapsed from the time of excavation up to creep failure by dilatancy depends on the relative distance between the point representing the secondary stress after excavation and the short term failure surface. If this distance is very small (as in the case shown in Figure 8.17) creep failure may be expected in hours or days. This time interval is longer if the point representing the stress state is farther on from the short-term failure surface. If this point is close to the compressibility/dilatancy boundary the time to creep failure increases to infinity (Cristescu [1986, 1989a, 1993b]).

For all the above cases shown in Figure 8.17 σ_v was obtained from $\sigma_v = \gamma h$, with $\gamma = 0.025$ Mpa m^{-1} and $h = 1000$ m. It was assumed that $\sigma_{h2} = \sigma_v$ while σ_{h1} is related to σ_{h2} by the relations given above.

Thus, instantaneous failure followed by a small amount of creep, if any, is possible in the direction σ_{hmax} coinciding with τ_{min}.

Evolutive damage (preceded or not by instantaneous failure) is possible in the direction of σ_{hmin} coinciding with τ_{max}. Fast creep convergence is possible in the same direction. Evolutive damage is possible for any stress state in the dilatancy domain: it develops and spreads faster in those locations where stresses are closer to the instantaneous failure surface, and much slower in locations where stresses are farther from this surface. For compressible/dilatant rocks, stress states belonging to the compressibility domain cannot produce failure. However, for rocks which are compressible only (very high initial porosity) or/and are anisotropic, things may be different.

In brief, for compressible/dilatant rocks failure is possible in both directions σ_{hmin} and σ_{hmax}. Fracture by tensile stresses is possible in the direction of σ_{hmax}. Progressive failure (lose of cohesion or spalling) is possible in the direction of σ_{hmin}. Time-dependent stability of boreholes was also discussed by Horsrud et al. [1994].

Other approaches to borehole failure or closure are due to Zoback et al. [1985], Aoki et al. [1993], Brudy and Zoback [1993], Ewy [1993], Cuisiat and Hudson [1993], Cheatham [1993], Shao and Krazraei [1994], Pellegrino et al. [1994], Charlez [1994], Germanovich et al. [1994], Mokhel et al.[1996].

9 Creep, Closure and Damage of Horizontal Tunnels

Stress distribution around horizontal circular tunnels has been studied for some time. Various particular formulations have been given. For instance, a **plane stress solution** of elastic stress distribution around a circular hole was given by Kirsch as early as in 1898 (see Obert and Duvall [1967]). Most of the solutions, sometimes involving thermal stresses, deal with linear elastic rock, with the computation carried out generally with FEM or BEM, though some viscoelastic or viscoplastic models have occasionally also been used (Nonaka [1981], Fritz [1984], Detournay [1985], Detournay and Fairhurst [1987], Yamatomi *et al.* [1987], Jiayou [1987], Swoboda *et al.* [1987], Detournay and St. John [1988], Pan and Hudson [1988a], Pan and Dong [1991], Pan and Reed [1991], Pelli *et al.* [1991], Wang *et al.* [1992], Obara *et al.*, [1992], Ewy [1993], Gioda and Cividini [1994], Gioda *et al.* [1994], and Wong and Simionescu [1995]). Carefully examined case studies are due to Panet [1969], Sulem *et al.* [1987], and Nilsen [1993], Kie [1993], Kovari and Amstad [1993], and many others (see also Section 8.5). The influence of water on the stability of deep tunnels has been studied too (Lembo Fazio and Ribacchi [1984], Huergo and Nakhle [1988]).

Failure and stability of rocks surrounding a tunnel, mainly in the case of unequal far field stresses, have been extensively studied in the last few years, either for tunnels *in situ* or on models tested in the laboratory: Kaiser and Morgenstern [1981a,b, 1982], Kaiser *et al.* [1983], Zoback *et al.* [1985], Bandis *et al.* [1986, 1987], Cristescu [1986], Ewy and Cook [1990a,b], Whittaker and Reddish [1991], Indraratna [1993], Kaiser [1993], Lee and Haimson [1993], Maury [1993], Young and Martin [1993] (see also Section 8.5).

Below we consider the stress and displacement distribution around deep horizontal tunnels as a **plane strain problem** (Massier and Cristescu [1981], Cristescu [1985b,c, 1989a, 1994a, c, d, e]). Besides the stress distribution, we intend to discuss tunnel closure, i.e. creep of tunnel walls, short term failure and creep failure, propagation of damage in the rock mass, finding the location where the rock around the tunnel becomes dilatant and where it becomes compressible as a result of excavation.

We do not consider here the interaction between the creeping rock and a lining. For a theory of self-adjusting lining, i.e., lining which would yield to the pressure exerted by a creeping rock, see Cristescu [1988, 1989a,1996a], Cristescu and Medves [1988], Cristescu and Duda [1989], Cristescu *et al.* [1987, 1988].

9.1   FORMULATION OF THE PROBLEM

We assume that the tunnel is quite deep so that the influence of the Earth's surface will be disregarded. The depth will be taken into account in the analysis by the value of the primary stress state at that particular depth. We consider circular cylindrical tunnels so that cylindrical coordinates r, θ and z will be used, with the z axis directed along the axis of the tunnel. It will be assumed that the primary stress state at the considered location is known, and generally the components are distinct: σ_v is the vertical component, σ_{h1} and σ_{h2} are the horizontal ones, with σ_{h2} directed along the axis of the tunnel. Since their values vary smoothly with the depth and since the radius of a deep tunnel is much smaller than the depth of the tunnel location, we assume that in the immediate neighborhood of the tunnel these components are practically constant and equal to their values estimated at the depth of the tunnel axis. Thus, using cylindrical coordinates and assuming **plane strain state** (for cross-sections which are not too close to the tunnel face) one obtains quite easily the stress state just after excavation (the elastic solution) as

$$\sigma_{rr}^S = p\frac{a^2}{r^2} + \frac{1}{2}\left(\sigma_{h1} + \sigma_v\right)\left(1 - \frac{a^2}{r^2}\right) + \frac{1}{2}\left(\sigma_{h1} - \sigma_v\right)\left(1 - \frac{4a^2}{r^2} + \frac{3a^4}{r^4}\right)\cos 2\theta \ ,$$

$$\sigma_{\theta\theta}^S = -p\frac{a^2}{r^2} + \frac{1}{2}\left(\sigma_{h1} + \sigma_v\right)\left(1 + \frac{a^2}{r^2}\right) - \frac{1}{2}\left(\sigma_{h1} - \sigma_v\right)\left(1 + \frac{3a^4}{r^4}\right)\cos 2\theta \ ,$$

$$\sigma_{r\theta}^S = \frac{1}{2}\left(\sigma_{h1} - \sigma_v\right)\left(-1 - \frac{2a^2}{r^2} + \frac{3a^4}{r^4}\right)\sin 2\theta \ , \tag{9.1.1}$$

$$\sigma_{zz}^S = \sigma_{h2} - \nu\left(\sigma_{h1} - \sigma_v\right)\frac{2a^2}{r^2}\cos 2\theta \ ,$$

with σ_{h1} the horizontal component of the far field stress normal to the tunnel axis. Formulae (9.1.1) supply the secondary stresses. In (9.1.1) superscript S means "secondary" stress component, i.e., $\sigma^S = \sigma^R + \sigma^P$, where σ^P are the primary stresses before excavation, and σ^R the relative stresses, e.g., the stress variation due to the excavation. Also $\sigma^P(X)$ depends on the location only, while $\sigma^R(X,t)$ may vary also in time. The primary stresses are obtained from (9.1.1) for $r \to \infty$ as

$$\sigma_{rr}^P = \frac{1}{2}\left(\sigma_{h1} + \sigma_v\right) + \frac{1}{2}\left(\sigma_{h1} - \sigma_v\right)\cos 2\theta \ ,$$

$$\sigma_{\theta\theta}^P = \frac{1}{2}\left(\sigma_{h1} + \sigma_v\right) - \frac{1}{2}\left(\sigma_{h1} - \sigma_v\right)\cos 2\theta \ ,$$

$$\sigma_{r\theta}^P = \frac{1}{2}\left(\sigma_{h1} - \sigma_v\right)\sin 2\theta \ , \tag{9.1.2}$$

$$\sigma_{zz}^P = \sigma_{h2} \ .$$

For the derivation of these formulae see for instance Cristescu [1985b,c, 1989a, Chap. 12], where the corresponding strain components are also given. The relative displacements with respect to the state *in situ* before excavation are

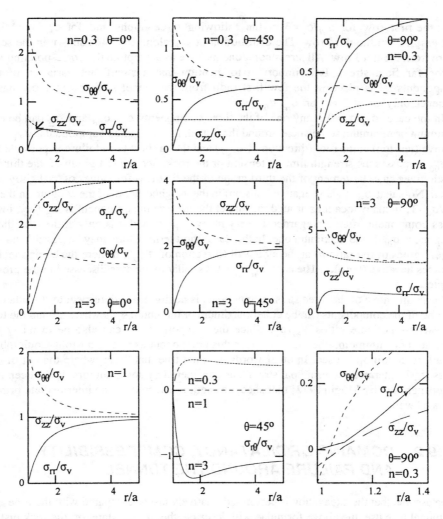

Figure 9.1 *The elastic stress distribution around a tunnel for three values of* $n = \sigma_{hl}/\sigma_v$ *and* $\sigma_v = $ *constant at depth h=500 m and at different directions* θ.

$$u_r^R = \frac{1+\nu}{E}\left\{-p\frac{a^2}{r} + \frac{1}{2}(\sigma_{hl}+\sigma_v)\frac{a^2}{r} + \frac{1}{2}(\sigma_{hl}-\sigma_v)\left[4(1-\nu)\frac{a^2}{r} - \frac{a^4}{r^3}\right]\cos 2\theta\right\}$$

$$u_\theta^R = -\frac{1+\nu}{E}\frac{1}{2}(\sigma_{hl}-\sigma_v)\left[2(1-2\nu)\frac{a^2}{r} + \frac{a^4}{r^3}\right]\sin 2\theta .$$

(9.1.3)

Figure 9.1 shows several examples of the initial stress distribution (just after excavation) of the normal components, according to (9.1.1) for the ratio $\sigma_{hl}/\sigma_v = 0.3$ in the three graphs from the first row, for $\sigma_{h1}/\sigma_v = 3$ in the following three graphs and for $\sigma_{h1}/\sigma_v = 1$ in the first graph from the third row. The distribution of $\sigma_{r\theta}$ along $\theta = \pi/4$ is shown in the second figure from the third row. All stress components have as asymptote the corresponding value of the far field stress. It is important to mention that a significant variation of these components takes place in the few first radii from the tunnel surface only. σ_{h2} has an influence only on σ_{zz}, not on $\sigma_{rr}, \sigma_{\theta\theta}, \sigma_{r\theta}$.

In the case $\sigma_{h1}/\sigma_v = 0.3$ any one of the three components $\sigma_{rr}, \sigma_{\theta\theta}$ and σ_{zz} can be either maximum or minimum somewhere around the tunnel. From all three components, $\sigma_{\theta\theta}$ can be positive (tension) at the roof (direction of σ^P_{max}) and thus a short term failure is possible there if $\sigma_{\theta\theta}$ surpasses the strength limit in tension of the rock. See the last graph of the third row which gives an enlargement of the third graph of the first row for values of r slightly larger than a. Note that σ_{zz} is also negative (tension) in the neighborhood of the contour. In the case $\sigma_{h1}/\sigma_v = 3$, in many locations around the tunnel the maximum component is σ_{zz} (i.e., just the stress component often disregarded in very many papers) and it is only at the roof that the component $\sigma_{\theta\theta}$ is the maximum one. In this case short term failure may be possible now (for larger values of the ratio n) at the side wall (direction of σ^P_{max}), if there the component $\sigma_{\theta\theta}$ becomes negative (tensile). The case $\sigma_{h1}/\sigma_v = 1$ is similar to the case discussed in the previous chapter.

Generally none of the three stress components is negligible with respect to the others and any one of the components can be either maximum or minimum somewhere around the tunnel for various possible ratios σ_{h1}/σ_v. Since the component σ_{zz} can also be in many cases maximum or minimum, one must consider this component as well if a Mohr-Coulomb type of failure condition is used. In other words, it is not true that "everywhere the out-of-plane stress σ_{zz} is intermediate principal stress", as considered by some authors. It has been found experimentally by Mogi [1972] that shear strength is affected by the intermediate principal stress as well.

9.2 DOMAINS OF DILATANCY, COMPRESSIBILITY AND FAILURE AROUND THE TUNNEL

If we assume that the excavation is performed relatively fast as compared with the time when the tunnel is in use, the above formulae will describe the initial state of the rock just after excavation. Thus, these formulae are used as "initial" data for the problem of tunnel closure and also to examine where around the tunnel the rock becomes dilatant and where compressible due to the excavation. To illustrate this last idea let us consider in the σ-τ plane the curve $X(\sigma,\tau) = 0$ representing the compressibility/dilatancy boundary, the short term failure surface $Y(\sigma,\tau) = 0$ and the initial yield surface $H(\sigma,\tau) = W^P$, together with the point representing the primary stress state (σ^P, τ^P). The secondary stress state (9.1.1) along the tunnel contour or in any other location in the neighborhood of the tunnel can also be represented on this plane. Thus this plane can be considered a **"stress map" of the model and of the initial stress state.** In order to plot this map the complete constitutive equation is not really needed. We need only the equations of the two surfaces $X=0$ and $Y=0$, which are obtained straightforwardly from tests, and the yield function $H(\sigma,\tau)$, which is obtained as shown in the previous chapters.

Figure 9.2 shows an example of such a map for the case $\sigma_h/\sigma_v = 0.3$ and $\sigma_{h1} = \sigma_{h2} = \sigma_h$.

The secondary stress state distributions along the contours $r=a$, $r=2a$ and $r=3a$ are shown by short-dashed lines. The stress state along this last contour is already quite close to the primary stress state shown as a diamond. The initial yield surface is shown as a long-dashed line, and the C/D boundary as a dash-dot line. Nearly all the stresses along the contour $r=a$ are in the dilatant region ($X<0$) with higher values of τ at the tunnel side wall and with the stresses at the crown in the elastic region (where stationary creep is still producing dilatancy: $(\dot{\varepsilon}^I_V)_s>0$ but $\dot{\varepsilon}^I_T=0$). It follows that creep failure is to be expected after some time at the side wall (direction of σ^P_{min}), while at the crown (direction of σ^P_{max}) a very small amount of rock will be involved in the short term failure ($Y\leq0$). This is shown as a single point (square) in Figure 9.3, which shows the "model state" in cross-section through the tunnel. Creep failure is predicted as described in the previous chapter. Figure 9.3 shows up to where the dilatancy region is spreading, where the compressibility region is located and where elastic unloading takes place. Again "unloading" refers to transient creep only ($\dot{\varepsilon}^I_T=0$) since stationary creep is present everywhere around the tunnel, with negligible effect there where τ has small values.

The presence of an inner pressure significantly changes the "map". For instance Figure 9.4 shows the stress distribution along the contour for the case when the tunnel is filled with brine so that the pressure exerted on the wall is that of a column of brine $p=\gamma h$, with $\gamma=0.01172$, p in MPa and h in meters. If Figures 9.2 and 9.4 are compared one finds that the possibility of creep failure at the side wall has significantly diminished since the value of τ at the side wall has decreased. However, at the crown the domain where short term failure is taking place has significantly increased. This is further shown in Figure 9.5 where the amount of the rock involved in short term failure is indicated by a full line and determined easily

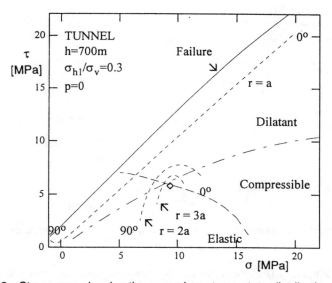

Figure 9.2 *Stress map showing the secondary stress state distribution (short-dashed lines) around a tunnel excavated in rock salt at h=700 m for σ_{hl}/σ_v =0.3; the full line is the short term failure surface, the dash-dot line is the C/D boundary and the long-dashed line is the initial yield surface.*

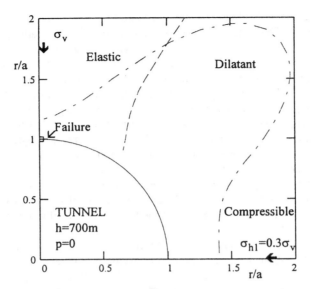

Figure 9.3 *Domains of dilatancy, compressibility, elasticity and failure (shown as a square at the crown) for p=0 and σ_{hl}/σ_v =0.3.*

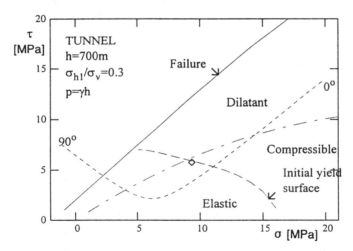

Figure 9.4 *Stress distribution along the tunnel contour in the case of presence of an internal pressure.*

from the condition $Y(\sigma,\tau) \le 0$. If Figures 9.3 and 9.5 are compared one can see important changes of the boundaries existing between the various domains due to the presence of an internal pressure: the domains of dilatancy have shrunk very much, while the elastic and compressibility domains have expanded.

The value 0.3 for the ratio σ_{hl}/σ_v was chosen to show that for values of this ratio that are

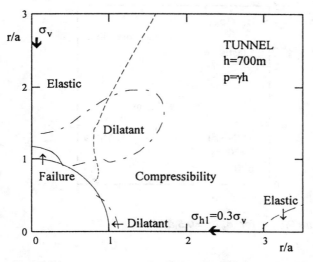

Figure 9.5 *Domains of dilatancy, compressibility, failure and elasticity around a tunnel for $\sigma_{hl}/\sigma_v = 0.3$ and internal pressure.*

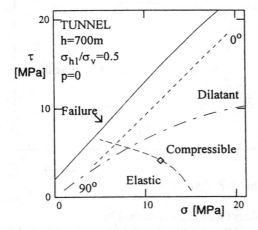

Figure 9.6 *As for Figure 9.4 but for $\sigma_{hl}/\sigma_v = 0.5$ and $p=0$.*

departing significantly from unity the tunnel may become unstable from the point of view of short term failure and of creep failure, and, as will be shown below, from the point of view of very fast closure in certain parts of the contour. If the ratio is closer to unity the tunnel is more stable. For instance Figure 9.6 shows the stress distribution along the tunnel contour for $\sigma_{hl}/\sigma_v = 0.5$ and $p=0$. In this case short term failure is no longer possible. However, the values of τ at the side wall are still quite large, and will ultimately result in creep failure. Figure 9.7 shows various domains existing around the tunnel in this case. The dilatancy domain is relatively small compared with Figure 9.3, but the octahedral shear stress along the

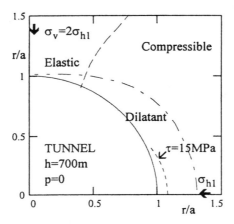

Figure 9.7 *As Figure 9.5 but for $\sigma_{hl}/\sigma_v=0.5$ and p=0.*

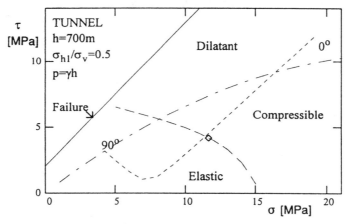

Figure 9.8 *As Figure 9.4 but for $\sigma_{hl}/\sigma_v=0.5$.*

side wall still reaches high values (see the short-dashed line).

An important improvement of the stability characteristics of the tunnel is obtained if an inner pressure due to a column of brine is applied on the tunnel wall (Figure 9.8). The stress distribution along the contour is nearly entirely in the compressibility domain, thus ensuring stability. Figure 9.9 shows the corresponding domains around the tunnel. The domain of dilatancy is very small and involving small values of τ, or, more precisely, the octahedral shear overstress above the C/D boundary is small. Creep failure is practically impossible. This is a very stable case.

For a ratio of the far field stresses close to unity the tunnel is quite stable. No other examples will be given to illustrate this case. We will rather discuss the case when this ratio is well above unity. Figure 9.10 shows the stress distribution along the contour for $\sigma_{hl}/\sigma_v=2$. Two cases were considered: a tunnel with no internal pressure and a tunnel filled with brine.

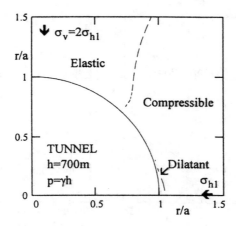

Figure 9.9 *As Figure 9.5 but for $\sigma_{hl}/\sigma_V =0.5$.*

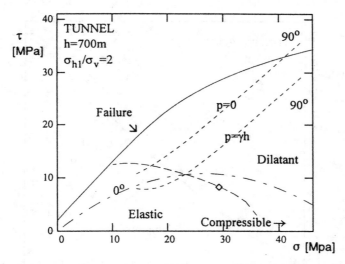

Figure 9.10   *Stress distribution along the tunnel contour for $\sigma_{hl}/\sigma_V =2$ and two internal pressures.*

The empty tunnel is an unstable case since at both the roof and floor a short term failure will take place, followed in the same direction (direction of σ^P_{min}) by creep failure which will enlarge the damaged area. The amount of rock involved in short term failure is shown in Figure 9.11a by a full line. In the same figure is shown the wide domain of dilatancy which completely surrounds the tunnel contour and expands primarily towards the vertical direction. Very high values of τ at the roof and the floor will produce a fast creep in that direction.

For the same ratio of far field stresses, the presence of an internal pressure greatly improves the stability conditions (Figure 9.11b). No short term failure is possible but at the

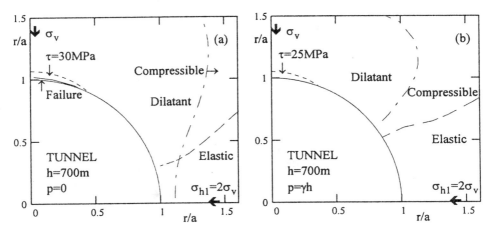

Figure 9.11 *Various domains around the tunnel for $\sigma_{hl}/\sigma_v = 2$ and a) no internal pressure; b) pressure of a column of brine.*

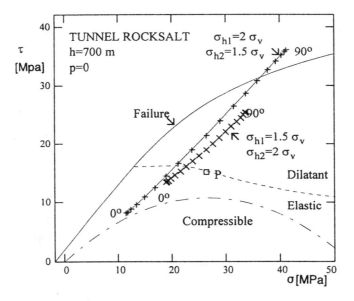

Figure 9.12 *Stress distribution along the tunnel wall contour for two possible far field stress distributions.*

roof and floor there are still high values of τ which will produce a fast creep quite soon after excavation. Figure 9.11b shows the various significant domains existing around the tunnel. The domain of dilatancy has shrunk compared with the previous case, but it is still spread significantly in the vertical direction, with high values of τ at the roof and floor.

As a conclusion we observe that short term failure is practically possible either in the direction of σ_{min}^P or σ_{max}^P but creep failure is possible in the direction of σ_{min}^P only.

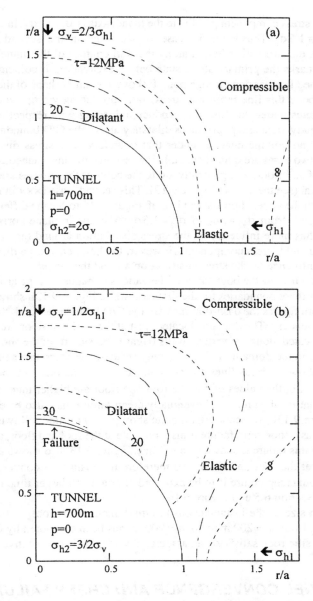

Figure 9.13 *Domains of significance around a tunnel and contours of τ = constant (short-dashed lines); the dash-dot line is the C/D boundary; the solid lines - show the boundary of the failure domain; the long-dashed line is the elastic boundary.*

Let us now discuss the case $\sigma_{h1} \neq \sigma_{h2}$. For this purpose let us consider a tunnel excavated in rock salt at depth $h = 700$ m. As possible far field stresses we assume two cases. In the first the horizontal in-plane far field stress σ_{h1} is related to σ_v by $\sigma_{h1} = 1.5\,\sigma_v$, while the horizontal far field stress component parallel to the tunnel axis is $\sigma_{h2} = 2\sigma_v$. In the second case $\sigma_{h1} = 2\sigma_v$ and $\sigma_{h2} = 1.5\sigma_v$. Thus in the first case $\sigma_{h1} < \sigma_{h2}$ and in the second one $\sigma_{h1} > \sigma_{h2}$. Figure 9.12 shows the stress distribution along the wall contour of the tunnel for these two cases. A square marks the primary stress state before excavation. It coincides for the two cases. A short-dashed line passing through point P shows the initial shape of the yield surface; thus, the points above this line represent stress states producing viscoplastic deformation, while the stress states under this line will produce unloading with respect to the transient creep. As ever, steady-state creep produces dilatancy above the C/D boundary (shown as a dash-dot line) but no volume change under that boundary. The stress distributions after excavation for the two cases are quite distinct. In the second case instantaneous failure is also possible at the roof and floor, and generally along the contour the stresses are spread out on a larger span interval (see the +++ in Figure 9.12). This second case (××× in Figure 9.12) is to be compared with the upper short-dashed line of Figure 9.10 to see the difference produced when σ_{h2} is changed from $\sigma_{h2} = 2\sigma_v$ to $\sigma_{h2} = 1.5\sigma_v$. Also the in-plane horizontal far field stress component has more influence on the stress distribution around the contour than the horizontal axial far field stress component. However, the latter cannot be disregarded since it has an obvious influence on the stress distribution around the tunnel.

Figure 9.13 a and b show the boundaries of the various domains existing around the tunnel. Figure 9.13a shows these boundaries for the first case while Figure 9.13b shows them for the second case. In both figures the dash-dot lines are the C/D boundaries, and the long-dashed line is the elastic boundary. The domains located just at the roof and floor are in the dilatant stages, and those located along the vertical walls (right-hand side from the short-dashed line) in the elastic stage. In the domain of elasticity only steady-state creep takes place. By short-dashed lines are shown various lines τ = constant. It is obvious that for the second case, shown in Figure 9.13b, the values of τ at the roof and floor are higher than in the first case (Figure 9.13a). That is why creep and eventually creep failure may also be expected in the second case (Figure 9.13b) to occur after a quite short period of time following excavation. Moreover, in this case short term failure is also possible at the roof and floor, and the amount of rock involved in this failure is shown by a full line. Figure 9.13 also shows that the highest values of τ occur at the roof and floor, and therefore the main wall convergence is in the vertical direction, but creep failure is to be expected after a much larger time interval than in the first case (see Section 6.5 and Figure 6.16).

The increase in size of the instantaneous failure domain at the roof, as the depth of the tunnel is increasing from $h = 200$ m up to $h = 900$ m, has been computed by Cristescu *et al.* [1985] for an andesitic rock satisfying an associated viscoplastic constitutive equation.

9.3   TUNNEL CONVERGENCE AND CREEP FAILURE

Let us discuss now the wall convergence of a tunnel. We would like to examine mainly where around the contour faster convergence is to be expected, and at what location and when a creep failure is initiated. For this purpose let us consider again a tunnel excavated in rock salt. As is well known for this rock, significant convergence is to be expected mainly at great depth. The first example considers a tunnel excavated in rock salt at a depth $h = 600$ m for the simplest case $\sigma_h = \sigma_v$ and $\sigma_{h1} = \sigma_{h2} = \sigma_h$. The starting of the tunnel convergence is

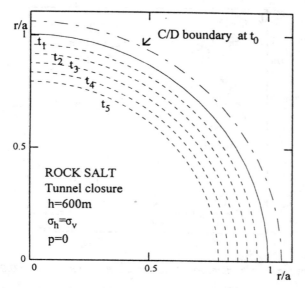

Figure 9.14 *Convergence of a tunnel excavated in rock salt at h = 600 m.*

described with a formula similar to the one used for vertical boreholes (see equation (8.2.1)). This formula does not take into account the stress variation during creep, but it can still show where along the circumference to expect the beginning of an accelerated convergence process.

An example of convergence at five successive dimensionless times is shown in Figure 9.14 for a depth $h=600$ m. The chosen five successive dimensionless times are $t_1 = k_S(t-t_o) = 1.5 \times 10^6$, $t_2 = 3 \times 10^6$, $t_3 = 4.5 \times 10^6$, $t_4 = 6 \times 10^6$ and $t_5 = 7.5 \times 10^6$. Some dimensionless times have been used for transient creep (assuming $k_T = k_S$), though most of the convergence is due to steady-state creep. The parameter from the constitutive equation which plays a fundamental role in the rate of the wall convergence is k_S. Since there is a certain uncertainty concerning the exact determination of this parameter, and in order to give more general solutions, in the examples given here a dimensionless time variable was used. The uncertainty is due to the fact that different types of rock salt show creep rates that can differ by a factor of 100 or even more; also reliable long term creep test data are scarce (see §3.1). The depth significantly influences the convergence. A dash-dot line shows where the C/D boundary is located at the initial time $t_o = 0$, just after excavation. Other examples for some other depths can be found in Cristescu [1994d].

How the creep failure progresses in the rock mass during creep is shown in Figures 9.15 and 9.16. The computation is done using the formula (8.2.1) given in Chapter 8. Also the spreading of dilatancy in the rock mass during creep is obtained using formula (5.3.40).

In the case $\sigma_h = \sigma_v$ all the boundaries shown in Figure 9.14 are symmetric (circles). If $\sigma_h \neq \sigma_v$ the boundaries are no longer symmetric and the tunnel contour is more unstable. For instance, Figure 9.15 shows the convergence of a tunnel excavated in rock salt at $h = 800$ m, for the ratio of the far field stresses $\sigma_h/\sigma_v = 0.9$, which is only a small deviation from 1. The convergence is faster in the horizontal direction, i.e. the direction of σ_{min}^P, which coincides with the direction where τ is maximum on the contour. The dimensionless times considered

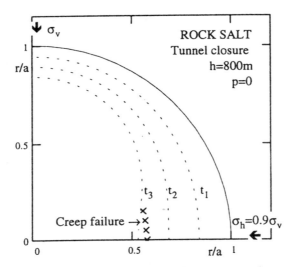

Figure 9.15 *Convergence and creep failure around a tunnel for* $\sigma_h = 0.9\,\sigma_v$.

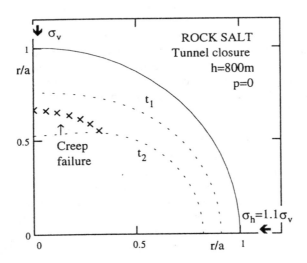

Figure 9.16 *As Figure 9.15 but for* $\sigma_h = 1.1\sigma_v$.

are $t_1 = k_s(t - t_o) = 1\times10^6$, $t_2 = 2\times10^6$ and $t_3 = 3\times10^6$. If we compare the results obtained for the case $\sigma_h = \sigma_v$ with those for the case $\sigma_h \neq \sigma_v$ we arrive at the conclusion that for $\sigma_h \neq \sigma_v$ the closure is faster in the direction σ_{min}^P, but slower in the direction σ_{max}^P. Creep failure at time t_3 is shown by xxx. We rerecall that creep failure is generally expected where τ reaches rather high values ($\tau_{max} = 17.17$ MPa at $\theta = 0°$).

Figure 9.16 shows the opposite case (obviously non-symmetric since for the same value of σ_v only σ_h has been changed) when $\sigma_h = 1.1\,\sigma_v$. This time the fastest convergence is in the vertical direction, i.e. the direction of the minimum far field stress and of the maximum value of τ. Convergence is shown for only two dimensionless times t_1 and t_2 (the same values as in the previous figure). Since in this case the maximum value of $\tau = 18.78$ MPa at $\theta = 90°$ is bigger than the highest values reached in the previous case, more convergence is taking place in this case. For the same reason the creep damage involves a greater amount of rock.

A more precise description of tunnel wall convergence has to take into account the stress variation during creep as described in the previous chapter for boreholes.

The spreading of creep failure into the rock mass around a tunnel has been considered in several papers. An example when both short-term failure and creep failure are possible in given by Cristescu [1986, 1989a] for a tunel excavated in andesitic rock at depth $h = 900$ m, for the case $\sigma_h/\sigma_v = 0.21$. In this case an instantaneous failure that occurs at the roof and floor as shown in Figure 9.17 is produced by tension stresses. In the vertical direction the rock is elastic (domain denoted D_e). Along the side walls the rock is dilatant (domain D_d). The maximum value of τ is at $\theta = 0°$, $r = a$. It is just there that creep-failure starts after 53 hours, and afterwards spreads in the rock mass. The short-dashed line shows how much the creep-failure has spread after 4 days and 16 hours. Thus, the evolutive failure, taking place in the direction of the minimum component of the far field stress is progressive, i.e., takes place in time.

For a similar study of possible evolutive damage and short-term failure, in the case of tunnels excavated in coal, see Cristescu [1989a, e]. For coal, damage evolution in time and its slowly spreading into the rock mass takes place at even small depths. The local (old mines have very many tunnels and shafts at small distances from one another) rate of the far field stress can be quite distinct from unity.

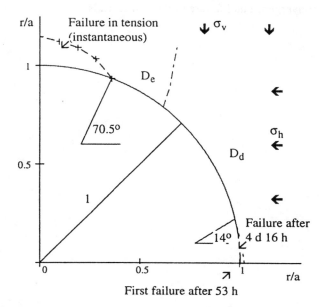

Figure 9.17   *Short term failure and creep failure for a tunnel excavated in andesite at h = 900 m (Cristescu [1986]).*

Experimental evidence. The short term failure and creep failure around a cylindrical opening have been studied both in the laboratory and by in-field measurements. Breakouts aligned in the direction of the least far field stress component acting in the cross-section plane have been found by Kaiser and Morgenstern [1982], Zoback *et al.* [1985], Bandis *et al.* [1986, 1987], Ewy and Cook [1990b], Lee and Haimson [1993], Herrick and Haimson [1994], Aoki *et al.* [1994] , Martin *et al.* [1994], Martin and Read [1996], Castro *et al.* [1995], besides others. Some of these authors have found that the phenomenon is evolutive in time, as described theoretically above. It has also been found that dilatational processes take place in the damaged domain, i.e. a volume change takes place in the direction of σ_{min}^{P}, where a clustering of the AE/MS events has also been found (Young and Martin [1993]). Interesting tests or analytical (numerical) discussions of possible failure modes are due to Whittaker and Reddish [1991], Maury [1993], Indraratna [1993], Ewy and Cook [1990a], Zheng and Khodaverdian [1996]. The influence of grain rotations (size effect) involved in the case of small holes, is discussed by van den Hoek *et al.* [1994].

Conclusions. Generally:
- faster creep closure of the tunnel walls takes place in the direction of the smallest far field stress component (acting in the cross section);
- creep failure is to be expected in the same direction;
- in the case $\sigma_h \neq \sigma_v$ (when $\sigma_{h1} = \sigma_{h2}$) the rock is more unstable compared with the case $\sigma_h = \sigma_v$;
- the convergence strongly depends on the depth, ratio of the far field stresses (acting in the cross section), and internal pressure;
- the far field stress component oriented along the tunnel axis also plays a role in the convergence and failure of the tunnel walls.

10 Creep, Damage and Failure around Rectangular-like Galleries or Caverns

Underground excavations of some other cross-sectional shape than circular are quite often used in mining engineering (galleries in mines), petroleum engineering (caverns used for storage of petroleum products), and civil engineering (underground deposits, garages, railway stations, power stations, stadiums, commercial/institutional purposes, parking; see Berest and Nguyen-Minh [1983], Barla and Rabagliati [1992], Broch and Nilsen [1993], Cheng and Liu [1993], Hoek and Moy [1993], Moretto *et al.* [1993], Nordmark and Franzen [1993], Sterling [1993], Barton *et al.* [1994], Nelson [1996]). Quite often these caverns or galleries are either square-like or rectangular-like in cross-sectional shape (Dawson and Munson [1983], Borns and Stormont [1989], Munson and Devries [1993], Mraz *et al.* [1993], Frayne *et al.* [1996]), i.e., their cross-sections are either squares or rectangles, with more or less rounded corners, the third dimension being much larger than the dimensions of the cross-section. This allows us to consider the problem to be a **plane strain problem**. The design of such caverns is based usually on numerical methods, in most cases either FEM or BEM (Saari [1987], Honecker and Wulf [1988], Nguyen-Minh and Pouya [1992], Cravero *et al.* [1993], Del Greco *et al.* [1993], Kikuchi *et al.* [1993], Lee Petersen and Nelson [1993], Liang and Lindblom [1994], McCreath and Diederichs [1994], Lux and Schmidt [1996], Vouille *et al.* [1996]). For the model of the rock involved, in most cases Hooke's law has been used, though occasionally some rheological models have been used as well (Passaris [1979], Nguyen-Minh and Pouya [1992], Kikuchi *et al.* [1993], Vouille *et al.* [1996]). In a very good survey paper Maury [1993] presents a variety of possible failure around underground excavations, and many case histories. Creep and failure of coal around a long wall was studied by Cristescu *et al.* [1994], Ionescu and Sofonea [1993]; the coal was modeled by an elasto/viscoplastic model with internal state variables of the kind described in Chapters 4 and 5. The variation in time of stress and displacement around the excavation was found. A method to calculate stresses within an elasto/viscoplastic seam of Bingham type is due to Nawrocki [1993, 1995]. Closed-form solutions for the thermoplastic behavior of rock around tunnels are due to Wong and Simionescu [1997], and approximate viscoelastic one to Bogobowicz [1996].

In the present chapter we discuss the problem of failure, creep dilatancy and/or compressibility, creep damage, and stability of large underground rectangular-like galleries, in a relatively unfractured rock, using the constitutive equation described in Chapter 4. The description follows the papers by Cristescu and Paraschiv [1995a, b, 1996], Cristescu [1996c], where more details can be found.

First an exact elastic solution (the "Massier solution", see Massier [1989, 1995a, b, 1996]) describing the state of stress, strain, and displacement around a rectangular-like gallery is given. Further we examine where around the gallery failure is possible just after excavation, and we determine the amount and shape of rock involved in failure. Furthermore we determine where around the gallery the rock becomes dilatant by microcracking, since this will seriously jeopardize the safety of the cavern or gallery, which may possibly be used for a repository of radioactive wastes or for other hazardous materials. Sometimes the dilatancy domain spreads quite far from the excavation, while in some other cases this domain does not exist at all or is very small. The determination of dilatant domains around the contour has an obvious significance not only for the safety of the underground work, but also for its use as a cavern or gallery for storage of radioactive wastes or some other contaminants, or for gases or hydrocarbons, for instance.

Also the exact solution shows where around the opening the stress state satisfies a short term failure condition and where a stress concentration occurs. From the point of view of rock creep, it is important to find out where the octahedral shear stress τ reaches very high values. It is there that a very fast creep is to be expected and possibly an early creep failure. The creep failure is a progressive phenomenon initiated at a certain time after excavation, at a certain location around the contour, and spreading in time into the rock mass (see Cristescu [1986, 1989a, 1993b, 1996c, d]). For circular excavations (tunnels or boreholes) the initiation of creep failure is governed mainly by the ratio of the far field stresses σ_h/σ_v: the creep failure initiates always in the direction of the minimum far field stress component σ_{min}^P since in that direction the octahedral shear stress τ reaches the highest values (see the above-mentioned papers). The spreading in time of the creep damage takes place both along the contour and into the rock mass. Internal pressure and depth have an obvious influence too.

With rectangular-like excavations, and generally with a non-circular cross-section of an excavation, the stress distribution around the cavern is much more involved. Each "corner" will introduce a stress concentration. Also, elongated cross-sections introduce a stress concentration. But a stress concentration is due also to σ_{min}^P. Thus one has to estimate where the initiation of the creep failure will take place as a result of the combined influence of non-circular cross-section, the ratio σ_h/σ_v, a possibly existing internal pressure, and obviously the depth. The same factors will influence the creep of the walls. Thus we would like to see where around the contour of the excavation the fastest creep convergence is to be expected. If the location of fastest creep convergence and that of initiation of creep failure are determined, and if the amount of rock involved in dilatancy is also determined, one can suggest where a support is necessary to keep the cavern reasonable stable, and the amount of this support.

After developing an exact solution which allows us to analyze in detail the influence of the various parameters involved, and thus to get a good understanding of how a cavern is behaving, one can develop a FEM method which has to reproduce the same results if particularized to the same problem. Further a FEM approach will allow the consideration of the galleries or caverns of various other cross-sectional shapes, or of the influence of stress relaxation, etc.

10.1 FORMULATION OF THE PROBLEM

The elastic stress state around a rectangular-like cross-section cavern or gallery can be obtained either with a complex variable method (see Muskhelishvili [1953], Savin [1961], Jaeger and Cook [1979], Gerçek [1993]) or with a method suggested by Greenspan [1944]. We use the latter, but change it from a plane-stress problem as Greenspan has considered to a **plane-strain problem**, as should be the case for a long underground cavern or gallery. Let us consider the map (see Greenspan [1944], and Obert and Duvall [1967]):

$$x = P\cos\beta + R\cos 3\beta \ ,$$
$$y = Q\sin\beta - R\sin 3\beta \ ,$$

(10.1.1)

where P, Q and R are constants and $\beta \in [0, 2\pi)$. We use the system of coordinates (α, β) defined by Massier [1995a, b]

$$x = \left(\bar{a}e^{\alpha} + \bar{b}e^{-\alpha}\right)\cos\beta + \bar{c}e^{-3\alpha}\cos 3\beta \ ,$$
$$y = \left(\bar{a}e^{\alpha} - \bar{b}e^{-\alpha}\right)\sin\beta - \bar{c}e^{-3\alpha}\sin 3\beta \ .$$

(10.1.2)

The coefficients in (10.1.1) are written as

$$P = \bar{a}e^{\alpha} + \bar{b}e^{-\alpha} \ , \quad Q = \bar{a}e^{\alpha} - \bar{b}e^{-\alpha} \ , \quad R = \bar{c}e^{-3\alpha} \ ,$$

(10.1.3)

where $\bar{a}, \bar{b}, \bar{c}$ are constants, α is a real number. α and β define a new curvilinear coordinate system. The contour of the cavern cross-section is obtained for a particular value $\alpha_o = 0$ of α. Various contours α=constant and β=constant are shown in Figure 10.1. By changing the parameter Q one can control mainly the height of the gallery (Figure 10.2b). In a similar way the parameter P controls mainly the horizontal width of the gallery. The parameter R is the one controlling mainly the rounding off of the corners of the gallery cross-section (Figure 10.2a), the values of P and Q being kept constant. For greater (algebraic) values of R the corners are more rounded. For $R = 0$ the contour becomes an ellipse if $P \ne Q$, while for $P = Q$ and $R = 0$ it becomes a circle. Thus by assigning various particular values to the parameters P, Q and R various shapes of the cavern contour can be obtained. Thus for the ratios $P/Q > 1$ we

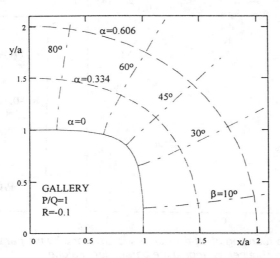

Figure 10.1 *Curvilinear coordinate lines* α = *const and* β = *const used in the analysis.*

obtain "wide" galleries, for $P/Q < 1$ "tall" galleries and for $P/Q = 1$ galleries of square- like cross-section. In all the figures, a is the initial horizontal size of half of the gallery.

The **boundary conditions** have to be formulated on the surface of the cavern and at infinity. We assume that a certain pressure may be exerted on the walls of the gallery, due either to the presence of a pressurized gas or to a liquid existing inside the gallery, or maybe due to the presence of a lining. One can consider also the case when this pressure may vary smoothly in time and we would like to examine the influence of an internal pressure on the gallery stability. Thus on the gallery cross-section contour we prescribe

$$\sigma_{\alpha\alpha}(\alpha,\beta) = p \quad , \quad \sigma_{\alpha\beta}(\alpha,\beta) = 0 \tag{10.1.4}$$

for $\beta \in [0, 2\pi)$, i.e., a pressure p is exerted on the walls only. We assume that at infinity the far field stresses can be defined by prescribing the values of the vertical component σ_v and the horizontal one σ_h. As usual we take

$$\sigma_v = \gamma h \quad , \quad \sigma_h = n\sigma_v \tag{10.1.5}$$

with the depth h in meters, σ_v in MPa and $0.3 \leq n \leq 4$, an interval quite often found from *in situ* measurements (see Chapter 7). For rock salt this interval is smaller, i.e., $0.5 \leq n \leq 2$. Several values for n will be used to illustrate various examples, since in different locations various far field stresses are to be found. Also, the ratio σ_h/σ_v is one of the main parameters influencing the stability of the gallery and for this reason different values of this parameter are considered in a parameter study concerning the gallery stability. In the examples given below we have used the value $\gamma = 0.025$. Thus, in cylindrical coordinates (the α,β coordinates become cylindrical at great distances, when $\alpha \to \infty$) the far field stresses are

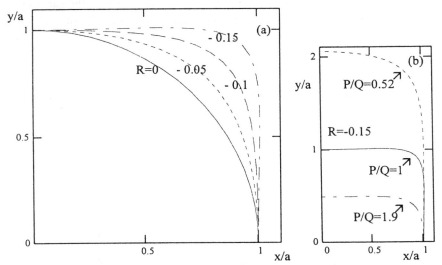

Figure 10.2   a) The rounding off of the corners is governed by the parameter R; b) Various shapes of the cavern obtained changing the parameter Q alone.

$$\lim_{\alpha \to \infty} \sigma_{\alpha\alpha} = \sigma_{\alpha\alpha}^{P} = \frac{1}{2}(\sigma_h + \sigma_v) + \frac{1}{2}(\sigma_h - \sigma_v)\cos 2\beta \ ,$$

$$\lim_{\alpha \to \infty} \sigma_{\beta\beta} = \sigma_{\beta\beta}^{P} = \frac{1}{2}(\sigma_h + \sigma_v) - \frac{1}{2}(\sigma_h - \sigma_v)\cos 2\beta \ ,$$

$$\lim_{\alpha \to \infty} \sigma_{\alpha\beta} = \sigma_{\alpha\beta}^{P} = -\frac{1}{2}(\sigma_h - \sigma_v)\sin 2\beta \ . \tag{10.1.6}$$

Just after a fast excavation the exact elastic solution has been obtained by Massier [1995a,b, 1996]:

$$\frac{\sigma_{\alpha\alpha}}{\bar{h}^4} = \left[\frac{1}{2}g_o'f_o' - 3g_1f_{c1} + \frac{1}{4}g_1'f_{c1}' - 12g_2f_{c2} + \frac{1}{4}g_2'f_{c2}'\right] + \left[\frac{1}{2}g_1'f_o' - \right.$$

$$\left. 4g_o f_{c1} + \frac{1}{2}g_o'f_{c1}' - 4g_2 f_{c1} + \frac{1}{4}g_2'f_{c1}' - 10g_1 f_{c2} + \frac{1}{4}g_1'f_{c2}'\right]\cos 2\beta$$

$$+ \left[\frac{1}{2}g_2'f_o' - g_1 f_{c1} + \frac{1}{4}g_1'f_{c1}' - 16g_o f_{c2} + \frac{1}{2}g_o'f_{c2}'\right]\cos 4\beta + \left[\frac{1}{4}g_2'f_{c1}'\right. \tag{10.1.7a}$$

$$\left. -6g_1 f_{c2} + \frac{1}{4}g_1'f_{c2}'\right]\cos 6\beta + \left(-4g_2 f_{c2} + \frac{1}{4}g_2'f_{c2}'\right)\cos 8\beta \ ,$$

$$\frac{\sigma_{\beta\beta}}{\bar{h}^4} = \left[g_o f_o'' - \frac{1}{2}g_o'f_o' + \frac{1}{2}g_1 f_{c1}'' + g_1 f_{c1} - \frac{1}{4}g_1'f_{c1}' - \frac{1}{2}g_2'f_{c2}'' + 4g_2 f_{c2} - \frac{1}{4}g_2'f_{c2}'\right]$$

$$+ \left[g_1 f_o'' - \frac{1}{2}g_1'f_o' + g_o f_{c1}'' - \frac{1}{2}g_o'f_{c1}' + \frac{1}{2}g_2 f_{c1}' + 2g_2 f_{c1} - \frac{1}{4}g_2'f_{c1}' + \frac{1}{2}g_1 f_{c2}''\right.$$

$$\left. + 2g_1 f_{c2} - \frac{1}{4}g_1'f_{c2}'\right]\cos 2\beta + \left[g_2 f_o'' - \frac{1}{2}g_2'f_o' + g_o f_{c2}'' - \frac{1}{2}g_o'f_{c2}' + \frac{1}{2}g_1 f_{c1}''\right. \tag{10.1.7b}$$

$$\left. - g_1 f_{c1} - \frac{1}{4}g_1'f_{c1}'\right]\cos 4\beta + \left[\frac{1}{2}g_2 f_{c1}'' - 2g_2 f_{c1} - \frac{1}{4}g_2'f_{c1}' + \frac{1}{2}g_1 f_{c2}''\right.$$

$$\left. - 2g_1 f_{c2} - \frac{1}{4}g_1'f_{c2}'\right]\cos 6\beta + \left(\frac{1}{2}g_2 f_{c2}'' - 4g_2 f_{c2} - \frac{1}{4}g_2'f_{c2}'\right)\cos 8\beta \ ,$$

$$\frac{\sigma_{\alpha\beta}}{\bar{h}^4} = \left[-g_1 f_o' + 2g_o f_{cl}' - g_o' f_{cl} - 2g_2 f_{cl}' + \frac{1}{2}g_2' f_{cl} + \frac{5}{2}g_1 f_{c2}' - g_1' f_{c2}\right]\sin 2\beta$$

$$+ \left[-2g_2 f_o' + \frac{1}{2}g_1 f_{cl}' - \frac{1}{2}g_1' f_{cl} + 4g_o f_{c2}' - 2g_o' f_{c2}\right]\sin 4\beta$$

$$+ \left[-\frac{1}{2}g_2' f_{cl} + \frac{3}{2}g_1 f_{c2}' - g_1' f_{c2}\right]\sin 6\beta + \left(g_2 f_{c2}' - g_2' f_{c2}\right)\sin 8\beta \ , \qquad (10.1.7c)$$

$$\sigma_{zz} = \sigma_h + \nu\left[\left(\sigma_{\alpha\alpha} - \sigma_{\alpha\alpha}{}^P\right) + \left(\sigma_{\beta\beta} - \sigma_{\beta\beta}{}^P\right)\right] \ ,$$

where the function $\bar{h}$ is defined by

$$\bar{h}^{-2}(\alpha,\beta) = g_o(\alpha) + g_1(\alpha)\cos 2\beta + g_2(\alpha)\cos 4\beta \ , \qquad (10.1.8)$$

with

$$g_o(\alpha) = \bar{a}^2 e^{2\alpha} + \bar{b}^2 e^{-2\alpha} + 9\bar{c}^2 e^{-6\alpha} \ ,$$
$$g_1(\alpha) = 6\bar{b}\bar{c}e^{-4\alpha} - 2\bar{a}\bar{b} \ , \qquad (10.1.9)$$
$$g_2(\alpha) = -6\bar{a}\bar{c}e^{-2\alpha}$$

and

$$f_o(\alpha) = F_{o2}\alpha + \bar{a}C_{21}\frac{e^{2\alpha}}{4} + \bar{b}(C_{12} - C_{11})\frac{e^{-2\alpha}}{4} + 3\bar{c}(C_{22} - C_{01})\frac{e^{-6\alpha}}{36}$$

$$f_{cl}(\alpha) = F_{cl1}e^{2\alpha} + F_{cl2}e^{-2\alpha} + \frac{1}{4}\left[\bar{b}C_{21} + \bar{a}(C_{12} - C_{11})\right] +$$

$$\left[3\bar{c}(C_{12} - C_{11}) + \bar{b}(C_{22} - C_{01})\right]\frac{e^{-4\alpha}}{12}$$

$$(10.1.10)$$

$$f_{c2}(\alpha) = \left[\alpha(C_{22} - C_{01}) + 3\bar{c}C_{21}\right]\frac{e^{-2\alpha}}{12} \ .$$

In the above formulae primes and double primes indicate first and second derivatives with respect to α. The constants involved in the above formulae are determined from the boundary conditions and are

$$F_{o2} = \left(\bar{a}^2 + \bar{b}^2 \frac{\bar{a}+\bar{c}}{\bar{a}-\bar{c}} + 3\bar{c}^2 \right) \left(p - \frac{\sigma_h + \sigma_v}{2} \right) + 2 \frac{\bar{a}^2 \bar{b}}{\bar{a}-\bar{c}} \frac{\sigma_h - \sigma_v}{2} ,$$

$$F_{c11} = -\frac{\bar{a}^2}{2} \frac{\sigma_h - \sigma_v}{2} ,$$

$$F_{c12} = -\frac{2\bar{a}\bar{b}\bar{c}e^{-2\alpha}}{\bar{a}-\bar{c}e^{-4\alpha}} \left(p - \frac{\sigma_h + \sigma_v}{2} \right) - \frac{\bar{a}^2 (\bar{a} + 3\bar{c}e^{-4\alpha})e^{2\alpha}}{2(\bar{a}-\bar{c}e^{-4\alpha})} \frac{\sigma_h - \sigma_v}{2} ,$$

$$C_{21} = \bar{a}(\sigma_h + \sigma_v) , \quad C_{22} - C_{o1} = 6\bar{c} \left(2p - \frac{\sigma_h + \sigma_v}{2} \right) , \qquad (10.1.11)$$

$$C_{12} - C_{11} = \frac{4\bar{a}\bar{b}p}{\bar{a}-\bar{c}e^{-4\alpha}} - 2\bar{b} \frac{\bar{a}+\bar{c}e^{-4\alpha}}{\bar{a}-\bar{c}e^{-4\alpha}} \frac{\sigma_h + \sigma_v}{2} + \frac{4\bar{a}^2 e^{2\alpha}}{\bar{a}-\bar{c}e^{-4\alpha}} \frac{\sigma_h - \sigma_v}{2} .$$

The elastic strain components corresponding to the stress state (10.1.7) are

$$\varepsilon_{\alpha\alpha}^E = \frac{1+\nu}{E} \left[(1-\nu)\left(\sigma_{\alpha\alpha} - \sigma_{\alpha\alpha}^P\right) - \nu\left(\sigma_{\beta\beta} - \sigma_{\beta\beta}^P\right) \right] ,$$

$$\varepsilon_{\beta\beta}^E = \frac{1+\nu}{E} \left[(1-\nu)\left(\sigma_{\beta\beta} - \sigma_{\beta\beta}^P\right) - \nu\left(\sigma_{\alpha\alpha} - \sigma_{\alpha\alpha}^P\right) \right] , \qquad (10.1.12)$$

$$\varepsilon_{\alpha\beta}^E = \frac{1+\nu}{E} \left(\sigma_{\alpha\beta} - \sigma_{\alpha\beta}^P \right) ,$$

while the other components are zero. Here, as in the previous chapters, the superscript P means "primary" component so that $\sigma - \sigma^P$ is the stress variation due to the excavation. The strains (10.1.12) are just due to this stress variation. Thus the elastic strain components (10.1.12) are relative. The viscoplastic components obtained during creep, if we assume the stresses remain constant, are similar to (5.3.36); i.e., for $\varepsilon_{\beta\beta}$

$$\varepsilon_{\beta\beta}^{VP} = H \left\langle 1 - \frac{W^P}{H} \right\rangle \frac{\partial F/\partial \sigma_{\beta\beta}}{\frac{\partial F}{\partial \sigma} \cdot \sigma} \left\{ 1 - \exp\left[\frac{1}{H} k_T \frac{\partial F}{\partial \sigma} \cdot \sigma (t_o - t) \right] \right\} + k_S \frac{\partial S}{\partial \sigma_{\beta\beta}} (t - t_o) , \qquad (10.1.13)$$

where t_o is the time of excavation, when creep starts.

The strain components are related to the displacements by the formulae

$$\varepsilon_{\alpha\alpha} = \bar{h}\frac{\partial u_\alpha}{\partial \alpha} - \frac{\partial \bar{h}}{\partial \beta}u_\beta \quad , \quad \varepsilon_{\beta\beta} = \bar{h}\frac{\partial u_\beta}{\partial \beta} - \frac{\partial \bar{h}}{\partial \alpha}u_\alpha \,,$$

$$2\varepsilon_{\alpha\beta} = \frac{\partial(\bar{h}u_\beta)}{\partial \alpha} + \frac{\partial(\bar{h}u_\alpha)}{\partial \beta} \,. \tag{10.1.14}$$

Since we consider here the beginning of the convergence process only, we disregard the component u_β since in most cases the wall convergence is essentially due to the component $\varepsilon_{\beta\beta}$ and u_α. Thus this displacement component is obtained from

$$u_\alpha = -\left(\varepsilon_{\beta\beta}^E + \varepsilon_{\beta\beta}^{VP}\right)\frac{1}{\partial \bar{h}/\partial \alpha} \,, \tag{10.1.15}$$

where

$$\frac{1}{\partial \bar{h}/\partial \alpha} = \frac{-2\left(g_o(\alpha) + g_1(\alpha)\cos 2\beta + g_2(\alpha)\cos 4\beta\right)^{3/2}}{g_o'(\alpha) + g_1'(\alpha)\cos 2\beta + g_2'(\alpha)\cos 4\beta} \,, \tag{10.1.16}$$

with $\bar{h}$ defined by (10.1.8). After finding the displacement u_α, the displacement in cartesian coordinates follows from (10.1.2) as

$$u_1 = \bar{h}\left(u_\alpha\frac{\partial x_1}{\partial \alpha} + u_\beta\frac{\partial x_1}{\partial \beta}\right) \quad , \quad u_2 = \bar{h}\left(u_\alpha\frac{\partial x_2}{\partial \alpha} + u_\beta\frac{\partial x_2}{\partial \beta}\right) \,, \tag{10.1.17}$$

while the convergence of the gallery walls follows from $x = X - u$, i.e.,

$$x_1 = X_1 - \bar{h}\,u_\alpha\frac{\partial x_1}{\partial \alpha} \quad , \quad x_2 = X_2 - \bar{h}\,u_\alpha\frac{\partial x_2}{\partial \alpha} \tag{10.1.18}$$

if u_β is disregarded. Here X are the initial coordinates of the points of the gallery wall, and x the present coordinates of the same points.

10.2   DILATANCY, COMPRESSIBILITY AND FAILURE AROUND A GALLERY

In order to study where around the gallery the rock becomes dilatant, where compressible and where failure is possible (with a possible estimation of the volume and shape of the region where the rock is involved in failure) we use the procedure described in the previous two chapters. Thus we intend to discuss the state just after excavation, which will be considered "initial data" for the study of creep, creep convergence and creep failure. For this purpose we do not need a full constitutive equation but only a few concepts involved in the constitutive equation. These are:

a) The **compressibility/dilatancy boundary**, which is the locus in the $\sigma - \tau$ plane

$$X(\sigma,\tau) = 0 \ , \tag{10.2.1}$$

making precise whether for a certain stress state considered the rock becomes compressible $X(\sigma,\tau) > 0$ or dilatant $X(\sigma,\tau) < 0$ (see the dash-dot line in Figure 10.3).

   b) The short-term **failure surface**

$$Y(\sigma,\tau) = 0 \ , \tag{10.2.2}$$

which is obtained straightforwardly from tests as the maxima on the stress-strain curves obtained in triaxial tests (solid line in Figure 10.3).

   c) The initial **yield surface** (shown as dashed line in Figure 10.3) is obtained assuming that the initial primary stress state σ^P (shown as a diamond in Figure 10.3) is a point located on the initial stabilization boundary $H(\sigma^P) = W^P$ for transient creep. However, the constitutive equation also describes steady-state creep, which may be due either to a stress variation produced by an excavation, or to a slow tectonic motion produced by the primary stress state. For this reason steady-state creep in Figure 10.3 produces volumetric dilatancy for stress states corresponding to points belonging to the domain labeled [DE], while from the point of view of the transient creep in the same domain an "unloading" is taking place. Similarly [CE] is an "elastic" domain for transient creep, but an "incompressible" one for steady-state creep. If no excavation is performed, a primary stress state represented by a point located in a domain [C] or [CE] will produce a steady-state creep (a slow tectonic motion, or land slides, for example) with no irreversible volumetric changes, while a primary stress state located in the domains [D]

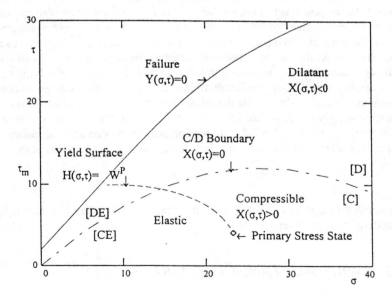

Figure 10.3   *Possible stress states around a cavern. The dashed line represents the initial yield surface and the diamond a possible primary stress belonging to this surface. The dash-dot line is the compressibility/dilatancy boundary.*

or [DE] will result in a steady-state creep producing irreversible dilatancy besides a change in shape.

d) The response of the rock to any sudden loading or unloading is **elastic**. As already mentioned, "loading" means a stress change $\sigma(t_o) \to \sigma(t)$, with $\sigma(t) \neq \sigma(t_o)$ for which $H(\sigma(t)) > H(\sigma(t_o))$, while "unloading" means $H(\sigma(t)) < H(\sigma(t_o))$. The "neutral" loading, i.e., $H(\sigma(t)) = H(\sigma(t_o))$ does not play an important role. If the yield function $H(\sigma)$ is determined from triaxial tests done in the laboratory, W^P follows from $H(\sigma^P) = W^P$ if the initial primary stress state is a state on the stabilization boundary.

The procedure we use is the following. First we choose a contour of the cross -section of the gallery by assigning specific values to the constants (10.1.3) (see Figure 10.4a). Then, for a chosen depth h and a certain ratio σ_h / σ_v of the far field stresses, one can plot a "map" in the σ-τ plane, showing all the boundaries and regions defined by the concepts a) - d) described above (see Figure 10.3). If we assume that the excavation is performed quite fast, the stress distribution which results just after excavation is obtained from (10.1.7). One can show on this map the stress state existing along the gallery contour and, if necessary, the stress state in the rock neighboring the gallery. Further we plot a second "map" in the x-y plane, showing the cavern contour as well as the boundaries of various significant regions (see further Figures 10.11, for instance): the boundary of the region where failure is to be expected just after excavation (solid line), the boundary between the region where the rock becomes dilatant due to excavation, and the region where it becomes compressible (dash-dotted line), the region where an "unloading" for transient creep is taking place due to excavation (long-dashed line). As already mentioned, the unloading is affecting the transient creep only, since steady-state creep exists everywhere though a significant effect of steady-state creep is felt only there where the octahedral shear stress reaches high values. To find out where a fast wall convergence by creep is to be expected, lines of constant values for τ are shown on several figures as short-dashed lines. Thus it is shown where around the contour τ reaches very high values and where these values are low. As mentioned in previous chapters, most steady-state creep laws as used for various rocks are generally expressed in terms of power functions involving τ. It follows that when τ reaches high values, the irreversible rate of deformation components are several orders of magnitude higher than in the domains where τ has much smaller values.

The examples given below are for rock salt (see Chapter 5). The model used is that described at Sections 5.2.3 and 5.3.2 with the compressibility/dilatancy boundary (4.3.5) and short term failure surface (4.4.2) which is supplemented with limit cut-off planes

$$\sigma_i = -\sigma_{it} \quad (i = 1, 2, 3) \tag{10.2.3}$$

for tensile stress states. Here σ_i are the principal components of the stress tensor, and σ_{it} the limit strength of the rock in tensile tests.

10.3 WIDE RECTANGULAR-LIKE GALLERY

First let us consider wide caverns, i.e., caverns, or galleries, with rectangular cross- section and more or less rounded corners. The width of the caverns (in the horizontal direction) is assumed to be larger than the height (vertical), and both are assumed to be much smaller than the length of the cavern. In these conditions the plane strain assumption is quite justified. In the first examples given we consider the width to be twice as large as the height. Several such cross-sections are shown in Figure 10.4a. Various values for the parameter R controlling the rounding off of the corners were considered. The starting idea is to choose the most appropriate cross-section (value of R), from the point of view of failure and stability (meaning that nowhere around the cavity does a domain of fast closure exist and where afterwards, in a quite short time interval, a creep failure may also be expected). In order to give

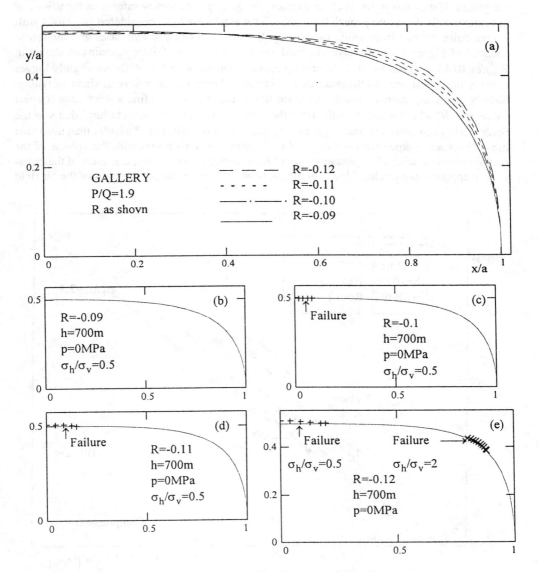

Figure 10.4 *a) Various possible contours for wide rectangular-like cross-sections; b) -d) possible failure zones for two ratios of the far field stresses: $\sigma_h/\sigma_v = 0.5$; e) the same for $\sigma_h/\sigma_v = 0.5$ and 2.*

an example, we choose the depth $h = 700$ m. If the gallery is to be used for storage of a gas under pressure, for example, we will try to find out the interval in which the internal pressure can be varied without jeopardizing the gallery stability. Since the most unstable case is an empty gallery with no internal pressure, Figure 10.4 considers just this case. For each gallery shape shown in Figure 10.4a, and for several possible values of the ratio of the far field stresses, we have examined if just after excavation, a short term failure is possible somewhere around the gallery. If the ratio σ_h/σ_v is close to unity, the gallery is generally safer than for values of this ratio which depart very much from unity. That is why we have considered two cases with values quite distinct from unity, i.e., $\sigma_h/\sigma_v = 0.5$ and 2. For these two cases and the cross-sections of Figure 10.4, we have obtained possible short term failure domains as shown in Figures 10.4 b-e. It follows that the cross- section obtained with $R = -0.09$ (or slightly bigger ones) is the safest one. All the other cross-sections will involve one or several kinds of failure. Shortly, too sharp corners result in unsafe stress state, but we can find a limit case (largest value for R) which would be safe, from the point of view of short term failure (that was the first criterion considered in analyzing cavern stability). For values of R smaller than this value the corners are sharper but the danger of short term failure increases with the volume of the rock involved in failure. For greater values of R the corners are safer, but the shape of the cross-section approaches a circle. Also, if at the location of a future gallery the ratio of the far field

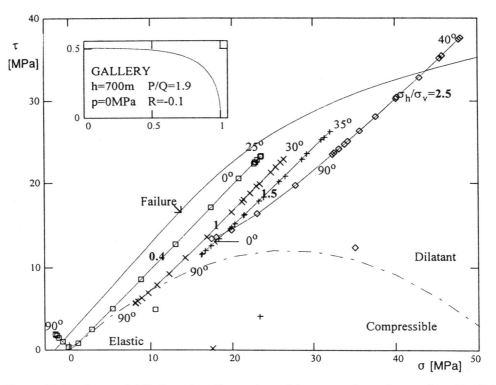

Figure 10.5 *Stress distribution along the contour of the cavern for various ratios of far field stresses. Isolated symbols mark the corresponding primary stress state.*

stresses is known, one can find out very precisely which would be the optimal shape of the cross-section, in order either to completely avoid failure or to reduce the amount of rock involved in failure to a minimum (thus to reduce the amount of necessary support and to know precisely where it is to be applied, and how to design a minimal but still safe support). Salt is used here as an example. In salt it is unusual, in general, to reinforce the walls. Spalling is removed periodically, instead, thus forming an improved shape.

The above discussion is a preliminary one. We must further examine if for a particular depth, the just considered shape of the cross-section will perform reasonably well if the ratio of the far field stresses departs from unity (i.e., is either bigger or smaller) and/or if the gallery is subjected to an internal pressure p. Figure 10.5 shows the stress distribution along the wall of the gallery for various values of the ratio of the far field stresses, with the shape of the cavern shown in the insert. The stress states at various points located along the contour are marked from $\beta = 0°$ up to $\beta = 90$ (at the roof). As expected, the "extreme" values for σ_h/σ_v considered, i.e., 0.4 and 2.5, produce instabilities. The corresponding primary stress states are shown by isolated symbols (i.e., by squares, x, +, and diamonds). The full line is the short term failure condition, while the dash-dotted line is the compressibility/dilatancy boundary. From Figure 10.5 it follows that for $\sigma_h/\sigma_v = 0.4$ a short term failure is expected at the crown. Here the condition (10.2.3) for failure is satisfied. The amount of rock involved in failure is shown in Figure 10.6a. Figure 10.6a shows by long-dashed lines the convergence of the gallery at the dimensionless time $t = 2$. The initiation of creep damage is expected at time $t = 0.458$ and location $\beta = 19°$ (shown by x). Afterwards the creep damage spreads along the contour so that for $t = 2.95$ it comprises the portion $0° \leq \beta \leq 31°$ of the contour. The extremity of this interval is marked by a square in Figure 10.6a. The initiation of the creep failure takes place at a point on the contour located between the corner and the horizontal axis of the gallery cross-section (the direction from where the smallest far field stress component σ_{min}^P acts).

Figure 10.5 also shows that the gallery is more stable for $\sigma_h/\sigma_v = 1$ and 1.5. However, since in the neighborhood of $\beta = 30°$ and $\beta = 35°$ respectively, τ reaches very high values, a fast closure by creep is to be expected in the neighborhood of these locations, where afterwards, a creep failure may also be expected. Indeed, Figure 10.7 shows the convergence of the gallery for $\sigma_h/\sigma_v = 1.5$. The initiation of creep failure takes place very early, after $t = 0.474$ in the region of the contour where $\beta = 36°$ (a square in Figure 10.7; see Figure 10.5 too). Afterwards the creep failure spreads out along the contour and the damaged length of the contour at time $t = 8 \times 10^4$ lies between the two x shown in Figure 10.7, while the shape of the contour at that same time is shown by a dash-dot line.

For the case $\sigma_h/\sigma_v = 2.5$ (see Figure 10.5) besides instantaneous failure around $\beta = 40°$ we have to expect an additional amount of rock to be involved in long term failure during creep dilatancy of the rock, at those locations along the contour where τ reaches very high values, close to the short term failure surface. Thus, for $\sigma_h/\sigma_v = 2.5$ it will be very difficult to design at that depth a gallery that is reasonably stable without internal pressure or without reinforcing the walls or removing the spalling periodically.

Let us further examine the stability of the cavern for $\sigma_h/\sigma_v = 2.5$ if an internal pressure is applied. Figure 10.8 shows the stress distribution along the contour of the gallery cross-section, for various internal applied pressures. The long-dashed line is the initial shape of the "yield surface", so that all points under this line are stress states which are "elastic" from the point of view of transient creep (unloading); steady-state creep is present everywhere and in the

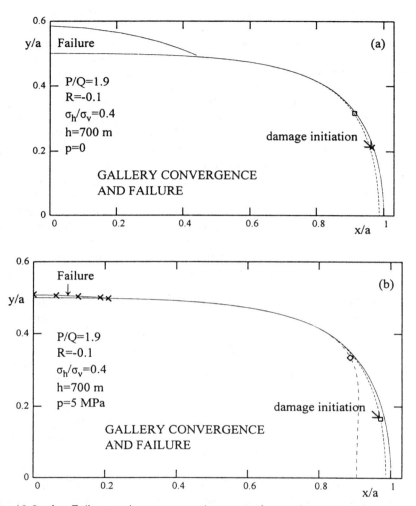

Figure 10.6 a,b *Failure and creep around a cavern for two internal pressures shown; location of creep failure is shown by an arrow; long-dashed line and short-dashed lines show successive shapes of gallery cross-section due to creep.*

dilatant domain (above the dash-dotted line) steady-state creep will produce dilatancy. From Figure 10.8 it follows that an applied pressure of 10 to 15 MPa will improve the stability of the cavern (neither short term failure nor fast creep closure is expected for a forceable time interval). However, if the pressure is too high, for instance $p=20$ MPa, failure is again possible, this time at the walls (around $\beta=0°$). Thus for relatively high ratios σ_h/σ_v one can still improve gallery stability if the internal pressure is kept within well determined limits. However, for high ratios σ_h/σ_v and great depth, the domain of dilatancy spreads up to great distances around the gallery. That means that the volume of the rock damaged by microcracking spreads out greatly. Also, since for the cases shown in Figure 10.8 the value of the octahedral stress τ reaches very high values mainly at the roof and floor, a fast closure

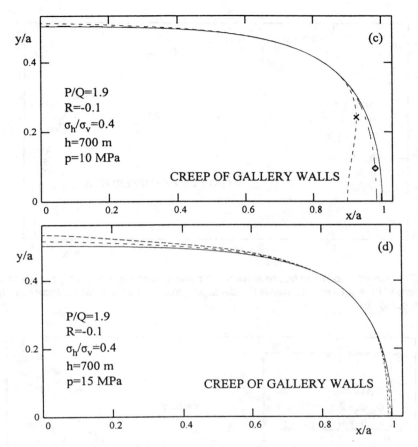

Figure 10.6 c,d   *Failure and creep around a cavern for another two internal pressures shown.*

by creep is to be expected there, and after a long time interval creep failure as well. Figure 10.9 shows how the creep damage initiates and spreads along the contour. For a small internal pressure of $p = 5$ MPa two successive positions of the walls during creep are shown in Figure 10.9a. As expected (see Figure 10.8), for $\beta = 45°$ a very early convergence takes place; the short-dashed line shows the contour at time 0.107 with the square marking the location of the creep failure initiation. Up to the time $t = 8$ (dash-dot line) the roof and floor are creeping very much and the creep failure domain spreads very much from $\beta = 45°$ mainly towards the roof $30° \leq \beta \leq 73°$, i.e., towards that part of the contour subjected to the smallest far field stresses σ_{min}^{P}. For a higher pressure $p = 10$ MPa creep failure initiation is found to be located around $\beta = 46°$ at time $t = 0.474$ (Figure 10.9 b). The dash-dot line contour of the cavern corresponds to $t = 25$. Thus, if we compare this case with the previous one, we find that an increased pressure delays both cavern convergence and creep failure. The domain damaged by creep failure is less spread along the contour, and comprises the interval $35° \leq \beta \leq 66°$. This is further shown in Figure 10.9 c, corresponding to $p = 15$ MPa, where the initiation of creep failure occurs for $\beta = 48°$ after $t = 2.324$. The short-dashed line shows the

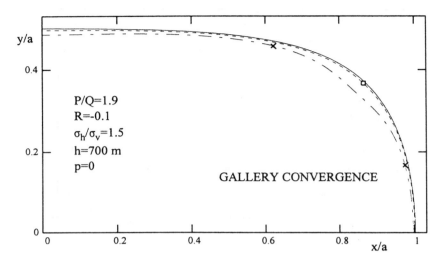

Figure10.7   *Creep and creep failure around the cavern (first expected after t = 0.474 at the location marked by a square: afterwards the damaged portion of the contour is located between the two crosses).*

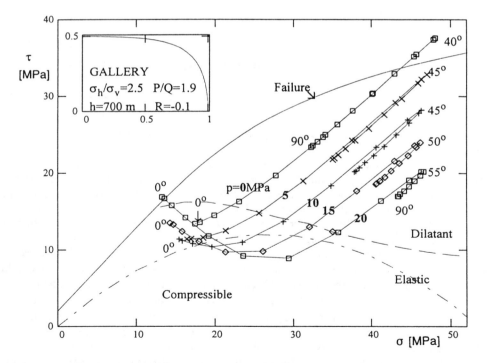

Figure 10.8   *Stress distribution along the contour for $\sigma_h/\sigma_v = 2.5$ and various internal pressures.*

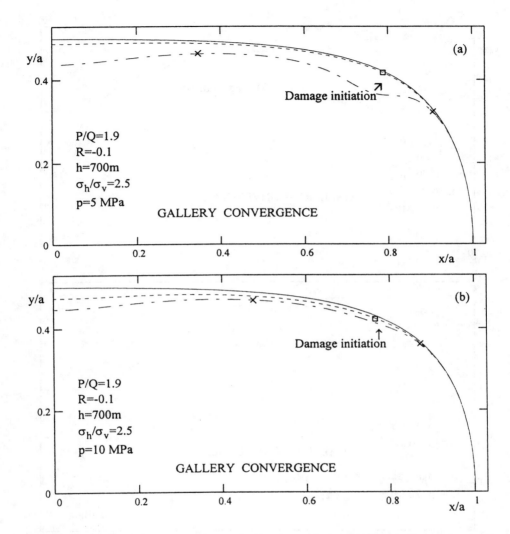

Fig.10.9 a,b *Convergence of cavern with two different internal pressures. Creep damage is marked by figures on the contours. First damage is at the location marked by a square; afterwards the damaged portion of the contour is located between the two stars.*

contour at that time while the dash-dot line shows the contour after $t = 6.5 \times 10^3$. For an even higher pressure, of $p = 20$ MPa, the initiation of creep failure is delayed to $t = 2.665 \times 10^3$ and located at $\beta = 52°$ (Figure 10.9 d); the dash-dot line corresponds to the contour at $t = 1.2 \times 10^4$, while the damaged portion of the contour is $42° \leq \beta \leq 82°$. In the horizontal direction the portion $0° \leq \beta \leq 82°$ of the contour is involved in short term failure. However, the amount of rock involved is so small that it could even not be shown in Fig.10.9 d. Shortly, as the pressure is increasing, the gallery becomes slightly more stable, i.e., any creep instability is to

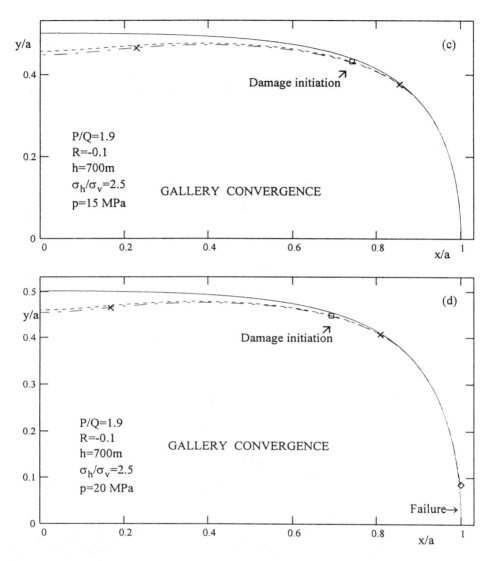

Fig.10.9 c,d *Same as in Fig.10.9 a,b but for other internal pressures. Creep damage is marked by figures on the contours. The diamond on Fig.10.9 d shows up to where the vertical wall failed instantly.*

be expected only after longer time interval. Also, as the stress state along the contour of the gallery is located farther from the instantaneous failure surface, the time up to creep failure initiation increases very much. In fact this time increases to infinity when the stress state approaches the compressibility/dilatancy boundary. For practical purposes any stress state above the compressibility/dilatancy boundary can produce creep damage, but the stress states closer to this boundary than to the short term failure surface can be considered safe for quite a long time. For these cases, the volume of rock damaged by dilatancy around the gallery

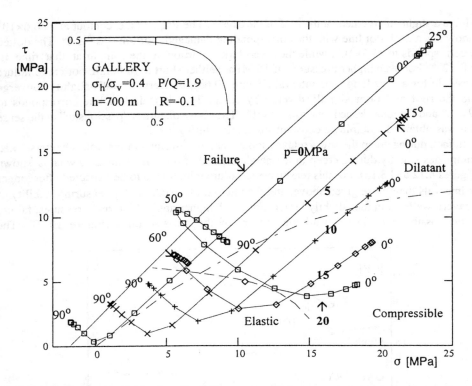

Figure 10.10 *Stress distribution along the contour of the cavern for $\sigma_h / \sigma_v = 0.4$ and various internal pressures. The long-dashed line represents the initial yield surface, and the diamond located on it marks the initial stress state.*

is small and the damage itself (dilatant volumetric strain) is also small.

For small values of σ_h/σ_v and the same depth, the stress distribution around the gallery is completely different (see Figure 10.10). An isolated diamond again marks the primary stress state, and the long-dashed line on which the diamond is situated is the initial yield surface. If the pressure is very small, or even zero, failure is expected at the roof (crown), since then condition (10.2.3) will be satisfied. For a moderate pressure (10 MPa, say) the gallery becomes relatively stable: neither failure nor fast creep closure is expected. Figures 10.6 a-d show the evolution of failure and creep failure when the pressure is increased. For $p = 0$ (Figure 10.6a) creep failure initiation is expected at $\beta = 19°$ and time $t = 0.458$. The contour at time $t = 2.95$ is shown as a long-dashed line, while the square marks up to where along the contour creep failure has spread out $0° \le \beta \le 29°$. A small pressure (5 MPa, say; Figure 10.6b) does not avoid short term failure at the roof but reduces the amount of rock involved, while creep failure of the vertical walls is also to be expected after some time interval. The initiation of creep failure takes place at time $t = 8.105 \times 10^4$ at $\beta = 14°$. The shape of the contour at time $t = 10^5$ is shown as a long-dashed line, while the creep failed portion is $0° \le \beta \le 21°$. At time $t = 8.5 \times 10^5$ the short-dashed line represents the contour shape and the damaged portion is $0° \le \beta \le 32°$. For $p = 10$ MPa (Figure 10.6c) no instantaneous failure is possible, while at the vertical walls creep failure may be expected after $t = 6.14 \times 10^5$ (a few years). For this pressure the creep of the walls produces a slight enlargement at the roof of the cavern. The shape of the contour at $t = 6.6 \times 10^5$ is shown as a dash-dot line with the corresponding damaged portion $0° \le \beta \le 8°$. The dotted line corresponds to $t = 5 \times 10^6$, while the creep-failed portion of the contour at that time is $0° \le \beta \le 23°$. For even higher pressures of 15 MPa, neither short term failure nor creep failure is possible for a very long time interval (Figure 10.6d); the vertical walls slightly converge while the roof and floor slightly diverge by creep. The short-dashed line corresponds to $t = 2 \times 10^6$ and the long-dashed one to $t = 5 \times 10^6$. The above picture is "pessimistic" in the sense that it was obtained assuming a constant stress state during creep.

In fact, during creep the stress relaxes somewhat, as described in previous chapters, i.e., τ diminishes slightly during creep in the neighborhood of the wall (similar to what is shown in Figures 8.12 and 8.13). For this reason creep failure initiation is to be expected after longer time intervals than those given above, and the interval of variation of p, ensuring stability of the cavern walls, is in fact slightly larger. For much higher applied internal pressures, failure is again possible, this time toward the corners of the cross-sections (Figure 10.10). The

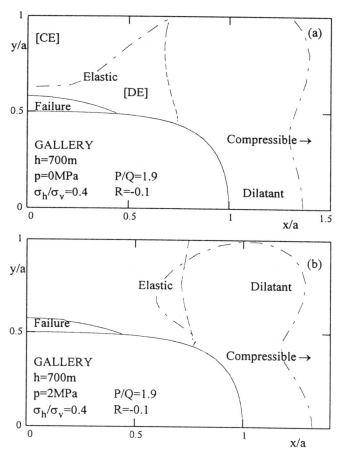

Figure 10.11 a,b   Domains of compressibility, dilatancy and failure around a gallery for $\sigma_h/\sigma_v = 0.4$ and pressures shown.

conclusion would be that for relatively small ratios of the far field stresses, to ensure gallery stability a certain internal pressure is to be applied and its value must be kept between well established limits. These limits can be made precise for each depth and magnitude of σ_h/σ_v.

A similar analysis can be developed for other values of the ratio σ_h/σ_v. As already mentioned, values of this ratio close to unity generally lead to favorable stress states with little, if any, danger of losing, stability. Values of σ_h/σ_v much smaller or much greater than unity are those which may lead to various hazardous cases. One can also examine what happens in the rock mass surrounding the gallery and situated at a certain distance from the contour. Let us give some examples. The case $\sigma_h/\sigma_v = 0.4$ was already discussed in conjunction with Figures 10.6 and 10.10. If we examine where around the gallery the rock becomes dilatant and where compressible we obtain the results shown in Figure 10.11 for four internal pressures. The amount of rock involved in short term failure is also shown (condition (10.2.3) is

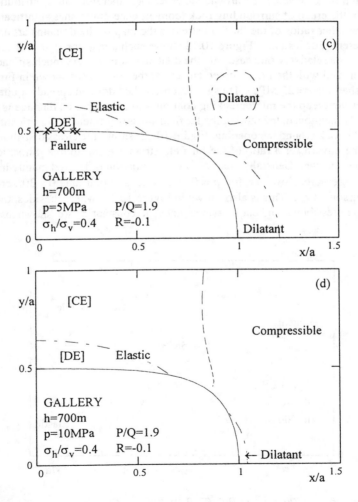

Figure 10.11 c,d *As in Figure 10.11 a,b.*

satisfied) by solid lines in Figures 10.11a,b, and by crosses in Figure 10.11c. Most of the rock in the lateral side (horizontal) of the cavern is in a dilatant state, while the rock at the roof and floor is in an "elastic" state, i.e., due to the excavation, "unloading" has taken place, with respect to the yield surface existing *in situ* before excavation (thus if by excavation $\sigma(t_o) = \sigma^P \to \sigma(t)$ we have $H(\sigma(t)) < H(\sigma^P)$). Therefore, in this domain no transient creep will take place, while steady-state creep is still present, producing changes in shape and dilatancy in the domain labeled [DE], but no volumetric changes in [CE]. It is interesting to follow the change of the shape of the dilatancy domain as the inner pressure is increased (Figures 10.11a to d). As this pressure is increased, this domain shrinks, again showing improved gallery performances under high internal pressures. As a general trend, a small internal pressure shrinks the short-term failure domains and the domains of dilatancy (Figure 10.11b,c,d), i.e., the domains where the rock may be damaged by microcracking. Also, a moderate applied pressure generally diminishes the value of τ around the cavern and therefore the contribution of steady-state creep in the creep of surrounding rock (convergence diminishes significantly).

For some other ratios of the far field stresses the map of the domains around the gallery is quite different. For instance, Figure 10.12 shows such a map for $\sigma_h/\sigma_v = 2$ and no internal pressure. This is a relatively safe case with the dilatancy domain very much spread towards the roof and floor, and with the fastest creep closure at the corners (as shown in Figure 10.12 by the short-dashed line $\tau = 25$ MPa). Thus the microcracked domain spreads quite far from the walls, and during creep the microcracking continues to develop, i.e., damage is progressing.

As already mentioned, the ratio of the far field stresses influences significantly the overall behavior of the rock around the opening. Figure 10.13 shows the stress distribution along the contour of the cavern for a ratio of the far field stresses equal to one. Short term failure is nowhere to be expected. Generally the creep failure initiation is delayed mainly in the presence of an internal pressure. However, for $p = 0$ and values of β around $\beta = 30°$, creep failure is possible after some time. This is also shown in Figure 10.14, where besides the domains of dilatancy, compressibility, and elasticity, the lines of constant τ are shown as short-dashed

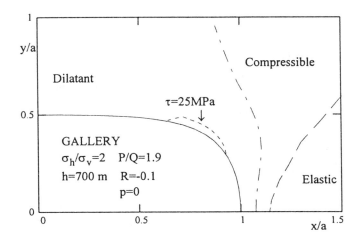

Figure 10.12   *As in Figure 10.11 but for $\sigma_h/\sigma_v = 2$.*

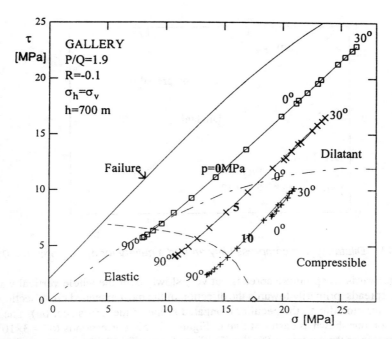

Figure 10.13   *Stress distribution along the contour of the gallery for $\sigma_h = \sigma_v$ and various internal pressures.*

lines. The highest values of τ along the contour are to be found around $\beta = 30°$. Recalling that all steady-state creep laws are expressed as power functions of τ (for rock salt the exponent is found to be somewhere between five and seven; see Chapters 2 and 3, Carter and Hansen [1983], Hunsche [1994a, 1996], Hunsche and Schulze [1994], Cristescu and Hunsche [1996], Haupt *et al.* [1996]) we obtain the result that along the contour around $\beta = 30°$ the order of magnitude of the irreversible rate of strain due to the steady-state creep is of the order of magnitude $10^9 \, s^{-1}$, while at the roof, this order of magnitude is $10^3 \, s^{-1}$. That is why the horizontal closure of the walls is in this case much faster than the vertical roof-floor closure. Creep and creep failure are further analyzed in Figures 10.15a,b. Figure 10.15a shows the beginning of the convergence for $p = 0$. Creep failure initiation occurs first at $t = 500$ at the location on the contour $\beta = 29°$ (shown as a square in Figure 10.15a). Afterwards the damaged region spreads out along the contour and at $t = 2 \times 10^5$ it spreads over the whole interval $0° \leq \beta \leq 46°$ (interrupted line). The spreading of damage along the contour occurs mainly towards the direction of higher elongation of the cross-section.The extremity of this interval is shown as a cross in Figure 10.15a. A faster closure takes place in the horizontal direction since higher values of τ are to be found just there (see also Figure 10.13). At the roof and floor the only mechanism producing convergence is steady-state creep (see Figures 10.13 and 10.14, from where it follows that at the roof an "unloading" takes place with respect to transient creep and the stress existing *in situ* before excavation). For higher internal pressures $p = 5$ MPa (see Figure 10.15b) the initiation of creep failure is significantly delayed and occurs at the point of the contour $\beta = 29°$ at time $t = 1.038 \times 10^5$. As compared with the case $p = 0$, the location of creep failure initiation is now shifted clock wise towards the biggest elongation of the cross

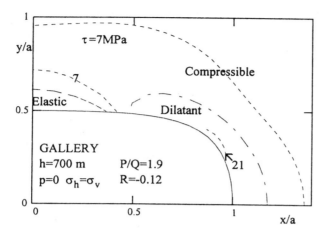

Figure 10.14 *Dilatancy and compressibility around a gallery for $\sigma_h = \sigma_v$ and $p = 0$.*

section. Afterwards creep failure spreads out very slowly over the whole vertical wall; again the damage spreads primarily towards the direction of the most extended cross-sectional shape (stress concentration due to the peculiar elongated shape of the cross-section). The position of the wall contour shown as dash-dot line in Figure 10.15b corresponds to $t = 3 \times 10^6$ and the damaged portion of the contour is $0° \le \beta \le 47°$. Since for $p = 5$ MPa the time up to the initiation of creep failure is three orders of magnitude higher than in the case $p = 0$, it follows that even a small internal pressure significantly improves the cavern stability. Furthermore, for even higher pressures $p = 10$ MPa the cavern is completely stable with no evolutive damage and slow creep convergence of the walls (see Figure 10.13). Figure 10.13 also shows where a reinforcement (by anchors, for instance) is necessary if we would like to delay the creep closure.

For higher values of the ratio σ_h / σ_v the gallery is quite unstable (see Figures 10.7 and 10.8).

Very wide horizontal galleries or caverns are sometimes excavated at shallow depth in various rocks, to become stadiums, railway stations, garages, power stations, commercial/institutional facilities, underground storage facilities, etc. (see Broch and Nilsen [1993], Cheng and Liu [1993], Hoek and Moy [1993], Moretto *et al.* [1993], Nordmark and Franzen [1993], Sterling [1993], Barton *et al.* [1994]). In order to give an example let us consider a cavern, again excavated in rock salt, at 100 m depth and with the ratio of horizontal diameter of the cross-section to the vertical one equal to $P/Q = 3.73$. Three far field ratios have been considered, the ones which may most probably be found at that depth: $\sigma_h / \sigma_v = 1$, or 2, or 4. The stress distribution along the contour of the cavern is shown in Figure 10.16. The shape of the gallery is shown in the insert. It follows from this figure that for all three ratios σ_h / σ_v the cavern is quite stable at the considered depth. Three isolated diamonds show the primary stress states for these three cases, respectively. Since all the stress states are very close to, or even under the compressibility/dilatancy boundary, and since everywhere along the contour, τ reaches quite small values, no creep failure is possible for several hundred years. For the case $\sigma_h / \sigma_v = 1$ creep failure is not expected at $\beta = 0°$ but only after several thousand years. Thus, for all three cases the cavern is extremely stable.

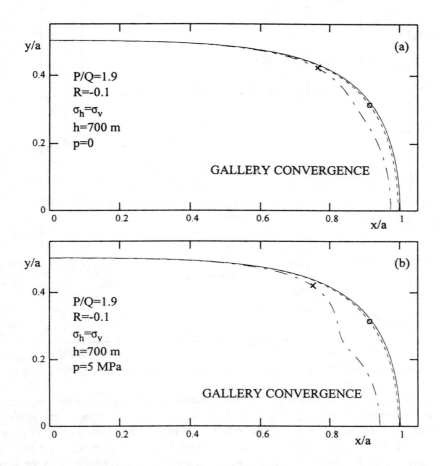

Figure 10.15   *Gallery convergence and creep failure for $\sigma_h = \sigma_v$ and two values of p shown.*

This is also shown in Figure 10.17. For $\sigma_h/\sigma_v = 1$ convergence takes place in the horizontal direction only (direction of the stress concentration, i.e. of relatively highest values of τ along the contour)   but is extremely slow: the long-dashed line in Figure 10.17a shows the convergence after $t = 2 \times 10^8$, and the short-dated line the same contour after $t = 1.1 \times 10^9$. Whatever values we choose for k_T and k_S, the time intervals involved are of the order of many thousand of years! The square on the last line shows the limit up to where creep failure has spread along the vertical walls of the contour.

For $\sigma_h/\sigma_v = 2$ convergence is possible in all directions. While in the horizontal direction a stress concentration exists (higher values of τ) due to the elongated shape of the cross-section in that direction (and is causing convergence), it is in the vertical direction that is acting the minimum primary far field stress acts. That is why in the vertical direction there can be faster convergence. However, due to the overall small values of τ involved (see Figure 10.16) the convergence is extremely slow: the long-dashed line in Figure 10.17b corresponds to $t = 10^9$

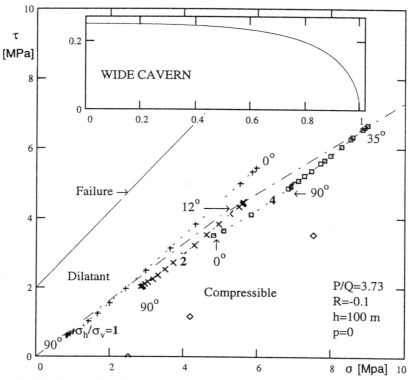

Figure 10.16 *Stress distribution along the walls of a wide horizontal gallery.*

and the short-dashed line to $t = 2 \times 10^9$, and thus involve exceedingly long time intervals (of the order of several thousand of years). Therefore in this case the cavern is very stable. For even higher ratios, i.e. $\sigma_h / \sigma_v = 4$, we again obtain a stable case. The convergence is shown in Figure 10.17c. Since higher values of τ are now found on the portion $35° \le \beta \le 90°$ of the contour (see Figure 10.16), more convergence is expected to occur primarily in the vertical direction (direction of σ_{min}^P). This is shown in Figure 10.17c. Higher values of τ mean faster creep so that the long-dashed line in Figure 10.17c corresponds to $t = 5 \times 10^6$ and the short-dashed line to $t = 2 \times 10^7$, which are somewhat smaller than in the previous case, but still long time intervals.

For $\sigma_h = \sigma_v$ the stress concentration is at $\beta = 0°$, i.e., in the direction of the largest dimension of the cross-section. If $\sigma_h > \sigma_v$ the stress concentration slides anticlockwise along the contour towards the direction of $\sigma_{min}^P = \sigma_v$. For $\sigma_h = 2\sigma_v$, the stress concentration is at $\beta = 12°$, while for $\sigma_h = 4\sigma_v$ it slides further on up to $\beta = 35°$. Thus, contrary to what happens for circular tunnels, where the stress concentration and location of damage initiation are governed solely by the value of the far field stresses, in the case of elongated cross-sections, there are two reasons for stress concentration: the elongated shape of the cross-section and the far-field stresses ratio with $\sigma_h \ne \sigma_v$. Due to the elongated shape in the horizontal direction the stress concentration should be in that direction, but due to the fact that $\sigma_{min}^P = \sigma_h < \sigma_v$ the stress concentration shifts anticlockwise along the contour toward σ_{min}^P. Thus, for this case the two reasons for stress concentration and damage initiation balance each other somewhat, and as a

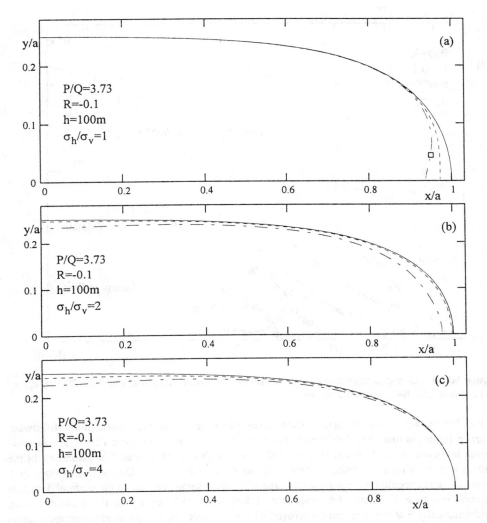

Figure 10.17 *Convergence of very wide caverns for depth h = 100 m and values of σ_h/σ_v shown.*

result the case of an elongated cross-section with σ_{min}^P acting along the smallest dimension of the cross-section is a more stable case than the case $\sigma_h = \sigma_v$ (contrary to what happens for circular cross-sections).

For greater depths the results are different. While generally, very wide caverns are not expected to be excavated at great depths, to show the influence of the depth on the overall cavern stability, let us give an example of very wide caverns excavated at a depth $h = 700$ m. For the case $\sigma_h = \sigma_v$, more likely to be found at that depth, the stress distribution along the contour is shown in Figure 10.18 for several internal pressures. For $p = 0$ the cavern is unstable in the horizontal direction (stress concentration involving high values of τ due to the elongated

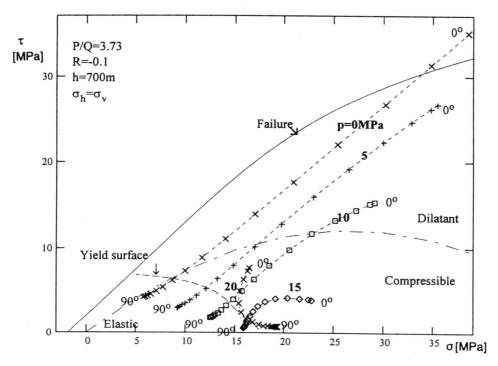

Figure 10.18 *Stress distribution along the contour of a wide cavern at depth h=700 m, $\sigma_h = \sigma_v$ and several internal pressures shown.*

shape of the cavern): a short term failure takes place just after excavation and is followed by creep failure in that same horizontal direction. This is shown in Figure 10.19. The amount of rock involved in the short term failure is shown by a short-dashed line. The boundary of the domain where the rock becomes dilatant is shown by a dash-dot line. Due to high values of τ creep failure is expected quite soon afterwards in the very same domain. In the vertical direction the rock response is elastic, i.e., unloading takes place with respect to the transient creep (mathematically that is expressed by $H(\sigma(t)) < H(\sigma^P)$, where $\sigma(t)$ is the stress after excavation and σ^P is the primary stress in that same material point). However, in the same region, a small closure in the vertical direction is expected due to steady-state creep.

Going back to Figure 10.18 one can see that for $p = 5$ MPa the gallery or cavern is expected to be more stable. The convergence of the walls in this case is shown in Figure 10.20. The convergence at the roof and floor is very small, mainly due to steady-state creep. As expected, as a consequence of stress concentration, the horizontal convergence is more pronounced; the long-dashed line corresponds to the shape of the cross section at $t = 0.7$, and the short-dashed line to $t = 3 \times 10^5$. Creep failure initiation takes place in the horizontal direction (highest value of τ, see Figure 10.18) and afterwards along the contour, the x on the dash-dot line shows up to where creep failure has spread by time $t = 3 \times 10^5$. Thus even for $p = 5$ MPa creep failure produces an instability after a certain time interval elapsed from the moment of excavation. However, Figure 10.18 also shows that for higher internal pressures $p = 10$ MPa, $p = 15$ MPa

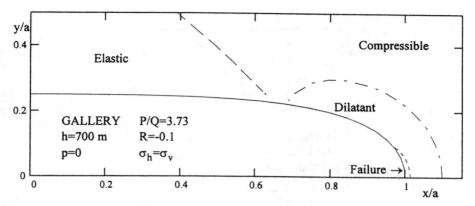

Figure 10.19 *Domain of failure, dilatancy, compressibility, and elasticity around a wide cavern.*

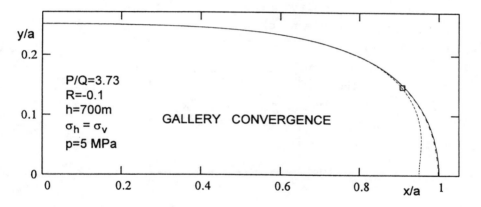

Figure 10.20 *Convergence of the wide cavern for $\sigma_h = \sigma_v$ and $p = 5$ MPa.*

and $p = 20$ MPa, for example, even at that depth the cavern is perfectly stable; creep convergence produces compressibility, i.e., closure of microcracks, and no creep damage is possible.

The **location of creep failure initiation** and the **time when this initiation starts** depend on the stress concentration around the gallery or cavern contour, i.e., on the location where τ reaches very high values. In turn, this stress concentration depends mainly on three factors: concentration induced by the elongated shape of the cross-section, concentration induced by the ratio of the in-plane far field stresses, and concentration induced by the possibly existing "corners" of the cross-section of the contour. If these three factors produce around the contour a combined stress concentration in the very same location, then there is a big danger of very early creep failure. In order to avoid this superposition of effects we can either change the shape of the cross-section or adapt the shape of the cavity to the ratio σ_h/σ_v (rounding the corners, or choosing less elongated cross-sections) if possible, or try to accommodate to a more favorable ratio of the in-plane far field stresses. The latter can be envisaged in those locations

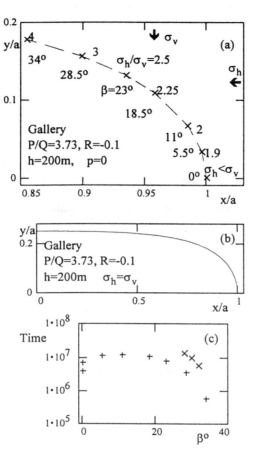

Figure 10.21 *Location of the initiation of creep failure (see text).*

where the horizontal far field stresses are different in various horizontal directions. Then, if the design of the cavern project allows it, one can orient the excavation of the gallery in the most favorable direction with respect to the stress concentration produced by the ratio of the in-plane far field stresses. Thus we can contemplate choosing an appropriate shape of the cross-section and an appropriate orientation of the gallery axis with respect to the existing horizontal far field stresses, in order to reduce the accumulation in the same location (or neighboring locations) of the effects of stress concentration due to several factors involved. Let us discuss now the **influence of the ratio** σ_h/σ_v **on the stability** of a horizontal, elongated cavern or gallery.

Figure 10.21 shows the case of such a cavern, excavated in rock salt, with an elongated cross-section in the horizontal direction, at depth $h = 200$ m. The shape of the cavern is shown in Figure 10.21b and is the same as in Figure 10.17. For $p=0$ the location of creep failure initiation is shown in Figure 10.21a, for various possible ratios σ_h/σ_v. Thus if $\sigma_h \leq \sigma_v$ the stress concentration is always at the tip $x = a$ of the contour, i.e., the stress concentrations induced by the ratio σ_h/σ_v and by the elongated shape of the cross-sections combine and are located at the very same place on the contour. If the ratio σ_h/σ_v takes larger values, this location shifts counterclock-wise along the contour as shown in Figure 10.21a. Figure 10.21c shows by +++ the time up to initiation of creep failure, as a function of the location of creep failure initiation β. Thus when $\beta = 0°$ and $\sigma_h/\sigma_v \le 1$ this time is shorter. When the locations of the induced stress concentration due to elongated cross-sectional shape and far field stresses are slightly separate, the time to creep failure initiation increases. For very high values of σ_h/σ_v (i.e., for $\sigma_h/\sigma_v > 3$) this time starts decreasing again since the stress concentration induced by far field stresses starts to become dominant.

If a small internal pressure of $p = 1$ MPa is added, the location of creep failure initiation shifts counterclock-wise compared with the previous case $p=0$ (see Figure 10.21c). The time up to creep failure initiation increases very much, as expected, and is shown in Figure 10.21c by multiplication signs. Only the cases where a creep failure initiation is still possible are shown; for other values of β, even a small added internal pressure increases the gallery stability. Generally, the cases just discussed are quite stable.

Let us give a specific example of the "decoupling" of stress concentrations when σ_h/σ_v is changed. We consider the same case as in Figure 10.17, but for $\sigma_h/\sigma_v = 1.8$ and $\sigma_h/\sigma_v = 2.65$. For this case cavern stability must be improved, since the stress concentration due to the elongated shape of the cross-section and to the ratio σ_h/σ_v do not completely coincide. Figure 10.22 shows the stress distribution along the contour of the same wide cavern as in Figure 10.17. For $\sigma_h \le \sigma_v$ the stress concentration due to the far field stresses is located at $x = a$ and coincides with the stress concentration due to the elongated shape of the cross-section. For $\sigma_h > \sigma_v$, as in the case in Figure 10.22, this coincidence no longer occurs. The stress concentration due to the ratio $\sigma_h/\sigma_v > 1$ shifts from $x = a$ towards the direction of $\sigma^P_{min} = \sigma_v$.

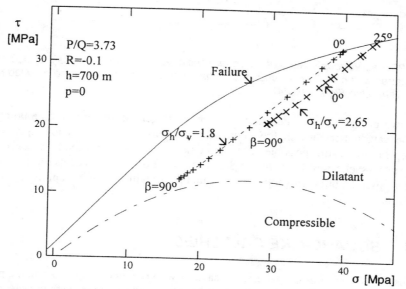

Figure 10.22 *Stress distribution along the contour of a wide cavern for $\sigma_h/\sigma_v = 1.8$ and $\sigma_h/\sigma_v = 2.65$.*

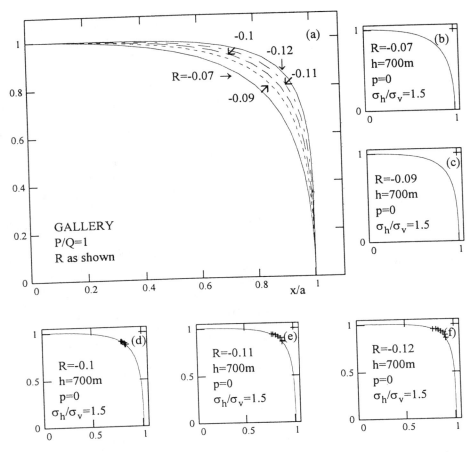

Figure 10.23 *a : Various shapes of the square-like cavern with more or less rounded corners; b and c: shapes with no failure around; d to f: shapes with increasing volume of failed rock (in that order) (see also Figure 10.24).*

Since the two causes of stress concentration are no longer located at the same point of the contour, the cavern becomes slightly more stable: short-term failure is no longer possible, while creep failure is still possible after some time in the neighborhood of $x = a$. Figure 10.22 shows two such cases for $\sigma_h / \sigma_v = 1.8$ and $\sigma_h / \sigma_v = 2.65$. In both cases short term failure is no longer possible, but high values of τ around $\beta = 0°$ will lead quite rapidly to initiation of creep failure.

10.4 SQUARE-LIKE GALLERIES

Let us consider now square-like galleries or caverns with rounded corners. As in the previous section we would like to find out how much we have to round off the corners in order to avoid failure and to ensure stability (again with the meaning that creep failure is not to be expected for a time interval which is of significance for the mining problem considered). In order to give an example and to discuss the problem, we have chosen a depth $h = 700$ m, and a ratio of the far field stresses $\sigma_h/\sigma_v = 1.5$. Several shapes of square-like cross- sections are shown in Figure 10.23a. Several values of the parameter R were used. For the same values of R, Figures 10.23 b to f have been obtained. For the depth and ratio σ_h/σ_v considered, and $p = 0$, we check for which of these five cases short term failure is possible in the surrounding rock. We find that for $R = -0.07$ and -0.09 no short time failure is possible, while for the other three cases, i.e., $R = -0.1, -0.11$, and -0.12 a short-term failure domain is present with an increasing size in that order. The conclusion would be that for sharper corners (closer to a right angle), the domain of short term failure is more pronounced. That is why we have chosen for further analysis the contour obtained with $R = -0.09$ (the sharpest corner which would not yet involve instantaneous failure).

We have to check how this contour performs for other possible values of the far field stress ratio and various internal pressures. For this purpose we have used $\sigma_h/\sigma_v = 0.3, 0.7, 1$, and 2, values quite often met in mining engineering practice. The stress distribution along the contour of the gallery for these cases is shown in Figure 10.24 for $p=0$. As is obvious from this figure, even in the absence of an internal pressure this cavern shape performs reasonably well for all ranges of the far field stresses considered. This is further shown in Figure 10.25, where short- and long-dashed lines correspond to various successive positions of the cavern wall for various ratios σ_h/σ_v shown. On each of these lines the interval between symbols indicates the portion of the contour where creep failure occured. It was shown in the previous chapters that with circular excavations (tunnels or boreholes) the **location of**

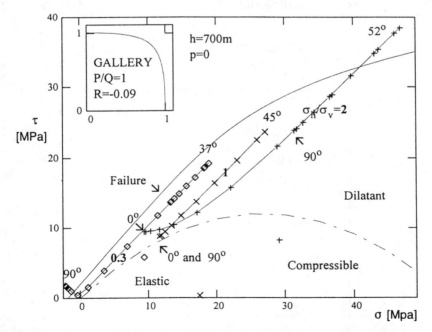

Figure 10.24 *Stress distribution along the contour of the square-like gallery for various ratios σ_h/σ_v and $p=0$.*

incipient creep failure is generally found on the contour in the direction of σ^P_{min} since from all possible directions around the contour of the tunnel τ reaches its highest values just in that direction. With square-like caverns the location of incipient creep failure is much more difficult to find since the corners induce a stress concentration which combines with the stress concentration due to far field stresses. Both are influenced by internal pressure so that the location of the initiation of creep failure and the direction in which it spreads out in time depend at each depth primarily on the stress concentration induced by the corner, by the ratio

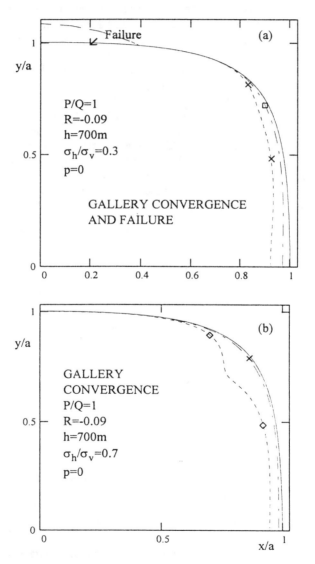

Fig.10.25 a,b *Convernence and creep failure around square-like cavern for two ratios* σ_h / σ_v.

σ_h/σ_v, and by the influence of the internal pressure.

For $\sigma_h/\sigma_v = 0.3$ (Figure 10.25a) short term failure (condition (10.2.3) is satisfied there) takes place at the roof and the floor. The amount of rock involved in failure is shown in Figure 10.25a. From Figure 10.24 it follows that the rock becomes dilatant mainly along the vertical

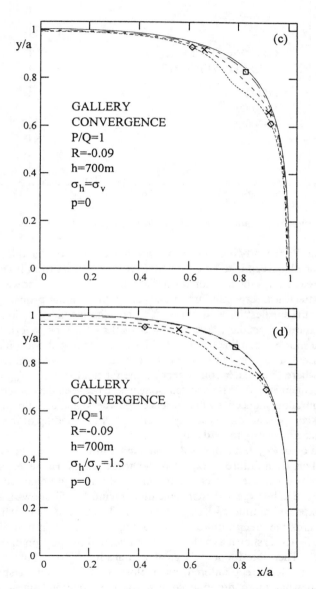

Fig.10.25 c,d   *Convergence and creep failure around square-like gallery for two other ratios* σ_h/σ_v.

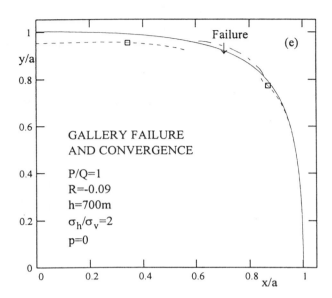

Figure 10.25 e *Convergence and creep failure around square-like gallery for* $\sigma_h/\sigma_v=2$.

portion of the contour. Particularly high values of τ are found for $\beta<45°$ and it is there that an important convergence by creep is to be expected. Indeed it follows from Figure 10.25a that incipient creep failure occurs at the location $\beta = 37°$ at time $t = 2.5\times10^4$ (shown as a square on the dash-dot line). Afterwards creep failure spreads along the contour primarily towards the direction from where the smallest far field stress σ_{min}^P acts (horizontal direction). Thus at time $t = 2.2\times10^5$ creep failure affects the whole portion $22° \leq \beta \leq 44°$ of the cavern contour shown in Figure 10.25a as a short-dashed line (the extremities of the damaged portion are marked on this line by two crosses). Thus for small values of the ratio σ_h/σ_v short term failure occurs in the direction from where the greater component of the far field stresses σ_{max}^P acts, while the location of the incipient creep failure is expected somewhere between the corners (the direction of the stress concentration induced by the presence of corners) and the direction from where the smallest far field component σ_{min}^P is acting. Afterwards the spreading of the damage along the contour takes place primarily towards σ_{min}^P too.

For higher values of σ_h/σ_v, but still smaller than unity, things change. From Figure 10.24 it follows that no short term failure is expected for these cases. However, since there are regions around the cavern where τ reaches quite high values, significant convergence and maybe creep failure are to be expected after some time. Figure 10.25b shows that the initiation of creep failure is expected at time $t = 1.393\times10^4$ at the location $\beta = 41°$, which is close to the corner. Afterwards the creep failure spreads along the contour in both directions; the contour shape at $t = 6\times10^5$ is shown as a short-dashed line, and the portion damaged by creep failure is $22° \leq \beta \leq 55°$. Therefore for slightly greater values for σ_h/σ_v, but where $\sigma_h/\sigma_v<1$ still, the location of incipient creep failure is much closer to $\beta = 45°$ and the spreading of the damage is only somewhat more pronounced towards the direction from where σ_{min}^P acts. The main convergence takes place in the horizontal direction (the direction of σ_{min}^P).

An interesting case, and one most likely to be found in many locations (at that particular

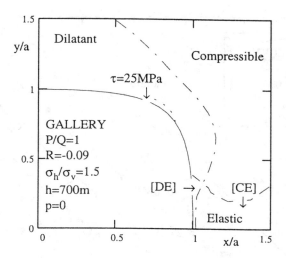

Figure 10.26 *Domains of dilatancy, compressibility and elasticity around the gallery. The region of maximum* τ *is shown as a short-dashed line.*

depth), is $\sigma_h = \sigma_v$. The portion of the contour where fast wall convergence as well as creep failure are to be expected, is influenced by the stress concentration due to the presence of the corners only (which in turn is influenced by depth and internal pressure). The stress distribution along the contour is shown in Figure 10.24. After excavation the rock around the contour of the cavern becomes dilatant, and for $\beta = 45°$ (at the corner) τ reaches quite high values. Figure 10.25c gives details. This time, convergence and spreading of creep failure are symmetric with respect to the diagonals of the square cross-section. The initiation of creep failure occurs at $\beta = 45°$ at time $t = 1.1$, and the new shape of the cross-section is shown as a dash-dot line. Afterwards, the creep damage spreads out symmetrically in both directions along the contour as shown in Figure 10.25c. The shape of the contour at time $t = 8 \times 10^4$ is shown as a long-dashed line and the failed portion of the contour $33° \leq \beta \leq 57°$ is marked by crosses. A successive position of the contour at $t = 2 \times 10^5$ is shown as short-dashed line, while the damaged portion of the contour is $29° \leq \beta \leq 61°$. Thus, instability takes place in the neighborhood of the corners only. In order to stop or at least to delay this instability one has to install a proper support at the corners. Outside the neighborhood of the corners wall convergence is quite small.

Let us consider now the case $\sigma_h > \sigma_v$. Since at a certain depth the magnitude of σ_v is determined from formula (10.1.5)$_1$, the two cases $\sigma_h = 2\sigma_v$ and $2\sigma_h = \sigma_v$, for instance, are obviously non-symmetric. If $\sigma_h > \sigma_v$ the location of creep failure initiation is found around $\beta = 50° \pm 10°$. Figure 10.25d shows that the initial location on the contour where creep failure is first to be expected at time $t = 0.0744$ is $\beta = 50°$ (a square marked on the dash-dot line). Due to convergence, the contour shape after time $t = 4 \times 10^3$ is shown as a long-dashed line, and the portion $38° \leq \beta \leq 64°$ of the contour has failed by creep. Afterwards, at time $t = 3 \times 10^4$ the contour shape is represented as a short-dashed line and the creep failure is found at $34° \leq \beta \leq 71°$. Therefore if $\sigma_h > \sigma_v$ creep failure initiates somewhere at $\beta > 45°$, i.e., shifts from the corner $\beta = 45°$ (where it was located for $\sigma_h = \sigma_v$) towards σ_{min}^P, i.e., a location where

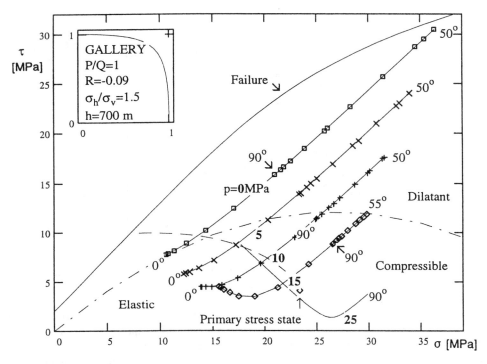

Figure 10.27 *Stress distribution along the contour for four internal pressures and*
$\sigma_h/\sigma_v = 1.5$.

$\beta > 45°$, and spreads in time mainly towards the direction of σ_{min}^P, which this time is just σ_v. Figure 10.26 shows where around the contour the rock becomes dilatant and where it becomes compressible. A domain of high values of τ is shown at the corner by a short-dashed line.

Finally the case $\sigma_h = 2\sigma_v$ is completely unstable, i.e. instantaneous failure takes place in the neighborhood of $\beta = 52°$ on the contour (see Figure 10.24). The amount of rock involved in failure is shown in Figure 10.25e as a dash-dot line. Further creep failure spreads out along the contour starting from this location. The spreading of creep failure takes place mainly towards σ_{min}^P. An estimation of the position at time $t = 0.5$ of the not yet failed contour is shown as a short-dashed line. The segment of the contour between the two squares, i.e., $40° \le \beta \le 76°$ no longer exists at that time. This case is totally unstable.

It follows from Figures 10.24 and 10.25 that creep failure starts from $\beta = 45°$ if $\sigma_h = \sigma_v$ and spreads equally in both directions along the contour. If $\sigma_h < \sigma_v$ the location of creep failure initiation shifts towards $\sigma_{min}^P = \sigma_h$ and the time-dependent evolution of damage along the contour progresses faster in that direction σ_{min}^P, too. The reversible for $\sigma_h > \sigma_v$, when $\sigma_{min}^P = \sigma_v$. Thus for such a cavern reinforcement is primarily needed along the portion of the contour comprising the corners and extended more or less towards the direction of σ_{min}^P (depending on the magnitude of σ_{min}^P compared with σ_{max}^P).

The above analysis has been developed for $p = 0$. If an internal pressure is present, the cavern is much more stable. Figure 10.27 shows for relatively small values of p and the case

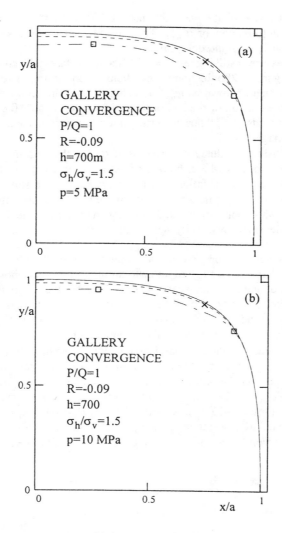

Fiure 10.28 *Cavern convergence with internal pressures: a) p = 5 MPa; b) p = 10 MPa.*

$\sigma_h = 1.5\,\sigma_v$ the stress distribution along the contour. The initiation of creep failure starts on the contour from $\beta = 50°$ (i.e., an internal pressure only slightly shifts this location towards σ_{min}^P). However, for higher internal pressures $p > 15$ MPa, for example, the cavern is completely stable. The long-dashed line in Figure 10.27 represents the initial yield surface and the isolated diamond situated on this curve represents the primary stress state. Thus, an obvious unstable case ($\sigma_h = 1.5\,\sigma_v$ with $p = 0$) becomes a stable one if the internal pressure p reaches values around 15 MPa. For smaller pressures, for instance $p = 5$ MPa, an instability is still possible as shown in Figure 10.28a. The same features as obtained for the case $p = 0$ are found here too, with the difference that creep and creep failure are delayed by the presence of internal pressure: creep initiation starts quite fast at $t = 700$ at location $\beta = 50°$, and spreads afterwards so that at $t = 2 \times 10^5$ it covers the interval $36° \le \beta \le 81°$. The main convergence is in the vertical direction (direction of σ^P_{min}). Thus an added internal pressure slightly shifts the damaged region towards σ^P_{min} and delays it.

If the internal pressure is still higher, $p = 10$ MPa, for instance, the convergence and creep failure is shown in Figure 10.28b. Incipient creep failure starts at time $t = 3.52 \times 10^4$ at location $\beta = 53°$ (shifted still farther on towards σ^P_{min}). The short-dashed line shows the wall convergence at $t = 5 \times 10^5$ when the portion of the contour failed by creep is $40° \le \beta \le 79°$, i.e., damage is progressing faster towards σ^P_{min}. Thus increasing internal pressure means less creep damage and delay of that damage.

A **general conclusion** for square-like caverns: if $\sigma_h = \sigma_v$ the most unstable regions are the neighborhoods of the corners where the rate-of-deformation components are very high and thus fast convergence and creep failure are to be expected there soon after excavation. If $\sigma_h < \sigma_v$ or $\sigma_h > \sigma_v$ the domain of instability shifts more or less towards the direction of σ^P_{min} and spreads in time towards some direction depending on the magnitude of the ratio $\sigma^P_{min} / \sigma^P_{max}$. An internal pressure delays the instability and slightly shifts the domain of instability even more towards σ^P_{min}.

As another general conclusion we can discuss the time necessary for a square-like gallery to reach by evolutive damage the stage of ultimate failure. It has been shown in Chapter 6 that any stress state of the dilatancy domain can sooner or later produce initiation of creep failure. Figure 10.29 (Cristescu [1996d]) shows the time to creep failure initiation for three constant mean pressures corresponding to three depths. The octahedral overstress above the C/D boundary means the magnitude of τ minus the magnitude of τ taken on the C/D boundary for the same mean stress. This last quantity was denoted by $\tau_{C/D}$, so that the overstress is $\tau - \tau_{C/D}$. For high values of $\tau - \tau_{C/D}$ the time to creep failure is very small, tending towards zero if τ tends towards its value on the short term failure surface (again for the same σ). As the overstress $\tau - \tau_{C/D}$ decreases, the time to failure increases very much. The case shown in Figure 10.29 corresponds to a square-like gallery with $P/Q = 1$, $R = -0.09$, $\sigma_h = \sigma_v$, $h = 700$ m, where the initiation of creep failure is just at the corner $\beta = 45°$. Thus, any stress state in the dilatancy domain can produce creep failure: for stress states close to the failure surface that happens

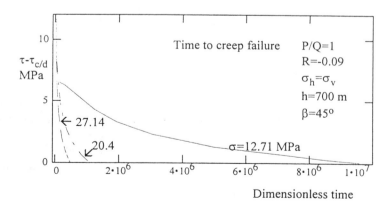

Figure10.29 *Dimensionless time to creep failure versus octahedral overstress above the C/D boundary for various mean stresses.*

after a short period of time, while for those which are farther from this surface, it happens after a long time.

At greater depths (greater σ), and for the same $\tau - \tau_{CID}$, the time to failure is obviously shorter. The results are similar to those for circular tunnels or boreholes. The above discussion is only indicative, since during creep significant stress relaxation takes place. This stress relaxation increases the time to initiation of creep failure very much. However, the trend is the same: the time to initiation of creep failure depends on the octahedral overstress $\tau - \tau_{CID}$.

10.5   TALL RECTANGULAR-LIKE GALLERIES

An interesting case is that of tall rectangular-like caverns, very often used in mining engineering (see Hoek and Moy [1993]) or for underground deposits of petroleum products. We use here the same procedure as used for the other two cases discussed above: we start by considering several contours for tall rectangular-like caverns with more or less rounded corners. Thus a contour chosen for further analysis is that for which at the depth considered and $p = 0$ no instantaneous failure will occur. The depth chosen is $h = 300$ m, where many such caverns exist and are stable for a long time. For instance, at the Slanic-Prahova salt mine in Romania there are caverns of up to 100 m high, 60 to 70 m wide and several hundred meters long, and these caverns are stable for more than half a century. The parameters describing the contour (10.1.3) were chosen as $P/Q=0.52$, and several values for R have been tested. Values for R smaller than -0.12, for example, produce instantaneous failure. Thus a safe value was considered to be $R=-0.1$.

Figure 10.30 shows the stress distribution around the contour of the cavern for $p=0$ and several ratios $\sigma_h/\sigma_v = 0.4, 1, 2$, and 3, which can be found at that depth. The shape of the cavern is given in the insert, and the lower graph gives details of the upper graph corresponding to small stresses. It follows that for ratios σ_h/σ_v greater than 2, for instance, instantaneous failure is to be expected at the vertical walls (around $\beta = 0°$). For smaller values of this ratio no instantaneous failure can occur. Also a wide domain of the rock around the cavern is in a dilatant state (see Figure 10.31c obtained for $\sigma_h/\sigma_v= 2$). As shown in this figure, the compressibility domain touches the contour on quite a small portion. It follows that this case can be expected to be unstable. Let us recall that Figure 10.31 is obtained from the exact elastic solution and the equation of the dilatancy/compressibility boundary (10.2.1) (which follows straightforwardly from tests).

If we examine the convergence of the gallery, it follows from Figure 10.30 that the fastest convergence around the contour is expected where τ reaches very high values. For instance, for $\sigma_h/\sigma_v = 0.4$, that happens in the neighborhood of $\beta = 49°$ (see also Figure 10.32a). The isolated x in Figure 10.30 locates the corresponding primary stress. Thus, due to the overall small values of τ no creep failure is to be expected, but may occur for exceedingly long time intervals. The various domains existing around the cavern are shown in Figure 10.31a; the dilatancy domain is quite small and involves small values of τ only. No significant closure is expected, while creep damage is practically impossible. Figure 10.32a shows the closure by creep if stresses are assumed to remain constant. The closure takes place mainly in the horizontal direction, i.e., the direction of σ_{min}^P; the dash-dot line corresponds to the contour shape at time $t = 7 \times 10^7$. Since this is a very long time interval, the cavern can be considered to be quite stable, if we also take into account that stresses relax during creep.

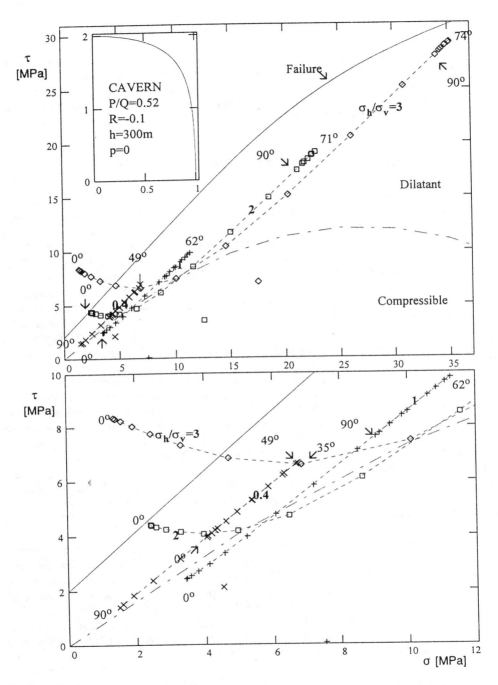

Figure 10.30 *Stress distribution along the contour of the cavern for various ratios σ_h/σ_v and no internal pressure.*

For $\sigma_h = \sigma_v$ the stress distribution along the gallery is shown by +++ in Figure 10.30. The highest values of τ (i.e., stress concentration) are found at $\beta = 62°$. The dilatancy domain is now at the roof and floor (Figure 10.31b), but involves a relatively small amount of rock. No significant convergence is expected here either. This convergence is shown in Figure 10.32b. The dash-dot line corresponds to time $t = 7.3 \times 10^6$ and the square on this line shows the location of the initiation of creep failure (at $\beta = 61°$). The short-dashed line shows the cavern contour after $t = 3 \times 10^7$, and the portion of the contour damaged by creep failure is $49° \leq \beta \leq 81°$. Thus, the damage initiation is found on the contour somewhere between the corner and the roof. Also the spreading in time of the damage takes place primarily from the initiation location towards the roof (and the floor respectively), i.e., in the direction of stress concentration where τ reaches high values.

For $\sigma_h/\sigma_v = 2$, high values of τ can be found at the roof and floor (see Figure 10.30) and an elastic domain is present in the horizontal direction. No short term failure is found though the stress state is very close to the failure surface at $\beta = 0°$. The creep convergence is nearly entirely in the vertical direction (in this case the direction of σ_{min}^P and that of the stress concentration are not far from each other) and takes place much faster than for the case $\sigma_h/\sigma_v = 1$. Thus, assuming constant stresses in Figure 10.32c, the dash-dot line shows the contour at time $t = 4.295 \times 10^4$ when the creep failure initiation starts at $\beta = 70°$ (marked by a square). The short-dashed line shows the contour at $t = 3 \times 10^5$, when the whole roof (and floor) $55° \leq \beta \leq 90°$ is damaged by creep failure. The time intervals involved are orders of magnitude smaller. Thus, this is a quite unstable case due to the high values of τ involved. These high values are due to two converging reasons: in the vertical direction we have the stress concentration due to the elongated shape of the cavern, and in the same direction σ_{min}^P also acts. For such caverns a support is absolutely necessary at the roof and floor if the cavern is to be used for a longer time period.

The case $\sigma_h/\sigma_v = 3$ is also shown in Figure 10.30. This time short-term failure takes place in the horizontal direction (direction of σ_{max}^P) and very high values of τ are found at the roof (direction of σ_{min}^P and stress concentration). Therefore, besides the instantaneous failure in the horizontal direction, one has to expect creep failure initiation quite soon at the roof and floor. Thus for $\sigma_h/\sigma_v = 3$ the cavern is totally unstable.

Figure 10.31d shows the domain of dilatancy for the same case as in Figure 10.31b, but for the depth $h = 700$ m. In this case the domain of dilatancy spreads out around the upper (and lower) part of the contour. The C/D boundary is marked by a dash-dot line, and the two short-dashed lines are lines ε_v = constant, showing where dilatancy is more advanced inside this region. The region of maximum damage is obviously at the very top (and bottom) of the cross-section.

In conclusion, such tall caverns are reasonably stable for quite long time intervals (tens or hundreds of years) for ratios in the range $0.4 < \sigma_h/\sigma_v < 1.5$. for instance, if no internal pressure is applied. For high values of σ_h/σ_v short term failure is expected in the direction of σ_{max}^P and quite early creep failure in the direction of σ_{min}^P. The location of creep failure initiation depends on the ratio σ_h/σ_v and on the stress concentration due to the elongated shape of the cross-section: if the directions of stress concentration due to the peculiar shape of the cross-section and that of σ_{min}^P are not too far apart, creep failure initiation is to be expected after excavation and afterwards the damage progresses quite fast in that same direction.

Even a small internal pressure changes the whole picture of the stress distribution around the cavern (see Figure 10.33 and compare it with Figure 10.30). For $\sigma_h/\sigma_v = 0.4$ short term

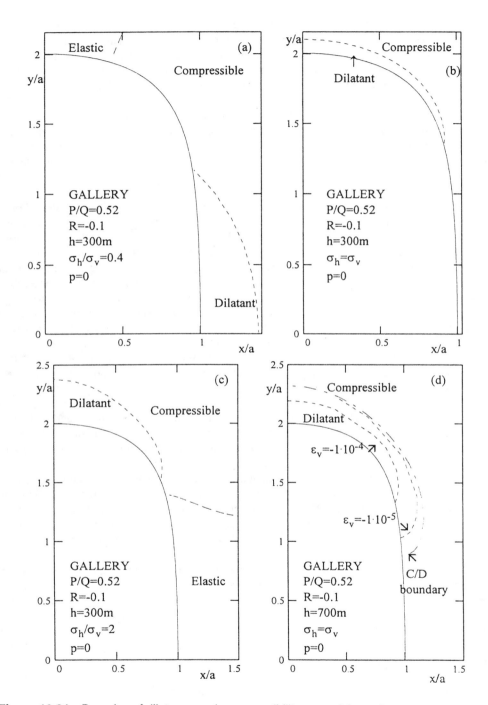

Figure 10.31   *Domains of dilatancy and compressibility around the gallery.*

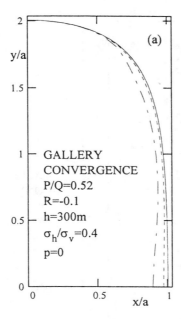

Figure 10.32a *Convergence of tall caverns for $\sigma_h/\sigma_v = 0.4$ and no internal pressure.*

failure takes place at the roof and floor (direction of σ_{min}^{P}). For $\sigma_h = \sigma_v$ the cavern is perfectly stable. Comparing the stress distributions shown in Figures 10.30 and 10.33 for $\sigma_h/\sigma_v=2$ one can conclude that for $p = 10$ MPa the cavern becomes quite stable, and creep failure is practically not possible. For higher values of $\sigma_h/\sigma_v=3$, for instance, short term failure is possible in the horizontal direction (direction of σ_{max}^{P}), though the stress state is quite close to the short term failure surface (failure may be expected during stress relaxation), while at the roof and floor the octahedral shear stress reaches moderate high values implying that creep failure could be expected some time after excavation. In conclusion one can say that an internal applied pressure somewhat improves cavern stability for $\sigma_h/\sigma_v = 2$, for instance, but has an unfavorable effect in the case $\sigma_h/\sigma_v = 4$.

The **location of creep failure initiation** and the **time when this initiation starts** can be discussed as for the other cases previously considered. Depending on the shape of the cross section and on the far field stresses, we have to find the location of stress concentration induced either by cross-sectional shape or by the ratio σ_h/σ_v. Figure 10.34a shows for a tall gallery excavated in rock salt at depth $h = 300$ m, with $P/Q = 0.27$ and $R = -0.1$, the location of creep damage initiation. The shape of the gallery is shown in Figure 10.34c. If $\sigma_h/\sigma_v \geq 0.5$, this location induced both by the elongated shape of cross-section and by the far field stresses is at the tip of the vertical axis. As the ratio σ_h/σ_v is smaller, the location of stress concentration induced by the far field stresses has a tendency to shift the location of stress concentration clock wise (Figure 10.34a). It is a tendency to separate the stress concentration induced by the far field stresses from the stress concentration location induced by the elongated shape of the cross- section. In this way the gallery tends to become more stable. That is shown in Figure 10.34b: the time up to the initiation of creep failure has the smallest value

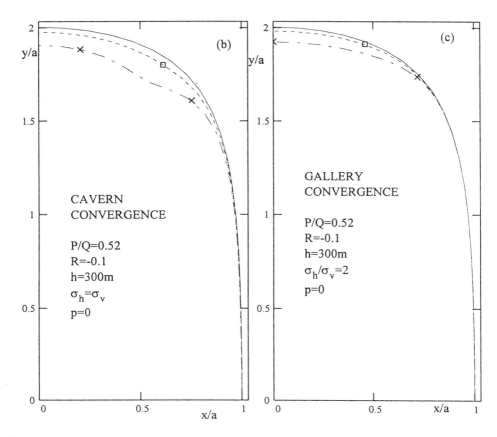

Figure 10.32 b,c *Convergence of tall caverns for two other ratios of far field stresses and no internal pressure.*

for the highest values of σ_h/σ_v, i.e., in those cases when the location of the stress concentration induced by the elongated shape of the cross-section and that induced by the far field stresses coincide (at the vertical ends of the axis of the cross-section).

10.6 CONCLUSIONS

In the case of rectangular-like caverns or galleries, the location of incipient creep failure depends mainly on the shape of the cross-section (the presence of corners or elongated shapes) and on the orientation of the far field stresses (of $\sigma^P_{h\,min}$ and $\sigma^P_{h\,max}$), and also on depth. As an additional example, Figure 10.35 shows for square-like cross sections ($P/Q=1$), $p=0$ and $h=700$ m, the dependency of this location on the ratio σ_h/σ_v of the far field stresses. In fact, since at a fixed depth the value of σ_v is fixed, it is question of orienting the axis of the gallery with respect to $\sigma^P_{h\,min}$ and $\sigma^P_{h\,max}$ to get optimal stability, i.e., to separate as much as possible

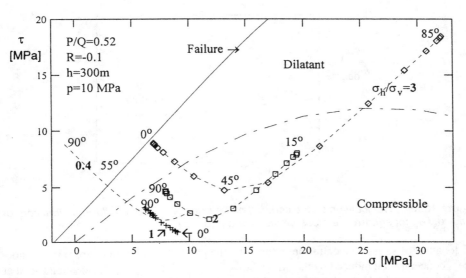

Figure 10.33 *Stress distribution along the contour of tall cavern for p = 10 MPa and several values of σ_h/σ_v.*

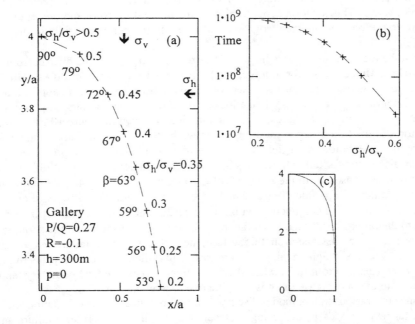

Figure 10.34 *a) Location and initiation of creep failure around a tall gallery, b) time to creep failure initiation; c) shape of the cross-section of the gallery.*

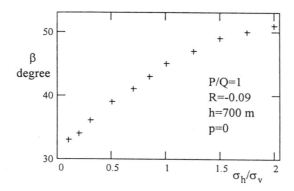

Figure 10.35 *The location on the contour, of the initiation of creep failure, as function of the ratio* σ_h/σ_v *for square-like cross-sections and p=0.*

the location of stress concentration induced by the presence of corners, from the location of stress concentration induced by the orientation of $\sigma_{h\,min}$. For $\sigma_h = \sigma_v$ the location is just at the corner and entirely due to the presence of the corner. However, for $\sigma_h \neq \sigma_v$ this location shifts more or less towards σ^P_{min} (either σ_h or σ_v), depending on how much the ratio σ_h/σ_v departs from unity. For $\sigma_h < \sigma_v$ this location shifts towards the horizontal axis (direction of $\sigma^P_{min} = \sigma_h$) and *vice versa*. This location depends also on the internal pressure and on depth. After creep failure initiation the damage progresses faster along the contour mainly in the direction of σ^P_{min}.

If at the location where the gallery is to be excavated, $\sigma_{h\,min}$ is quite distinct from $\sigma_{h\,max}$, one can **orient the gallery** (for storage, etc.) in order **to get maximum stability**, which is governed by geometry of the cross-section, on the ratio σ_h/σ_v (with σ_h appropriately chosen in the interval $\sigma_{h\,min} < \sigma_h < \sigma_{h\,max}$ at the location), on depth, and on internal pressure.

For wide caverns or galleries, a similar situation is shown in Figure 10.21. In this case the location of initiation of creep failure is governed by the elongated shape of the cross-section and by the orientation of far field stresses, besides depth and internal pressure, obviously. A similar discussion for tall galleries has been presented in conjunction with Figure 10.35.

The time elapsed from the moment of excavation up to creep failure initiation depends primarily on the magnitude of τ and on how close the stress state is to the short term failure surface. As shown in conjunction with Figure 10.30, one finds out that if the stress state is close to the failure surface the creep failure initiation is to be expected quite soon. If this stress state is at a greater distance from this surface, the time up to creep failure initiation increases, while for values of τ approaching the compressibility/dilatancy boundary this time increases very much (tending to infinity when the "overstress" above the C/D boundary tends towards zero). Note that a decrease of τ is possible due to a relaxation.

Since the various failure modes discussed above are related to the orientation and magnitude of the far field stresses, by observing the location and timing of short term failure and of creep failure initiation around the contour, and by following the analysis given above, one can obtain information concerning the orientation and relative magnitude of the in-plane far field stress components σ^P_{min} and σ^P_{max}, if these were not known *a priori*.

Once the location of loss of stability (short term failure, creep failure, fast creep) is determined and the amount sf rock involved is also estimated, one can use this information for the design of an optimal economical support (rock-bolts, lining, etc.).

References[‡]

Addis, M.A., Last, N.C., and Yassir, N.A. (1994). The estimation of horizontal stresses at depth in faulted regions and their relationship to pore presure variations. *Rock Mechanics in Petroleum Engineering*. Balkema, Rotterdam, 887-895. [Ch.7]

Ahrens, T.J. (Editor) (1995) *Handbook of Physical Constants*, Vol. 1, 2, 3. American Geophysical Union, Washington. [Ch.1-3]

Akai, K., Yamamoto, K. and Arioka, M. (1970) Experimentelle Forschung über anisotropische Eigenschaften von kristallinen Schiefern. *Proc. 2nd Int. Congr. on Rock Mechanics*, Belgrade, Vol. II, 181. [Ch.5]

Albrecht, H., and Hunsche, U. (1980) Geomechanical aspects of radioactive waste disposal in salt domes with emphasis on the flow of rock salt (in German). *Fortschr. Miner.* **58**(2). 212-247. [Ch.2]

Albrecht, H., Hunsche, U., Plischke, I., and Schulze, O. (1990). Mapping of the mechanical properties of a salt dome. *Proc. 6th Congr. of the Int Assoc. of Engineering Geology (IAEG)*. Vol.1, Ed. D.G. Price. Balkema, Rotterdam, 23-29. [Ch.2]

Albrecht, H., Hunsche, U., and Schulze, O. (1993) Results from the application of the laboratory test program for mapping homogeneous parts in the Gorleben salt dome. *Geotechnik-Sonderheft 1993*. Glückauf, Essen, 152-155. [Ch.2]

Alheid, H.-J., and Rummel, F. (1976) Energieregelung bei Druckversuchen an Gesteinen. Sonderforschungsbereich 77 - Felsmechanik. Univ. Karlsruhe. Jahresbericht 1975. Karlsruhe. [Ch.6]

Alheid, H.-J., and Knecht, M. (1996) Distribution of seismic velocities around underground drifts. *Proc. Int. Conf. on Deep Geological Disposal of Radioactive Waste*, Winnipeg, Canada. [Ch.2]

Alheid, H.-J. (1997) Private Communication. [Ch.2]

Allemandou, X., and Dusseault, M.B. (1996) Procedures for cyclic creep testing of salt rock, results and discussion. *The Mechanical Behavior of Salt. Proc. of the 3rd Conf.*,Eds. M.Ghoreychi, P. Berest, H.R. Hardy, Jr., and M. Langer. Trans Tech Publ. 207-220. [Ch.4, 6]

Allirot, D., and Boehler, J.-P. (1979) Évolution des propriétés mécanique d'une roche stratifiée sous pression de confinement. *Proc. 4th Int. Soc. Rock Mechanics*, Montreux, Balkema, Rotterdam, Vol.1, 15-22. [Ch.5]

Amadei, B. (1993) *Rock Anisotropy and the Theory of Stress Measurements*. Springer-Verlag, Berlin-Heidelberg. [Ch.5]

Andersen, M.A., Foged, N., and Pedersen, H.F. (1992) The rate-type compaction of a weak North Sea chalk. *Rock Mechanics. Proc. 33rd US Symp.* Eds. J.R.Tillerson, and W.R. Wawersik, Balkema, 253-261. [Ch.4]

Aoki, T., Tan, C.P., and Bamford, W.E. (1993) Effects of deformation and strength anisotropy on borehole failures in saturated shales. *Rock Mechanics in the 1990s*, Ed. B.C. Haimson, *Int. J. Rock Mech. Min. Sci. & Geomech. Abstr.* **30**, 1031-1034. [Ch.8]

[‡] At the end of each reference, the chapters where it is mentioned, are given in brackets.

Aoki, T., Tan, C.P., and Bamford, W.E. (1994) Stability analysis of inclined wellbores in saturated anisotropic shales. *Computer Methods and Advances in Geomechanics,* Eds. H.J. Siriwardane and M.M. Zaman. Balkema, Rotterdam, 2025-2030. [Ch.9]

Arjang, B. (1989) Pre-mining stresses at some hard rock mines in the Canadian Shield. *Rock Mechanics as a Guide for Efficient Utilization of Natural Resources. Proc. 30th US Symp. Rock Mechanics.* Balkema, Rotterdam, 545-551. [Ch.7]

Ashley, M.F., Gandhi, C., and Taplin, D.M. R. (1979). Overview No.3. Fracture mechanism maps and their construction for F.C.C. metals and alloys. *Acta Met.* **27**, 699-729 [Ch.1-3]

Attewell, P.B., and Sandford, M.R. (1974) Intrinsic shear strength of a brittle anisotropic rock -I. Experimental and mechanical interpretation. *Int. J. Rock Mech. Min. Sci. & Geomech. Abstr.,* **11**, 423-430. [Ch.5]

Aubertin, M., Gill, D.E., and Ladanyi, B. (1991) An internal variable model for the creep of rocksalt. *Rock Mechanics and Rock Engineering.* **24**, 81-97. [Ch.5]

Aubertin, M., Sgaoula, J., and Gill, D.E. (1993a) A Damage Model for Rocksalt: Application to Tertiary Creep. *Proc.7th Symp. on Salt.* Elsevier,Amsterdam, 117-125. [Ch.6]

Aubertin, M., Servant, S., and Gill, D.E. (1993b) Relations constitutives pour le comportement mécanique du sel gemme: quelques développements récents et tendances actuelles. *Proc. 46th Canadian Geotech. Conf., Saskatoon,* 117-126. [Ch.5]

Aubertin, M., Gill, D.E. and Ladanyi, B. (1994) Constitutive equations with internal state variables for the inelastic behavior of soft rocks. *Appl. Mech. Reviews, ASME* **47**(6-2), S97-S101. [Ch.5]

Aubertin, M. (1996) On the physical origin and modelling of kinematic and isotropic hardening of salt. *The Mechanical Behavior of Salt. Proc. 3rd Conf.* Eds. M. Ghoreychi, P. Berest, H.R. Hardy, Jr., and M. Langer. Trans Tech Publ., Clausthal-Zellerfeld. 3-17 . [Ch.2 , 5]

Aubertin, M., Gill, D.E., and Servant, S. (1996a) Preliminary determination of constants for an updated version of the SUVIC Model. *The Mechanical Behavior of Salt. Proc.3rd Conf.* Eds. M. Ghoreychi, P. Berest, H.R. Hardy, Jr., and M. Langer. Trans Tech Publ. Clausthal-Zellerfeld. 19-30 .[Ch.2]

Aubertin, M., Sgaoula, J., Servant, S., Julien, M.R., Gill, D.E., and Ladanyi, B. (1996b) An up-to-date version of SUVIC-D for modeling the behavior of salt. *The Mechanical Behavior of Salt. Proc. 4th Conf.* Eds. P. Habib, H.R. Hardy, B. Ladanyi, and M. Langer. Trans Tech Publ., Clausthal-Zellerfeld (In press). [Ch.5]

Aubertin, M., Simon, R. (1997) A damage initiation criterion for low porosity rocks. *Int. J. Rock Mech. Min. Sci. & Geomech. Abstr.* **34** (In press). [Ch.5]

Ayling, M.R., Meredith, P.G., and Murrel, S.A.F. (1995) Microcracking during triaxial deformation of porous rock monitored by changes in rock physical properties, I. Elastic wave propagation measurements on dry rocks. *Tectonophysics.* **245**, 205-221. [Ch.2]

Bandis, S.C., and Barton, N. (1986) Failure modes of deep boreholes. *Proc. 27th US Rock Mechanics Symp.* 599-605. [Ch.8]

Bandis, S.C., Nadim, F., Lindmann, J., and Barton, N. (1986) Stability investigations of modelled rock caverns at great depth. *Large Rock Caverns,* Ed. K.H.O. Saari. Pergamon Press, Oxford, Vol.2, 1161-1170 [Ch.9]

Bandis, S.C., Lindman, J., and Barton, N. (1987) Three-dimensional stress state and fracturing around cavities in overstresses weak rock. *Proc. Int. Cong. on Rock Mechanics.* Eds. G. Herget and S. Vongpaisal. Balkema, Rotterdam, 769-775. [Ch.9]

Barber, D.J., and Meredith, P.G. (Eds.) (1990) *Deformation Processes in Minerals, Ceramics and Rocks.* Unwin, Hyman. [Ch.3]

Barla, G., and Rabagliati, U. (1992) Observed cavern performance of a large underground storage facility in weak jointed chalk. *Rock Mechanics. Proc. 33rd US Symposium,* Eds. J.R. Tillerson and W.R. Wawersik. Balkema, Rotterdam, 409-419. [Ch.10]

Barrett, C.R., and Nix, W.D. (1965) A model for steady-state creep based on the motion of jogged screw dislocations. *Acta Metall.* **13**, 1247. [Ch.3]

Barton, N., By, T.L., Chryssanthakis, P., Tunbridge, L., Kristiansen, J., Løset, F., Bhasin, R.K., Westerdahl, H., and Vik, G. (1994) Predicted and measured performance of the 62m span Norwegian Olympic Ice Hockey Cavern at Gjøvik. *Int. J. Rock Mech. Min. Sci. & Geomech. Abstr.* **31**, 597-616 [Ch.10].

Baumgärtner, J., Healy, J.H., Rummel, F., and Zoback, M.D. (1993) Deep hydraulic fracturing stress measurements in the KTB (Germany) and CAJON PASS (USA) scientific drilling projects - A summary. *Proc. 7th Int. Congr. on Rock Mechanics,* Balkema, Rotterdam, 1685-1690. [Ch.7]

Bayuk, E.I. (1966) Velocities of waves in specimens of eruptive and metamorphic rocks at pressures up to 4000 kg/cm^2. *Trudy Inst. Fiz. Zemli.* No.37 (204), 16-36. [Ch.6]

Bell, J.F. (1973) *The Experimental Foundations of Solid Mechanics.* Handbuch der Physik, Vol.VI a/1, Springer-Verlag, Berlin. [Ch.1, 4, .6]

Berest, P., and Nguyen, M.D. (1983) Comportement mécanique des cavités profondes de stockage d'hydrocarbures dans le sel . *Proc. 5th Int. Congr. on Rock Mechanics,* Melbourne Preprints Section D, D227-D231. [Ch.10]

BGR - Bundesanstslt für Geowissenschaften und Rohstoffe, Hannover (1995). *Untersuchungen zum Kompaktionsverhalten von Salzgrus als Versatzmaterial für Endlagerbergwerke in Salz unter besonderer Berücksichtigung der Wechselwirkung zwischen Gebirge und Versatz.* Förderkennzeichen 02 E 8552 8, Hannover. [Ch.3]

Biberger, M., and Blum, W. (1989) On the natural law of steady state creep. *Scripta Metall.* **23**, 1419-1424. [Ch.2, 3]

Bieniawski, Z.T. (1967). Mechanics of brittle fracture of rock. Part I - Theory of the fracture process; Part II - Experimental studies; Part III - Fracture in tension and under long-term loading. *Int.J. Rock Mech. Min. Sci,* **4**, 365-430. [Ch.1, 4]

Blacic, J.D., and Christie, J.M. (1984) Plasticity and hydrolytic weakening of quartz single crystals. *J. Geophys. Res.* **89**, 4223-4239. [Ch.3]

Blum, W., and Pschenitzka, F. (1976) Durchführung von Zug-, Druck- and Spannungsrelaxationsverfahren bei erhohter Temperatur. *Z. Metallkunde,* **67**, Heft 1, 62-65. [Ch.1]

Blum, W. 1977. Dislocation models of plastic deformation of metals at elevated temperature. *Z. Metallkunde.* **68**, 484-492. [Ch.3]

Blum, W., and Finkel, A. (1982) New technique for evaluating long range internal back stresses. *Acta Metall.* **30**, 1705-1715. [Ch.2]

Blum, W., and Weckert, E. (1987) On the interpretation of the "internal back stress" determined from dip tests during creep of Al-5 at % Mg. *Materials Science and Engineering.* **86**, 145-158. [Ch.2]

Blum, W., and Fleischmann, C. (1988) On the deformation mechanism map of rock salt. *The Mechanical Behavior of Salt, Proc. 2nd Conf.* Eds. H.R. Hardy, and M. Langer. Trans Tech Publ., Clausthal-Zellerfeld, 7-22. [Ch.2]

Blum, W. (1991) High temperature deformation and creep of crystalline solids. *Plastic Deformation and Fracture of Materials*. Ed. H. Mughrabi. VCH , Weinheim. 360-405. [Ch.2]

Blum, W., Straub, S., and Vogler, S. (1991). Creep of pure materials and alloys. *Proc. 9th Int. Conf. on Strength of Metals and Alloys (ICSMA9)*, Eds. D.G. Brandon, R. Chaim, and A. Rosen, Freund, London, 111-126. [Ch.2]

Bock, H.F. (1993) Measuring *in situ* rock stress by borehole Slotting. *Comprehensive Rock Engineering*. Ed-in-Chief J. A. Hudson., Vol.3, Rock Testing and Site Characterization. Pergamon Press ,433-444 [Ch.7].

Bodner, S.R. (1981) A procedure for including damage in constitutive equations for Elastic-viscoplastic work-hardening materials. *Proc. IUTAM Symp.Senlis*, Eds. J. Hult and J. Lemaitre. Springer-Verlag, Berlin, 21-28. [Ch.6]

Boehler, J.-P. (1975) Contributions théoriques et expérimentales à l'étude des milieux plastiques anisotropes. Thése de Doctorat ès Sciences, Grenoble. [Ch.5]

Boehler, J.-P. (1978) Lois de comportement anisotrope des milieux continus. *Journal de Mécanique*. **17**, 153-190. [Ch.5]

Boehler, J.-P. (1987) Yielding and failure of transversely isotropic solids. *Applications of tensor functions in solid mechanics*. Ed. J.-P. Boehler, Courses and Lectures No.292, CISM Udine, Springer Verlag, Berlin. 3-140. [Ch.5]

Bogobowicz, A. (1996) Non-Newtonian creep into a two-dimensional cavity of near-rectangular shape. *Trans. ASME, Journal of Applied Mechanics*. **63**, 1047-1051. [Ch.10]

Borns, D.J., and Stormont, J.C.(1989) The delineation of the disturbed rock zone surrounding excavations in salt. *Rock Mechanics as a Guide for Efficient Utilization of Natural Resources. Proc. 30th US Symp. Rock Mechanics*. Balkema, Rotterdam, 353-360. [Ch.10]

Brace, W.F. (1965). Some new measurements of linear compressibility of rocks. *J. Geophys. Res.* **70**, 2, 391-398. [Ch.4]

Brace, W.F., Paulding, B.W. Jr., and Scholz, C. (1966). Dilatancy in the fracture of crystalline rocks. *J. Geophys. Res.* **71**, 16, 3939-3953. [Ch . 1, 4]

Bridgman, P.W. (1949). Volume changes in the plastic stages of simple compression. *J. Applied Physics,* **20**, 1241-1251. [Ch.1, 4].

Brignoli, M., and Sartori, L. (1993) Incremental constitutive relations for the study of Wellbore failure. *Rock Mechanics in the 1990s*. Ed. B.C. Haimson. *Int. J. Rock Mech Min. Sci.& Geomech. Abstr.* **30**, 1319-1322. [Ch.4, 5, 8]

Broch, E., and Nilsen B. (1993) Recent development in rock engineering in Norway: Gas-tight rock caverns, subsea road tunnels, steel-fiber reinforced shotcrete. *Comprehensive Rock Engineering*. Ed-in-Chief J. A. Hudson., vol.5, Surface and Underground Project Case histories, Pergamon Press , Oxford 55-83. [Ch. 9, 10]

Brodsky, N.S., and Munson, D.E. (1991) The effect of brine on the creep of WIPP salt in laboratory tests. *Proc. 32nd US Symp. on Rock Mechanics as a Multidisciplinary Science*. Ed. J.-C. Roegiers. Balkema, Rotterdam, 703-712. [Ch.3]

Brudy, M., and Zoback, M.D. (1993) Compressive and tensile failure of boreholes arbitrarily-inclined to principal stress axes: application to the KTB boreholes, Germany. *Rock Mechanics in the 1990s*. Ed. B.C. Haimson, *Int. J. Rock Mech. Min. Sci. & Geomech. Abstr.* **30**, 1035-1038. [Ch.8]

Burlet, D., and Cornet, F.H. (1993) Stress measurements at great depth by hydraulic tests in boreholes. *Proc. 7th Int. Congr. on Rock Mechanics*. Balkema, Rotterdam, 1691-1697. [Ch.7]

Caillard, D., and Martin, J.L. (1987) New trends in creep microstructural models for pure metals. *Revue Phys. Appl.* **22**, 169-183. [Ch.2]

Carmichael, R.S. (1982) *Handbook of physical properties of rocks*. Vols.1 and 2. CRC Press, Boca Raton, FL. [Ch.1]

Carmichael, R.S. (1984) *Handbook of physical properties of rocks*. Vol.3. CRC Press, Boca Raton, FL. [Ch.1]

Carter, J.P. (1988) A Semi-analytical solution for swelling around a borehole. *Int.J. Numerical and Analytical Methods in Geomechanics,* **12**, 197-212. [Ch.8]

Carter, N.L., Anderson, D.A., Hansen, F.D., and Kranz, R.L. (1981) Creep and creep rupture of granitic rock. *Mechanical Behaviour of Crustal Rocks*. Geophysical Monograph **24**, American Geophysical Union, 61-82. [Ch.6]

Carter, N.L., Hansen, F.D., and Senseny, P.S. (1982) Stress magnitudes in natural rock salt. *J. Geophys. Res.* **87** (B11), 9289-9300. [Ch.2]

Carter, N.L., and Hansen, F.D. (1983) Creep of rocksalt. *Tectonophysics,* **92**, 275-333. [Ch. 2, 3, 5, 10]

Carter, N.L., Kronenberg, A.K., Ross, J.V., and Wiltschko, D.V. (1990), Control of fluids on deformation of rocks. *Deformation Mechanisms. Rheology and Tectonics*. Eds. R.J. Knipe and E.H. Rutter. Geological Society Special Publication No.54, 1-13. [Ch.3]

Castro, L.A.M., McCreath, D.R., and Kaiser, P.K. (1995) Rockmass strength determination from breakouts in tunnels and boreholes. *Proc. Int. Congr. on Rock Mechanics, Tokyo*. Balkema, Rotterdam, 531-536. [Ch.9]

Cazacu, O. (1995) Contribution à la modelisation elasto-viscoplastique d'une roche anisotrope. Ph.D, Dissertation, University of Science and Technology, Lille, France. [Ch.4]

Cazacu O. and Cristescu, N. (1995) Failure of anisotropic compressible shale. *Impact, Waves and Fracture*, Eds. R.C.Batra, A.K.Mal, and G.P.MacSithigh. ASME, 1-8. [Ch.5]

Cazacu, O., Cristescu, N. D., Shao, J. P., and Henry, J. P. (1996) Elastic/viscoplastic constitutive equation for anisotropic shale. *Rock Mechanics. Tools and Technique. Proc. 2nd NARMS Symp.* Eds. M. Aubertin, F. Hassani, and H. Mitri. Balkema, Rotterdam, 1683-1690. [Ch.6]

Cazacu, O., Jin, J., and Cristescu, N.D. (1997) A new constitutive model for alumina powder compaction. *KONA, Powder and Particle,* (In press) [Ch.4]

Chan, K.S., Bodner, S.R., Fossum, A.F., and Munson, D.E. (1992) A constitutive model for inelastic flow and damage evolution in solids under triaxial compression. *Mechanics of Materials.* **14**, 1-14. [Ch.4]

Charlez, Ph., Segal, A., Heugas, O., and Quenault, O. (1990) Development of a microstatistical model for deep borehole ovalization in brittle rocks. *Rock at Great Depth*. Eds. V. Maury, and D. Fourmaintraux. Balkema, Rotterdam, 1447-1454. [Ch.8]

Charlez, Ph.A. (1991) *Rock Mechanics, Vol.1, Theoretical Fundamentals*. Editions Techniq, Paris. [Ch.1-3]

Charlez, Ph., and Heugas, O. (1991) Evaluation of optimal mud weight in soft shale levels. *Rock Mechanics as a Multidisciplinary Science. Proc. 32nd US Symp.* Ed. J.-C. Roegiers, Balkema, Rotterdam, 1005-1014. [Ch.8]

Charlez, P.A. (1994) The impact of constitutive laws on wellbore stability: A general review. *Rock Mechanics in Petroleum Engineering. EUROCK '94.* Balkema, Rotterdam, 239-249. [Ch.8]

Cheatham, J.B.Jr. (1993) A new hypothesis to explain stability of borehole breakouts. *Rock Mechanics in the 1990s, The 34th US Symp. on Rock Mechanics* Ed. B.C. Haimson. *Int. J. Rock Mech. Min. Sci. & Geomech. Abstr.* **30**, 1095-1101. [Ch.8]

Chen, W.F. (1982) *Plasticity in Reinforced Concrete.* McGraw-Hill, New York. [Ch.1-3]

Cheng, H., and Dusseault, M.B. (1993) Deformation and diffusion behaviour in a solid experiencing damage: a continuous damage model and its numerical implementation. *Rock Mechanics in the 1990s, The 34th US Symp. on Rock Mechanics,* Ed. B.C. Haimson. *Int. J. Rock Mech. Min. Sci. & Geomech Abstr.* **30**, 1323-1331. [Ch.6]

Cheng, Y., and Liu, S.-C. (1993) Power caverns in Mingtan Pumped Storage Project, Taiwan. *Comprehensive Rock Engineering.* Ed-in-Chief J. A. Hudson. vol.5, Surface and Underground Project Case Histories, Pergamon Press, 111-131 [Ch.10]

Chia, J.H., and Desai, C.S. (1994) Constitutive modelling of thermoviscoplastic response of rock salt. *Computer Methods and Advances in Geomechanics. Proc. 8th Conf. IACMAG,* Eds. H.J. Siriwardane and M.M. Zaman. Balkema, Rotterdam, 555-560. [Ch.5]

Choi, S.O., Synn, J.H., and Lee, H.K. (1991) Influence of rift in granite and stress conditions on hydraulic fracturing. *Proc. 7th Int. Congr. Rock Mechanics.* Balkema, Rotterdam, 441-445. [Ch.8]

Clark, S.P.Jr. (1966) *Handbook of Physical Constants.* Geological Society of America, Memoir 97, New York. [Ch.1]

Cleja-Tigoiu, S. (1991) Elasto-viscoplastic constitutive equations for rock-type materials (finite deformation). *Int. J. Engng. Sci.,* **29**, 1531-1544. [Ch.4]

Cooling, C.M., Hudson, J.A., and Tunbridge, L.W. (1988) *In Situ* rock stresses and their measurements in the U.K. Part II. *Site experiments and stress field interpretation. Int. J. Rock Mech. Min. Sci. & Geomech. Abstr.* **25**, 6, 371-382. [Ch.7]

Constantinescu, M. (1981) Experimental formulation of constitutive equations for solid porous materials. Ph. D. Dissertation, University of Bucharest, Bucharest. [Ch.4, 5]

Cornet, F.H. (1993) Stresses in rock and rock masses. *Comprehensive Rock Engineering,* Ed-in-Chief J. A. Hudson. Vol.3, *Rock Testing and Site Characteri.* Pergamon Press, 41-66. [Ch.7]

Costin, L.S. (1983) A microcrack damage model for brittle rock. Sandia Report SAND83-1590, Sandia National Laboratories, Albuquerque, N.M. [Ch.6]

Costin, L.S. (1985) Damage mechanics in the post-failure regime. *Mech. of Materials.* **4**, 149-160. [Ch.6]

Costin, L.S. (1987) Time-Dependent Deformation and Failure. *Fracture Mechanics of Rock,* Academic Press, New York [Ch.6].

Cottrell, A.H. (1953) *Dislocations and Plastic Flow in Crystals.* Clarendon Press, Oxford [Ch.1-3]

Coussy, O., Charlez, Ph. and Heugas, O. (1991) Thermoporoelastic response of a borehole. *Proc. 7th Int. Congr. Rock Mechanics.* Balkema, Rotterdam, 699-702. [Ch.8]

Cravero, M., Garino, A,. and Innaurato, N. (1993) Analysis of the stability of an underground opening in Central Appenines. *Assessment and Prevention of Failure in Rock Engineering,* Eds. A. G.Paşamehmetoğlu, T. Kawamoto, B.N. Whittaker, and Ō. Aydan. Balkema, Rotterdam, 507-512 [Ch.10]

Cristescu, N. (1975) Rheological properties of some deep located rocks. University of Bucharest Report. [Ch.4, 6]

Cristescu, N. (1979) A viscoplastic constitutive equation for rocks. *Preprint Series in Mathematics,* No.49/1979, INCREST Bucharest. [Ch.4]

Cristescu, N. (1982) Rock Dilatancy in Uniaxial Tests. *Rock Mechanics,* 15, 133-144. [Ch.4]

Cristescu, N. and Suliciu, I. (1982) *Viscoplasticity.* Martinus Nijhoff Co. Hague - Editura Tehnica Bucuresti. [Ch.4]

Cristescu, N. (1983) *Mechanics of Rocks,* University of Bucharest, Bucharest (in Romanian). [Ch.6, 8]

Cristescu, N. (1984) *Rock Mechanics.* University of Bucharest, Bucharest (in Romanian). [Ch.4]

Cristescu, N. (1985a) Fluage, dilatance et/ou compressibilité des roches autour des puits verticaux et des forages pétroliers. *Revue Française de Geotechnique.* No.31, 11-22. [Ch.8]

Cristescu, N. (1985b) Viscoplastic creep of rocks around horizontal tunnels. *Int.J. Rock Mech. Min. Sci.& Geomech Abstr.* 22, 6, 453-459. [Ch.4, 9]

Cristescu, N. (1985c) Irreversible dilatancy or compressibility of viscoplastic rock-like materials and some applications. *Int. J. Plasticity.* 1, 3, 189-204. [Ch.8, 9]

Cristescu, N. (1985d) Plasticity of compressible/dilatant rock-like materials. *Int. J.Engng. Sci.* 23, 10, 1091-1100. [Ch.4]

Cristescu, N. (1985e) Rock plasticity. *Plasticity Today; Modelling, Methods and Applications.* Eds. A. Sawczuk and G. Bianchi. Elsevier Applied Science, London 643-655. [Ch.4]

Cristescu, N., Massier, D. Dobre-Pal, M., Nicolae, M., Cristescu, C., and Teodorescu, S. (1985) Determination of the creep of rocks surrounding a circular underground opening. *Mine, Petrol, Gaze,* 36 , 10, 509-517. (in Romanian). [Ch.9]

Cristescu, N. (1986) Damage and failure of viscoplastic rock-like materials. *Int. J. Plasticity.* 2, 2, 189-204. [Ch.4, 8, 9]

Cristescu, N. (1987). Elastic/viscoplastic constitutive equations for rock. *Int. J. Rock Mech. Min. Sci. & Geomech Abstr.* 24, 5, 271-282. [Ch.4]

Cristescu, N., Fota, D., and Medves, E. (1987) Tunnel support analysis incorporating rock creep, *Int. J. Rock Mech. Min. Sci.& Geomech Abstr.* 24, 6, 321-330. [Ch 7, .9]

Cristescu, N. (1988) Viscoplastic creep of rocks around a lined tunnel. *Int. J. Plasticity.* 4, 4, 393-412. [Ch.9]

Cristescu, N., Fota, D., and Medves, E. (1988) Rock-support interaction in lined tunnels. *Rock and Soil Rheology.* Lectures Notes in Earth Sciences 14, Springer Verlag, Berlin, 245-272. [Ch.7, 9]

Cristescu, N., and Medves, E. (1988) Wood compressibility in mining applications. *Mechanical Behavior of Wood.* Actes, Bordeaux, 461-470. [Ch.9]

Cristescu, N. (1989a) *Rock Rheology,* Kluwer Academic, Dordrecht .[Ch. 1, 4, 6, 7, .8, 9]

Cristescu, N. (1989b) Plasticity of Porous Materials. *Proc. Plasticity '89. The 2nd Int. Symp.on Plasticity and its Current Applications,* Eds. A.S. Khan, and M. Tokuda. Pergamon Press, Oxford, 11-14. [Ch.5]

Cristescu, N. (1989c) A mechanical - AE/MA correlation. *Proc 4th Conf. on Acoustic Emission/Microseismic Activity in Geologic Structures and Materials. Pennsylvania State University, Oct.1985,* Trans Tech Publ., Clausthal-Zellerfeld, 559-567. [Ch.6]

Cristescu, N. (1989d) Viscoplasticity and damage of porous materials. *Advances in Constitutive Laws for Engineering Materials. Proc. ICCLEM*, Eds. Fan Jinghong and S. Murakami, International Academic Publ., Vol.1, 31-36. [Ch.6]

Cristescu, N. (1989e) Rock Rheology. *Advances in Constitutive Laws for Engineering Materials. Proc.ICCLEM*, Eds. Fan Jinghong and S. Murakami, International Academic Publ., Vol.2, 906-919. [Ch.4, 9]

Cristescu, N., and Duda, I. (1989) A tunnel support analysis incorporating rock creep and the compressibility of a broken rock stratum. *Computers and Geotechnics.* 7, 239-254. [Ch.7, 9]

Cristescu, N., Nicolae, M., and Teodorescu, S. (1989) Coal Rheology *Inelastic Solids and Structures*, Pineridge Press, Swansea, 165-177. [Ch.5]

Cristescu, N.(1991a) Nonassociated elastic/viscoplastic constitutive equations for sand. *Int. J. Plasticity.* 7, 41-64. [Ch.4, 5]

Cristescu, N. (1991b) Constitutive equations for rock salt. *Anisotropy and Localization of Plastic Deformation. Proc. Plasticity '91 Symp.* Eds. J-P. Boehler, and A.S. Khan, Elsevier Applied Science. London, 201-204. [Ch.4]

Cristescu, N. (1992) Constitutive equation for rock salt and mining applications. *Proc 7th Int. Symp. on Salt, 1992, Kyoto, Japan*, Elsevier, Amsterdam, Vol.1, 105-115. [Ch.4,8]

Cristescu, N. and Florea, D. (1992) Temperature influence on the elastic/viscoplastic behaviour of bituminous concrete. *Rev.Roumaine de Mecanique Appliquee*, 37, 603-614. [Ch.4]

Cristescu, N., and Hunsche, U. (1992) Determination of a nonassociated constitutive equation for rock salt from experiments. *Finite Inelastic Deformations - Theory and Applications. IUTAM Symp., Hannover,.1991.* Eds. D. Besdo and E. Stein. Springer Verlag, Berlin-Heidelberg, 511-523. [Ch.4, 5]

Cristescu, N. (1993a) A general constitutive equation for transient and stationary creep of rock salt. *Int. J. Rock Mech. Min. Sci.& Geomech Abstr.* 30, 2, 125-140. [Ch.4, 5]

Cristescu, N (1993b) Failure and creep failure around an underground opening. *Assessment and Prevention of Failure Phenomena in Rock Engineering, Istanbul*, Eds. A.G. Paşamehmetoğlu, T. Kawamoto, B.N. Whittaker, Õ. Aydan. Balkema, Rotterdam, 205-210. [Ch.8]

Cristescu, N. (1993c) Rock Rheology. *Comprehensive Rock Engineering.* Ed.-in-Chief J.A. Hudson., Vol.1, *Rock Mechanics Principles*, Pergamon Press, Oxford 523-544. [Ch.4]

Cristescu, N. and Hunsche, U. (1993a) A constitutive equation for salt. *Proc. 7th Int. Congr. Rock Mechanics*, Balkema, Rotterdam, Vol. 3, 1821-1830. [Ch.4]

Cristescu, N. and Hunsche, U. (1993b) A viscoplastic model for stationary and transient creep of rock salt. *Preprint 4th Int. Symp. on Plasticity and Its Current Applications, Baltimore, USA, July 19-23.* [Ch.4, 5, 10]

Cristescu, N. (1994a) Viscoplasticity of Geomaterials. *Visco-Plastic Behaviour of Geo-materials*, Eds. N.D. Cristescu and G. Gioda, Springer-Verlag, Wien-New York, 103-207. [Ch.4, 5, 7, 8, 9]

Cristescu, N (1994b) A procedure to determine nonassociated constitutive equations for geomaterials. *Int. J. Plasticity.* 10, 2, 103-131. [Ch.4, 5]

Cristescu, N. (1994c) Rheological models for geomaterials. *Computer Methods and Advances in Geomechanics. Proc. 8th Conf. IACMAG.*, Eds. H.J. Siriwardane and M.M. Zaman. Balkema, Rotterdam, Vol.1, 103-115. [Ch.9]

Cristescu, N. (1994d) Time effect in rock surrounding a horizontal tunnel. *Rock Mechanics; Models and Measurements; Challenges from Industry. Proc. 1st NARMS Symp.* Eds. P.P. Nelson and S.E. Laubach. Balkema, Rotterdam, 657-664. [Ch.9]

Cristescu, N. (1994e) Failure of compressible/dilatant geomaterials. *Mechanics USA 1994.* Ed. A.S.Kobayashi, Appl. Mech. Rev. **47**, .6, part 2, S102-S106. [Ch.9]

Cristescu, N., Ionescu, I.R. and Rosca, I. (1994) A numerical analysis of the foot-floor interaction in long wall workings. *Int. J. Numerical and Analytical Methods in Geomechanics*. **18**, 641-652. [Ch.10]

Cristescu, N. and Cazacu, O. (1995) Viscoplasticity of anisotropic rock. *Proc. 5th Int. Symp. on Plasticity and its Current Applications,* Eds. S. Tanimura and A.S. Khan, Gordon and Breach, New York. 499-502. [Ch.4, 5]

Cristescu, N. and Paraschiv, I. (1995a) The optimal shape of rectangular-like caverns and galleries. *Int. J. Rock Mech. Min. Sci.& Geomech Abstr.* **32**, 285-300. [Ch.10]

Cristescu, N. and Paraschiv, I. (1995b) Optimum design of large cavern. *Proc. 8th Int. Congr.on Rock Mechanics.* Balkema, Rotterdam. 923-928. [Ch.10]

Cristescu, N. (1996a) Design of yieldable tunnel supports for creeping rocks. *Society for Mining, Metallurgy, and Exploration, Inc. Transactions,* Vol. 300, 1996. 47 - 54. [Ch.4, 7, 9]

Cristescu, N. (1996b) Plasticity of porous and particulate materials. Nadai Award Lecture. *Transactions of the ASME, Journal of Engineering Materials and Technology.* **118**, 145-156. [Ch.1, 4, 5, 6]

Cristescu, N. (1996c) Stability of large underground caverns in rock salt. *Rock Mechanics. Tools and Technique. Proc. 2nd NARMS Symp.* Eds. M. Aubertin, F. Hassani, and H. Mitri. Balkema, Rotterdam, 101-108. [Ch.10]

Cristescu, N. (1996d) Evolutive damage in rock salt. *Mechanical Behavior of Salt. Proc.of the 4th Conf.* Eds. P. Habib, H.R. Hardy, B. Ladanyi, and M. Langer. Trans Tech Publ., Clausthal-Zellerfeld (in press). [Ch. 6, 10]

Cristescu, N. and Hunsche, U. (1996) A comprehensive constitutive equation for rock salt: determination and application. *The Mechanical Behavior of Salt. Proc.3rd Conf.* Eds. M. Ghoreychi, P. Berest, H.R. Hardy, Jr., and M. Langer. Trans Tech Publ. Clausthal-Zellerfeld. 191-206 . [Ch.4, 8, 10]

Cristescu, N. and Paraschiv, I. (1996) Creep and creep damage around large rectangular-like caverns. *Mechamics of Cohesive-Frictional Materials,* **1,** 165-197.[Ch.10]

Cristescu, N.D., O. Cazacu, and J. Jin (1997) Constitutive equation for compaction of ceramic powders. *IUTAM Symp. on Mechanics of Granular and Porous Materials.* Ed. N. Fleck, Kluwer, Dordrecht. (In press). [Ch.4]

Cristescu, N.D., and Cazacu, O. (1997) Time-effects on Uniaxial Compaction of Powders. *Proc. 6th Int. Symp. on Plasticity and its Current Applications, Alaska.* Neat Press, Fulton, Maryland, 303-304. [Ch.4]

Cuisiat, F.D.E., and Hudson, J.A. (1993) The influence of rock anisotropy on borehole breakouts: a microstatistical approach. *Rock Mechanics in the 1990s. The 34th US Symp. on Rock Mechanics.* Ed. B.C.Haimson, *Int. J. Rock Mech. Min. Sci. & Geomech. Abstr.* **30**, 1077-1083. [Ch.8]

Cunha, A.P. (1990) Scale effects in rock mechanics. *Scale Effects in Rock Masses.* Ed. A. Pinto da Cunha. Balkema, Rotterdam. 3-27. [Ch.7]

Dahou, A. (1994) Contribution à l'étude du comportement élasto-viscoplastique d'une craie poreuse. These de Doctorat, Université des Sciences et Technologie de Lille, Laboratoire de Mecanique de Lille, URA 1441 CNRS. [Ch.4, 5]

Dahou, A., Shao, J.F., and Bederiat, M. (1995) Experimental and numerical investigations on transient creep of porous chalk. *Mechanics of Materials.* **21**, 147-158. [Ch.4]

Dawson, P.R., and Munson, D.E. (1983) Numerical simulation of creep deformations around a room in a deep potash mine. *Int. J. Rock Mech. Min. Sci. & Geomech. Abstr.* **20**, 33-42. [Ch.10]

Deklotz, E.J., Brown, J.W., and Stemler, O.A. (1966) Anisotropy of schistose gneiss. *Proc. 1st Congr. Society on Rock Mechanics, Lisbon,* Vol. 1, 465-470. [Ch.5]

Del Greco, O., Ferrero, A.M., Oggeri, C., and Peila, D. (1993) Analysis of exploitation mine room instability in weak rocks. *Assessment and Prevention of Failure in Rock Engineering.* Eds. A.G. Paşamehmetoğlu, T. Kawamoto, B.N. Whittaker, and Ō. Aydan. Balkema, Rotterdam, 441-447. [Ch.10]

Demiris, C.A. (1985) Surface de rupture et de limite de dilatance du marbre de Thassos soumis à des efforts de compression multiaxiaux. *C.R. Acad. Sci. Paris.* **301**, Série II, No.17, 1225-1230. [Ch.4]

den Brok, S.W.J., and Spiers, C.J. (1991) Experimental evidence for water weakening of quartzite by microcracking plus solution-precipitation creep. *Int. J. Geological Society, London* **148**, 541-548. [Ch.3]

Desai, C.S., and Varadarajan, A. (1987). A constitutive model for quasi-static behavior of rock salt. *J. Geophys. Res.,* **92**, B11, 11445-11456. [Ch.4]

Desai, C.S., and Zhang, D. (1987) Viscoplastic model for geologic materials with generalized flow rule. *Int. J. Numer. Anal. Meth. Geomech.* **11**, 603-620. [Ch.5]

Desai, C.S. (1990) Modelling and testing: implementation of numerical models and their application in practice. *Numerical Methods and Constitutive Modelling in Geomechanics.* Eds. C.S. Desai and G. Gioda, Springer-erlag, New York, 1-168. [Ch.5, 6]

Desai, C.S., and Ma, Y. (1992) Modelling of joints and interfaces using the disturbed-dtate concept. *Int. J. Num. Analyt. Meth. Geomech.* **16,** 623-653. [Ch.5]

Desai, C.S., Samtani, N.C., and Vulliet, L. (1995) Constitutive modeling and analysis of creeping slopes. *J. Geotechnical Engineering* **121**, 43-56. [Ch.5]

Detournay, E. (1985) Solution approximative de la zone plastique autour d'une galerie souterraine soumise à un champ de contraine non hydrostatique. *C. Acad. Sci. Paris,* **301**, Série II, No.12, 857-860. [Ch.9]

Detournay, E. (1986) An approximate statical solution of the elastoplastic interface for the problem of Galin with a cohesive-frictional material. *Int. J. Solids Structures* **22**, 1435-1454. [Ch.8]

Detournay, E., and Fairhurst, C. (1987) Two-dimensional elasticoplastic analysis of a long, cylindrical cavity under non-hydrostatic loading. *Int.J. Rock Mech. Min.Sci. & Geomech. Abstr.* **24**, 4, 197-211. [Ch.9].

Detournay, E., and Cheng, A.H-D. (1988) Poroelastic response of a borehole in a non-hydrostatic stress field. *Int. J. Rock Mech. Min. Sci. & Geomech. Abstr.* **25**, 171-182. [Ch.8]

Detournay, E., and St. John, C.M. (1988) Design charts for a deep circular tunnel under nonuniform loading. *Rock Mechanics and Rock Engineering* **21**, 119-137. [Ch.9]

Di Benedetto, H., and Hameury, O. (1991) Constitutive law for granular materials: Description of the anisotropic and viscous effects. *Computer Methods and Advances in Geomechanics*, Eds. G. Beer, J.R. Booker, and J.P. Carter. Balkema, Rotterdam, 599-604. [Ch.4]

Donath, F.A. (1961) Experimental study of shear failure in anisotropic rocks. *Geological Society of America, Bull.* **72**, 985. [Ch.5]

Donath, F.A. (1964) Strength variation and deformational behavior in anisotropic rock. *State of Stress in the Earth's Crust.* Ed. W.R. Judd, Elsevier, Amsterdam, 281-297. [Ch.5]

Donath, F.A. (1972) Effects of cohesion and granularity on deformational behavior of anisotropic rock. *Studies in mineralogy and precambian geology.* Vol.135. Eds. B.R. Doc, and D.K. Smith,, *Geological Society of America, Boulder*, 95-128. [Ch.5]

Dreyer, W. (1972) *The Science of Rock Mechanics.* Part 1: *The Strength Properties of Rocks.* Trans Tech Publ. Clausthal-Zellerfeld. [Ch.1]

Drucker, D. (1988) Conventional and unconventional plastic response and representation. *Appl.Mech. Review* **41**, 151-162. [Ch.5]

Dudley, J.W. II, Myers, M.T, Shew, R.D., and Arasteh, M.M. (1994) Measuring compaction and compressibility in unconsolidated reservoir materials via time-scaling creep. *Rock Mechanics in Petroleum Engineering. EUROCK '94.* Balkema, Rotterdam, 45-54. [Ch.4]

Durup, G. and Xu, J. (1996) Comparative study of certain constitutive law used to describe the rheological deformation of salts. *The Mechanical Behavior of Salt. Proc. 3rd Conf.* Eds. M. Ghoreychi, P. Berest, H.R. Hardy, Jr., and M. Langer. Trans Tech Publ. Clausthal-Zellerfeld, 75-84. [Ch.4]

Dusseault,M.B.,and Fordham, C.J. (1993) Time-dependent behavior of Rocks. *Comprehensive Rock Engineering.* Ed.-in-ChiefJ.A.Hudson, Vol.3, *Rock Testing and Site Characterization.* Pergamon Press, Oxford 119-149. [Ch.4]

Eggeler, G., and Blum, W. (1981) Coarsening of the dislocation structure after stress reduction during creep of NaCl single crystals. *Philos. Mag.* **A44**, 1065-1084. [Ch.2]

Ewy, R.T., and Cook, N.G.W. (1990a) Deformation and fracture around cylindrical openings in rock. - I. Observations and analysis of deformations. *Int.J.Rock Mech. Min. Sci. & Geomech. Abstr .27*, 387-407. [Ch.9]

Ewy, R.T., and Cook, N.G.W. (1990b) Deformation and fracture around cylindrical openings in rock. - II. Initiation, growth and interaction of fractures. *Int.J.Rock Mech. Min. Sci. & Geomech. Abstr.* **27**, 409-427. [Ch.9]

Ewy, R.T. (1993) Yield and closure of directional and horizontal wells. *Rock Mechanics in the 1990s. The 34th US Symp. on Rock Mechanics.* Ed. B.C. Haimson, *Int. J. Rock Mech. Min. Sci. & Geomech. Abstr.* **30**, 1061-1065. [Ch.9]

Fakhimi, A.A., and Fairhurst, C. (1994) A model for the time-dependent behavior of rock. *Int. J. Rock Mech. Min. Sci. & Geomech. Abstr.* **31**, 117-126. [Ch.4]

Fan, C.-H., and Jones, M.E. (1994). The impact of tropical weathering on the mechanical behaviour of mudrocks. *Rock Mechanics Models and Measurements. Challenges from Industry.* Eds. P.P. Nelson and S.E. Laubach. Balkema, Rotterdam. 767-773. [Ch.4]

Fischer, R.F., Light, B.D., and Paslay, P.R. (1992). Salt-cavern closure during and after formation. *Proc. 7th Int. Symp. on Salt. 1992, Kyoto*, Elsevier, Amsterdam. [Ch.8]

Flavigny, E., and Nova, R. (1989) Viscous properties of geomaterials. *Geomaterials. Constitutive Equations and Modelling.* Ed. F. Darve, Elsevier Applied Science, London, 27-54. [Ch.4]

Fota, D. (1983) Private communications. [Ch.6]

Frayne, M.A., Rothenburg, L., and Dusseault, M.B. (1996) Four case studies in saltrock. *.The Mechanical Behavior of Salt. Proc. 3rd Conf.* Eds. M. Ghoreychi, P. Berest, H.R. Hardy, Jr., and M. Langer. Trans Tech Publ. Clausthal-Zellerfeld, 471-482. [Ch.4, 8]

Fritz, P. (1984) An analytical solution for axisymmetric tunnel problems in elasto-visco-plastic media. *Int. J. Numer. Analyt. Meth. Geomech.* **8**,325-342. [Ch.9]

Frost, H.J., and Ashby, M.F. (1982) *Deformation Mechanism Maps. The Plasticity and Creep of Metals and Ceramics.* Pergamon Press, Oxford. [Ch.2, 3]

Geisler, G. (1985) Entwiklung einer Präzisionsapparatur zur Messung des Kriechverhaltens von Steinsalzproblem. Pd.D. Dissertation, Universität Hannover. [Ch.1]

Georgi, F., Menzel, T., and Schreiner, W. (1975) Zum geomechanischer Verhalten von Steinsalz vershiedener Steinsalz lagerstätten der DDR Teil I: Das Festigkeitsverhalten. *Neue Bergbautechnik* **5**,(9), 669-676. [Ch.1]

Gerçek, H. (1993) Qualitative prediction of failures around non-circular openings. *Assessment and Prevention of Failure in Rock Engineering,* Eds. A. G. Paşamehmetoğlu, T. Kawamoto, B.N. Whittaker, and Ö. Aydan. Balkema, Rotterdam, 727-732. [Ch.10]

Germanovich, L.N., Roegiers, J.-C., and Dyskin, A.V. (1994) A model for borehole breakouts in brittle rocks. *Rock Mechanics in Petroleum Engineering. EUROCK '94 .* Balkema, Rotterdam, 361-370. [Ch.8]

Gessler, K. (1983) Vergleich der einaxialen Zugfestigkeit mit der Drei-Punkt-Biegezugfestigkeit und unterschiedlichen Spaltzugfestigkeiten. *Kali und Steinsalz.* **8**(12), 416-423. [Ch.2]

Gevantman, L.H. (1981) *Physical Properties Data for Rock Salt.* NBS Monograph 167, US Department of Commerce, National Bureau of Standards. [Ch.1]

Ghoreychi, M., Berest, P., Hardy, H.R.,Jr., and Langer, M. (Eds) (1996) *The Mechanical Behavior of Salt. Proc. 3rd Conference 1993*, Trans Tech Publ., Clausthal-Zellerfeld. [Ch.1]

Gioda, G., and Cividini, A. (1994) Finite element analysis of time dependent effects in tunnels. *Visco-Plastic Behaviour of Geomaterials,* Eds. N.D. Cristescu and G. Gioda, Springer-Verlag, Wien-New York, 209-244. [Ch.9]

Gioda, G., Sterpi, D., and Locatelli, L. (1994) Some examples of finite element analysis of tunnels. *Computer Methods and Advances in Geomechanics. Proc. 8th Conf. IACMAG,* Eds. H.J. Siriwardane and M.M. Zaman. Balkema, Rotterdam 165-176. [Ch.9]

Glabisch, U. (1996) Stoffmodell für Grenzzustände im Salzgestein zur Berechnung von Gebirgshohlräumen. Ph. D. Dissertation, Technical University Carolo-Wilhelmina, Braunschweig, Germany. [Ch.8]

Glover, P.W.J., Meredith, P.G., Sammonds, P.R., and Murrell, S.A.F. (1994) Measurements of complex electrical conductivity and fluid permeability in porous rocks at raised confining pressures. *Rock Mechanics in Petroleum Engineering, EUROK '94.* Balkema, Rotterdam, 29-36. [Ch.4]

Goldenblat, I.I. and Kopnov, V.A. (1966) Strength of glass-reinforced plastics in the complex stress state. *Polymer Mechanics.* **1**, 54. [Ch.5]

Greenspan, M. (1944) Effect of a small hole on the stresses in a uniformly loaded plate. *Quart. J. Appl. Math.* **2**, 60-71. [Ch.10]

Griggs, D.T., Turner, F.J., and Heard, H.C. (1960) Deformation of rocks at 500° to 800°C. *Rock Deformation. Geological Society of America.* Mem. 79, 39-104. [Ch.2].

Griggs, D.T., and Blacic, J.D. (1965) Quartz: Anomalous weakness of synthetic crystals. *Science.* **147**, 292-295. [Ch.3]

Griggs, D.T. (1967) Hydrolytic weakening of quartz and other silicates. *Geophys J,* **14**, 19-31. [Ch.3]

Guéguen, Y., and Palciauskas, V. (1994) *Introduction to the Physics of Rocks.* Princeton University Press, Princeton. [Ch.1]

Guenot, A. (1989) Borehole breakouts and stress fields. *Int. J. Rock Mech. Min. Sci.& Geomech.Abstr.,* **26**, 185-195. [Ch.8]

Haasen, P. (1974) *Physikalische Metallkunde.* Springer-Verlag. Berlin. [Ch.2, 3]

Habib, P., Hardy, H.R. Jr., Ladanyi, B., and Langer, M..(Eds). (1997) *The Mechanical Behavior of Salt. Proc. 4th Conf. 1996.* Trans Tech Publ., Clausthal-Zellerfeld. (In press) [Ch.1]

Haimson, B.C., Lee, M., and Herrick, C. (1993a) Recent advances in *in-situ* stress measurements by hydraulic fracturing and borehole breakouts. *Proc. 7th Int. Congr. on Rock Mechanics.* Balkema, Rotterdam, 1737-1742. [Ch.7, 8]

Haimson, B.C., Lee, M., Chandler, N., and Martin, D. (1993b) Estimating the state of stress from subhorizontal hydraulic fractures at the underground research laboratory, Manitoba. *Rock Mechanics in the 1990s, The 34th US Symp. on Rock Mechanics.* Ed. B.C. Haimson. *Int. J. Rock Mech. Min. Sci.& Geomech. Abstr.* **30**, 959-964. [Ch.7]

Haimson, B.C., and Song, I. (1993) Laboratory study of borehole breakouts in Cordova Cream: a case of shear failure mechanism. *Rock Mechanics in the 1990s, The 34th US Symp. on Rock Mechanics.* Ed. B.C.Haimson, *Int. J. Rock Mech. Min. Sci.& Geomech. Abstr.* **30**, 1047-1056. [Ch.8]

Haimson, B.C., and Song, I. (1995) A new borehole failure criterion for estimating in situ stress from breakout span. *Proc. Int. Congr. on Rock Mechanics, Tokyo,* Balkema, Rotterdam, 341-345. [Ch.8]

Haimson, B.C., Lee, M.Y., and Chandler, N. (1996) Estimating the state of stresss at the underground research laboratory from hydraulic fracturing tests in three test holes of different plunges. *Rock Mechanics. Tools and Techniques. Proc. 2nd NARMS.* Eds. M Aubertin, F. Hassani, and H. Mitri. Balkema, Rotterdam, 913-920. [Ch.7]

Hamami, M., Tijani, S.M., and Vouille, G. (1996) A methodology for the identification of rock salt behavior using multi-steps creep tests. *The Mechanical Behavior of Salt. Proc. 3rd Conf.* Eds. M. Ghoreychi, P. Berest, H.R. Hardy, Jr., and M. Langer. Trans Tech Publ. Clausthal-Zellerfeld, 53-66. [Ch.2, 4]

Hambley, D.F., and Fordham, C.J. (1989) General failure criteria for saltrock. *Rock Mechanics as a Guide for Efficient Utilization of Natural Resources,* Ed. Khair, Balkema, Rotterdam, 91-98. [Ch.5]

Hampel, A., Hunsche, U., Weidinger, P., and Blum, W. (1996) Description of the creep of rock salt with the composite model - II. Steady-state creep. *Mechanical Behavior of Salt. Proc. 4th Conf.* Eds. P. Habib, H.R. Hardy, B. Ladanyi, and M. Langer. Trans Tech Publ. Clausthal-Zellerfeld (in press). [Ch.2, 3, 5]

Hampel, A. (1997) Private Communication. [Ch.3]

Hansen, F.D., and Mellegard, K.D. (1980) Creep of 50-mm diameter specimens of some salt from Avery Island, Louisiana. RE/SPEC ONWI-104. [Ch.5]

Hansen, F.D., Mellegard, K.D., and Senseny, P.E. (1984) Elasticity and strength of ten natural rock salt. *The Mechanical Behavior of Salt. Proc. 1st Conf.* Eds. H.R.Hardy, Jr., and M. Langer, Trans Tech Publ., Clausthal-Zellerfeld, 71-83. [Ch.4]

Hansen, F.D., Callahan, G.D., and Van Sambeek, L.L. (1996) Reconsolidation of salt as applied to permanent seals for the Waste Isolation Pilot Plant. *The Mechanical Behavior of Salt. Proc. 3ird Conf.* Eds. M. Ghoreychi, P. Berest, H.R. Hardy, Jr., and M. Langer. Trans Tech Publ. Clausthal-Zellerfeld, 323-336. [Ch. 2, 4]

Hardy, H.R., Jr. (1972) Application of acoustic emission technique to rock mechanics research. *Acoustic Emission.* Eds. R.G. Liptai, D.O. Harris, and C.A. Tatro. ASTM STP 505, Amer. Soc. for Testing and Materials, Philadelphia. 41-83. [Ch.6]

Hardy, H.R., Jr., and Langer, M.(Eds) (1984) *The Mechanical Behavior of Salt. Proc. 1st Conf. 1981.* Trans Tech Publ. Clausthal-Zellerfeld. [Ch.1]

Hardy, H.R., Jr., and Langer, M. (Eds) (1988) *The Mechanical Behavior of Salt. Proc. 2nd Conf. 1984.* Trans Tech Publ. Clausthal-Zellerfeld. [Ch.1]

Hardy, H.R., Jr., and Leighton, F.W. (Eds) (1977) *Proc. 1st Conf. on Acoustic Emission/Microseismic Activity in Geologic Structures and Materials.* Trans Tech Publ., Clausthal-Zellerfeld. [Ch.6]

Hardy, H.R., Jr., and Leighton, F.W. (Eds) (1980) *Proc. 2nd Conf. on Acoustic Emission/ Microseismic Activity in Geologic Structures and Materials.* Trans Tech Publ.,Clausthal-Zellerfeld. [Ch.6]

Hardy, H.R., Jr., and Leighton, F.W. (Eds.) (1984) *Proc. 3rd Conf. on Acoustic Emission/ Microseismic Activity in Geologic Structures and Materials.* Trans Tech Publ., Clausthal-Zellerfeld. [Ch.6]

Hardy, H.R., Jr., and Leighton, F.W. (Eds) (1989) *Proc. 4th Conf. on Acoustic Emission/ Microseismic Activity in Geologic Structures and Materials.* Trans Tech Publ., Clausthal-Zellerfeld. [Ch.6]

Hardy, H.R., Jr.(Editor) (1995) *Proc. 5th Conf. on Acoustic emission/microseismic activity in geologic structures and materials. 1991,* Trans Tech Publ., Clausthal-Zellerfeld. [Ch.1]

Hardy, H.R., Jr. (1996) Application of the Kaiser effect for the evaluation of in situ stresses in salt. *The Mechanical Behavior of Salt. Proc. 3rd Conf.* Eds. M. Ghoreychi, P. Berest, H.R. Hardy, Jr., and M. Langer. Trans Tech Publ. Clausthal-Zellerfeld,323-336. [Ch.7]

Haupt, M. (1988) Entwicklung eines Stoffgesetzes für Steinsalz auf der Basis von Kriech- und Relaxationsversuchen. *Veroffentlichungen des Inst. für Bodenmech. und Felsmech. der Univ. Karlsruhe* Eds. G. Gudehus and O. Natau, Vol. **110**, 138 p. [Ch.2]

Haupt, M. (1991) A constitutive law for rock salt based on creep and relaxation tests. *Rock Mechanics and Rock Engineering.* **24**, 179-206. [Ch.4]

Heard, H.C. (1960) Transition from brittle to ductile flow in Solenhofen limestone as a function of temperature, confining pressure and interstial fluid pressure. *Rock Deformation,* Eds. D. Griggs and J. Handin.. *Geological Society of America.* Memoir 79, 193-226. [Ch.2]

Heard, C.H. (1972) Steady-state flow in polycrystalline halite at pressure of 2 kilobars. *Geophysical Monograph Series,* Vol. *16, Flow and Fracture of Rocks.* Eds. H.C. Heard, I.Y.Borg, N.L. Carter, and C.B. Raleigh. AGU, Washington, 191-209. [Ch.3]

Herget, G. (1987) Stress assumptions for underground excavations in the Canadian Shield. *Int. J. Rock Mech. Min. Sci. & Geomech. Abstr.* **24**, 1, 95-97. [Ch.7]

Herget, G. (1993) Rock stresses and rock stress monitoring in Canada. *Comprehensive Rock Engineering.* Ed-in-Chief J. A. Hudson., vol.3, *Rock Testing and Site Characterization* Pergamon Press , Oxford. 473-497. [Ch.7]

Herrick, C.G., and Haimson, B.C. (1994) Modeling of episodic failure leading to borehole breakouts in Alabama limestone. *Rock Mechanics. Models and Measurements. Challenges from Industry. Proc. 1st NARMS Symp.*, Eds. P.P. Nelson, and S.E. Laubach. Balkema, Rotterdam, 217-224. [Ch.9]

Herrmann, W., Wawersik, W.R., and Lauson, H.S. (1980) Creep curves and fitting parameters for southeastern New Mexico bedded salt. Sandia Rep. SAND-80-0087 [Ch.5]

Herrmann, W., and Lauson, H.S. (1981) Analysis of creep data for various natural rock salts. Sandia Rep. SAND81-2567. [Ch.5]

Hertzberg, R.W. (1983) *Deformation and Fracture Mechanics of Engineering Materials.* John Wiley and Sons, New York. [Ch.2]

Hesler, G.J., Cook, Neville G.W., and Myer, L. (1996) Estimation of intrinsic and effective elastic properties of cracked media from seismic testing. *Rock Mechanics. Tools and Techniques. Proc. 2nd NARMS.* Eds. M. Aubertin, F. Hassani, and H. Mitri. Balkema, Rotterdam, 467-473. [Ch.6]

Hettler, A., Gudehus, G., and Vardoulakis, I. (1984) Stress-strain behaviour of sand in triaxial tests. *Results of Int. Workshop on Constitutive Relations for Soils. 1982, Grenoble.* Eds. G.Gudehus, F. Darve, and I. Vardoulakis,, Balkema, Rotterdam, 55-66. [Ch.4, 5]

Heusermann, S. (1996) Measurements of initial rock stress at the Asse Salt Mine. *The Mechanical Behavior of Salt. Proc. 3rd Conf.* Eds. M. Ghoreychi, P.Berest, H.R. Hardy, Jr., and M. Langer. Trans Tech Publ.Clausthal-Zellerfeld, 101-116. [Ch.7]

Hill, R. (1948) A theory of the yielding and plastic flow of anisotropic metals. *Proc. Royal Society of London.* **A 193**, 282-287. [Ch.5]

Hoek, E. (1964) Fracture of anisotropic rock. *J. S. Afr. Inst. Min. Metall.* **64**, 10, 501-518. [Ch.5]

Hoek, E., and Moy, D. (1993) Design of large powerhouse caverns in weak rock. *Comprehensive Rock Engineering.* Ed-in-Chief J. A. Hudson. vol.5, *Surface and Undergound Project Case Histories*, Pergamon Press, Oxford. 85-110. [Ch.10]

Hoffers, B., Engeser, B., and Rischmüller, H. (1994) Wellbore stability of a superdeep borehole in crystalline rock. *Rock Mechanics in Petroleum Engineering. EUROCK '94.* Balkema, Rotterdam, 371-378. [Ch.8]

Hofmann, U. and Blum, W. (1993) Modelling transient creep of pure materials after sudden changes in stress or temperature. *Proc. 7th JIM Int. Symp. on Aspects of High Temperature Deformation and Fracture in Crystalline Materials,* Eds. Y. Hosoi, H. Yoshinaga, H. Oikawa, and K. Maruyama. The Japan Inst. of Metals. 649-656. [Ch.2 , 3]

Holcomb, D.J. (1981) Memory, relaxation, and microfracturing in dilatant rock. *J. Geophys. Res.* **86**, B7, 6235-6248. [Ch.4]

Holcomb, D.J. (1991) Evolution of damage surfaces and the plastic potential in a limestone. *Proc. Spring Conf. on Experimental Mechanics.* The Society for Experimental Mechanics, Inc. [Ch.6]

Holcomb, D.J. (1993) Observations of the Kaiser effect under multiaxial stress states: implications for its use in determining *in-situ* stress. *Geophys. Res. Letters,* **20**, 19, 2119-2122. [Ch.6]

Honecker, A. and Wulf, A. (1988) The FE-system ansalt a dedicated tool for rock and salt-mechanics. *The Mechanical Behavior of Salt. Proc. 2nd Conf.* Eds. H.R. Hardy Jr., and M. Langer. Trans Tech Publ., Clausthal-Zellerfeld, 509-520. [Ch.10]

Horseman, S.T. (1988) Moisture content - a major uncertainty in storage cavity closure prediction. *The Mechanical Behavior of Salt . Proc. 2nd Conf.* Eds. H.R. Hardy Jr., and M. Langer. Trans Tech Publ., Clausthal-Zellerfeld, 53-68. [Ch.3]

Horseman, S.T., Russell, J.E., Handin, J. and Carter, N.L. (1993) Slow experimental deformation of Avery Island salt. *Proc. 7th Symp.on Salt, Kyoto.* Eds. H. Kakihana, H.R. Hardy Jr., T. Hoshi and T. Toyokura. Vol.1, Elsevier, Amsterdam, 67-74. [Ch.2, 3]

Horsrud, P., Holt, R.M., Sonstebo, F., Svano,G., and Bostrom, B. (1994). Time-dependent borehole stability: Laboratory studies and numerical simulation of different mechanisms in shale. *Rock Mechanics in Petroleum Engineering. EUROCK '94.* Balkema, Rotterdam, 259-266. [Ch.8]

Hudson, J.A., (Ed-in-Chief) (1993a) *Comprehensive Rock Engineering,* Pergamon Press.[Ch.4]

Hudson, J.A. (1993b) Rock Properties, Testing Methods and Site Characterization. *Comprehensive Rock Engineering.* Ed-in-Chief J. A. Hudson., vol.3, *Rock Testing and Site Characterization* Pergamon Press, Oxford. 1-39. [Ch.4, 7]

Huergo, P.J. and Nakhle, A. (1988) L'influence de l'eau sur la stabilité d'un tunnel profond. *Tunnels and Water.* Ed. J.M. Serrano, Balkema, Rotterdam, 177-184. [Ch.9]

Hunsche, U. (1978) Modellrechnungen zur Entstehung von Salzstockfamilien. *Geol. Jahrb. E 12, Hannover,* 53-107. [Ch..2, 3]

Hunsche, U. (1984a) Fracture experiments on cubic rock salt samples. *The Mechanical Behavior of Salt. Proc. 1st Conf.* Eds. H.R.Hardy, Jr., and M. Langer. Trans Tech Publ., Clausthal-Zellerfeld, 169-179. [Ch.1]

Hunsche, U. (1984b) Results and interpretation of creep experiments on rock salt. *The Mechanical Behavior of Salt. Proc. 1st Conf.* Eds. H.R.Hardy, Jr., and M. Langer, Trans Tech Publ., Clausthal-Zellerfeld, 159-167. [Ch.1]

Hunsche, U., and Plischke, I. (1985) In situ creep experiments under controlled stress in rock salt pillars - design, instrumentation and evaluation. *Design and Instrumentation of In Situ Experiments in Underground Laboratories for Radioactive Waste Disposal. Proc. Joint CEC-NEA Workshop, Brussels 1984.* Eds. B. Come, P. Johnston and A. Müller. Balkema, Rotterdam, 417-424. [Ch.1]

Hunsche, U., Plischke, I., Nipp. H.-K., and Albrecht, H. (1985) An in situ creep experiment using a large rock salt pillar. *Proc. 6th Int. Symp. on Salt, Toronto,* Eds. B.C. Schreiber and H.L. Harner. The Salt Institute, Alexandria, USA, Vol.1, 437-454. [Ch.1]

Hunsche, U. (1988) Measurements of creep in rock salt at small strain rates. *The Mechanical Behavior of Salt. Proc. 2nd Conf.* Eds. H.R.Hardy, Jr., and M. Langer, Trans Tech Publ., Clausthal-Zellerfeld, 187-196. [Ch.1, 2]

Hunsche, U. (1989) A failure criterion for natural polycristalline rock salt. *Advances in Constitutive Laws for Engineering Materials. Proc. ICCLEM,* Eds. Fan Jinghong , and S. Murakami. International Academic Publ., Vol.2, 1043-1046. [Ch.1]

Hunsche, U. (1990) On the fracture behavior of rock salt. *Constitutive Laws of Plastic Deformation and Fracture,* Eds. A.S. Krausz, J.I. Dickenson, J.-P.A. Immarigeon, and W. Wallace. Kluwer, Dordrecht. 155-163. [Ch.1, 4, 5]

Hunsche, U., and Albrecht, H. (1990). Results of true triaxial strength tests on rock salt. *Engineering Fracture Mechanics.* 35, 4/5, 867-877. [Ch.1, 2]

Hunsche, U. (1991) Volume change and energy dissipation in rock salt during triaxial failure tests. *Mechanics of Creep Brittle Materials 2, Sept. 1991, Leicester*, Eds. A.C.F. Cocks, and A.R.S. Ponter. Elsevier, London, 172-182. [Ch.1, 2, 4, .5]

Hunsche, U. (1992) True Triaxial Failure Tests on Cubic Rock Salt Samples - Experimental Methods and Results. *Finite Inelastic Deformations - Theory and Applications.IUTAM Symp.* Eds. D. Besdo and E. Stein. Springer-Verlag, Berlin-Heidelberg. 525-536. [Ch.1, 2, 4, 5]

Hunsche, U. (1993a) Failure Behaviour of Rock Salt Around Underground Cavities. *Proc. 7th Symp. on Salt, Kyoto.* Eds. H. Kakihana, H.R. Hardy Jr., T. Hoshi and T. Toyokura. Elsevier, Amsterdam, Vol. 1, 59-65. [Ch.1, 2, 4, 5]

Hunsche, U. (1993b) Strength of rock salt at low mean stresses. *Proc. 7th Int. Congr. on Rock Mechanics,* Balkema, Rotterdam, 160-163. [Ch.2, 5]

Hunsche, U. (1994a) Uniaxial and triaxial creep and failure tests on rock: Experimental technique and interpretation. *Visco-Plastic Behaviour of Geomaterials,* Eds. N.D. Cristescu, and G.Gioda, Springer-Verlag, Wien-New York, 1-53. [Ch.1, 2, 5]

Hunsche, U. (1994b) Ein- und dreiaxiale Kriechversuche an Gesteinsproben. Empfehlung Nr.16 des Arbeitskreises 19 - Versuchstechnik Fels - der DGGT. *Bautechnik,* 71, 500-505. [Ch.1, 2]

Hunsche, U. and Schulze, O. (1994) Das Kriechverhalten von Steinsalz. *Kali und Steinsalz,* 11, 238-255. [Ch.2, 3, 4]

Hunsche, U., Schulze, O., and Langer, M. (1994) Creep and failure behavior of rock salt around underground cavities. *The Mining Industry on the Threshhold of XXI Century. Proc. 16th World Mining Congress (WMC), Sofia.* Vol.5, 217-230. [Ch.1, 2]

Hunsche, U. (1995) Salzmechanische Laboruntersuchungen an Steinsalz aus den Schächten in Gorleben. *Proc. Symp. Gefrierschachte Gorleben, Bochum, Germany.* Ed. H.L. Jessberger. Balkema, Rotterdam. 251-262. [Ch.1, 2]

Hunsche, U. (1996) Determination of the dilatancy boundary and damage up to failure for four types of rock salt at different stress geometries. *The Mechanical Behavior of Salt Proc. 4th Conf.* Eds. P. Habib, H.R. Hardy, B. Ladanyi, and M. Langer. Trans Tech Publ., Clausthal-Zellerfeld (In press) [Ch.1, 2]

Hunsche, U., Mingerzahn, G., and Schulze, O. (1996) The influence of textural parameters and mineralogical composition on the creep behavior of rock salt. *The Mechanical Behavior of Salt. Proc. 3rd Conf.* Eds. M. Ghoreychi, P. Berest, H. Hardy, Jr. and M. Langer. Trans Tech Publications, Clausthal-Zellerfeld, 143-151. [Ch.2, 3]

Hunsche, U. and Schulze, O. (1996) Effect of humidity and confining pressure on creep of rock salt. *The Mechanical Behavior of Salt. Proc. 3rd Conf..* Eds. M. Ghoreychi, P. Berest, H.R. Hardy, Jr., and M. Langer. Trans Tech Publ. Clausthal-Zellerfeld, 237-248. [Ch.2, 3, 4]

Hyett, A. and Hudson, J.A. (1989) In situ stress for underground excavation design in a naturally fractured rock mass. *Rock Mechanics as a Guide for Efficient Utilization of Natural Resources. Proc. 30th US Symp. Rock Mechanics.* Balkema, Rotterdam, 293-300. [Ch.7]

Ilschner, B. (1973) *Hochtemperatur-Plastizitat.* Springer-Verlag, Berlin. [Ch.2, 3]

Indraratna, B. (1993) Effects of bolts on failure modes near tunnel openings in soft rock. *Geotechnique.* 43, 3, 433-442. [Ch.9]

Ionescu, I.R., and Sofonea, M. (1993) *Functional and Numerical Methods in Viscoplasticity.* Oxford University Press, Oxford-New York-Tokyo. [Ch.10]

Jaeger, J.C. (1960) Shear failure of anisotropic rocks. *Geol. Mag.* **97**, 65-72. [Ch.5]

Jaeger, J.C. and Cook, N.G.W. (1979) *Fundamentals of Rock Mechanics*, Chapman and Hall, London   [ch.1, 2, 10]

Jaoul, O. (1984) Sodium weakening of Heavitree quartzite: preliminary results. *J. Geophys. Res.* **89**, 4271-4280. [Ch.2]

Jaoul, O., Tullis, J., and Kronenberg, A. (1984) The effect of varying water content on the creep behaviour of Heavitree quartzite. *J. Geophys. Res.* **89**, 4298-4312. [Ch. 3]

Jeffery, R.I. and North, M.D. (1993) Review of recent hydrofracture stress measurements made in the carboniferous coal measures of England. *Proc. 7th Int. Congr. on Rock Mechanics.* Balkema, Rotterdam. 1699-1703. [Ch.7]

Jeremic, M.L. (1994) *Rock Mechanics in Salt Mining.* Balkem, Rotterdam. [Ch.4]

Jiayou, Lu (1987) The elastoplastic analysis of rock masses surrounding circular opening considering strain hardening. *Large Rock Caverns*, Ed. K.H.O. Saari. Pergamon Press, Oxford, Vol.2, 1329-1335. [Ch.9]

Jin, J., Cazacu, O. and Cristescu, N.D. (1996a) An elastic/viscoplastic consolidation model for alumina powder. *Proc. 5th World Congress of Chemical Engineering, July 14-18, 1996, San Diego*, Vol.5, pp.573-576. [Ch. 4]

Jin, J. and Cristescu, N. (1996). Experimental study on the consolidation of alumina powder. *Preprint ASME Mechanics and Materials Conf. The Johns Hopkins University*. 354. [Ch.4, .5]

Jin, J., Cristescu, N.D., and Hunsche, U. (1996b) A new elastic/viscoplastic model for rock salt. *The Mechanical Behavior of Salt. Proc.4th Conf.* Eds. P. Habib, H.R. Hardy, B. Ladanyi, and M. Langer. Trans Tech Publ., Clausthal-Zellerfeld (In press)   [Ch.5]

Jin, J., Cazacu, O. and Cristescu, N. (1997). Compaction of Ceramic Powders, *Applied Mechanics in the Americas.* Vol.4. *Mechanics and Dynamics of Solids.* Eds. L.A. Godoy, M. Rysz, and L.E. Suarez. The University of Iowa, 212-215. [Ch.4]

Jin, J. and Cristescu, N.D. (1997) An Elastic/Viscoplastic Model for Transient Creep of Rock Salt. *Int.J. Plasticity* (in press). [Ch.5, 8]

Johnston, J.E., Christensen, N.I. (1993) Compressional to Shear Velocity Ratios in Sedimentary Rocks. *Rock Mechanics in the 1990s, The 34th US Symp. on Rock Mechanics.* Ed. B.C.Haimson, *Int. J. Rock Mech. Min. Sci.& Geomech. Abstr.* **30**, 751-754. [Ch.6]

Julien, M.P., Foerch, R., Aubertin, M., and Cailletaud, G. (1996) Some aspects of the numerical implementation of SUVIC-D. *The Mechanical Behavior of Salt. Proc.4th Conf.* Eds. P. Habib, H.R. Hardy, B. Ladanyi, and M. Langer. Trans Tech Publ.,Clausthal-Zellerfeld (In press)   [Ch.1, 2]

Jumikis, A.R. (1983) *Rock Mechanics.* Trans Tech Publ. Clausthal-Zellerfeld. [Ch.1]

Kaar, D.G., Law, F.P., Cox, G.F.N., and Hoo Fatt, M. (1989) Asymptotic and quadratic failure criteria for anisotropic materials. *Int. J. Plasticity.* **7**, 303-336. [Ch.1, 5]

Kaiser, P.K., and Morgenstern, N.R. (1981a) Time-dependent deformation of small tunnels -I. Experimental facilities. *Int. J. Rock Mech. Min. Sci. & Geomech. Abstr.* **18**, 129-140. [Ch.9]

Kaiser, P.K., and Morgenstern, N.R. (1981b) Time-dependent deformation of small tunnels -II. Typical test data. *Int. J. Rock Mech. Min. Sci. & Geomech. Abstr.* **18**, 141-152.[Ch.9]

Kaiser, P.K., and Morgenstern, N.R. (1982) Time-dependent Deformation of Small Tunnels -III. Pre-failure Behaviour. *Int. J. Rock Mech. Min. Sci. & Geomech. Abstr.* **19**, 307-324. [Ch.9]

Kaiser, P.K., Maloney, S., and Morgenstern, N.R. (1983) Time-dependent Behaviour of Tunnels in Highly Stressed Rock. *Reprint 5th Int. Congr. on Rock Mechanics*, D329-D336. [Ch.9]

Kaiser, P.K., Guenot, A., and Morgenstern, N.R. (1985) Deformation of small tunnels - IV- Behaviour during failure. *Int. J. Rock Mech. Min. Sci. & Geomech. Abstr.*, **22**, 3, 141-152. [Ch.6]

Kaiser, P.K. (1993) Deformation monitoring for stability assessment of underground openings. *Comprehensive Rock Engineering*. Ed-in-Chief J. A. Hudson. Vol.4, *Excavation, Support and Monitoring*, Pergamon Press, Oxford. 607-629. [Ch.9]

Kanagawa, T., Hibino, S., Ishida, T., Hayashi, M., and Kitahara, Y. (1986). *In Situ* stress measurements in the Japanese Islands: Over-coring results from a multi-element gauge used at 23 sites. *Int. J. Rock Mech. Min. Sci. & Geomech. Abstr.* **23**, 29-39. [Ch.7]

Karato, S.I., and Toriumi, M. (Eds.) (1989) *Rheology of Solids and of the Earth*. Oxford Science Publications, Oxford. [Ch.3]

Kenter, C.J., Doig, C.J., Rogoar, H.P., Fokker, P.A., and Davies, D.R. (1990) Diffusion of brine through rock salt roof of caverns. *SMRI, Fall meeting, Oct.1990, Paris*. [Ch.1-3]

Kern, H., and Schmidt, R. (1990) Physical properties of KTB core samples at simulated *In-Situ* conditions. *Scientific Drilling*, **1**, 217-223. [Ch.7]

Kern, H., Schmidt, R., and Popp, T. (1991) The velocity and density structure of the 4000m crustal segment at the KTB drilling site and their relationship to lithiological and microstructural characteristics of the rocks: an experimental approach. *Scientific Drilling*, **2**, 130-145. [Ch.7]

Kern, H., Popp, T. and Schmidt, R. (1994) The effect of a deviatoric stress on physical rock properties - An experimental study simulating the *in-situ* stress field at KTB drilling site. *Surveys in Geophysics*. Kluwer, Dordrecht. Vol.15, 467-479. [Ch.6]

Kern, H., and Popp, T. (1997) Private communication. [Ch.2]

Kessels, W. (1989) Observation and interpretation of time-dependent behaviour of borehole stability in the Continental Deep Drilling pilot borehole. *Scientific Drilling*, **1**, 127-134. [Ch.8]

Kie, T.T. (1993) The importance of creep and time-dependent dilatancy, as revealed from case records in China. *Comprehensive Rock Engineering*. Ed.-in-Chief J.A. Hudson. Vol.3, *Rock Testing and Site Characterization*, Pergamon Press, Oxford, 709-744. [Ch.4,.9]

Kikuchi, S., Oka, K., and Tokunaga, K. (1993) Study of stability evaluation of the underground openings in the abandoned "Ohya Stone" quarries. *Assessment and Prevention of Failure in Rock Engineering*. Eds. A. G. Paşamehmetoğlu, T. Kawamoto, B.N. Whittaker,and Ö. Aydan. Balkema, Rotterdam. 457-463. [Ch.10]

King, M.S. (1966) Wave velocities in rocks as function of changes in overburden pressure and pore fluid saturants. *Geophysics* **31**, 1, 50-73. [Ch.4]

King, M.S. (1970) Static and dynamic elastic moduli of rocks under pressure. *Rock Mechanics - Theory and Practice. Proc. 11th Symp. on Rock Mechanics*. Ed. W.H. Somerton. Amer. Inst. Mining Metallurgical and Petroleum Engrs, New York, 329-351. [Ch.4, 6]

King, M.S., and Paulson, B.N.P. (1981) Acoustic velocities in heated block of granite subjected to uniaxial stress. *Geophys. Res. Lett.* **8**, 7, 699-702. [Ch.6]

Klee, G. and Rummel, F. (1993) Hydrofrac Stress Data for the European HDR Research Project Test Site Soultz-Sous-Forets. *Rock Mechanics in the 1990s, The 34th US Symp. on Rock Mechanics,* Ed. B.C. Haimson. *Int. J. Rock Mech. Min. Sci. & Geomech. Abstr.* **30**, 973-976. [Ch.7]

Klepaczko, J.R., Gary, G., and Barberis, P. (1991) Behaviour of rock salt in uniaxial compression at medium and high strain rates. *Arch. Mech.* **43**, 499-517. [Ch.2]

Knipe, R.J., and Rutter, E.H. (Eds) (1990) *Deformation Mechanisms. Rheology and Tectonics.* Geological Society Special Publication No.54. [Ch.3]

Kolsky, H. (1963) *Stress Waves in Solids.* Dover, New York. [Ch.6]

Korthaus, E. (1996) Measurement of crushed salt consolidation under hydrostatic and deviatoric stress conditions. *The Mechanical Behavior of Salt. Proc. 3rd Conf..* Eds. M. Ghoreychi, P. Berest, H.R. Hardy, Jr., and M. Langer. Trans TechPubl.,Clausthal-Zellerfeld 311-322. [Ch.2, 4]

Kovari, K., and Amstad, C. (1993) Decision making in tunneling based on field measurements. *Comprehensive Rock Engineering.* Ed-in-Chief J. A. Hudson. vol.4, *Excavation, Support and Monitoring,* Pergamon Press, Oxford, 571-606. [Ch.9]

Kranz, R.L., and Scholz, C.H. (1977) Critical dilatant volume of rocks at the onset of tertiary creep. *J. Geophys. Res.* **82,** 4893-4898. [Ch.6]

Kranz, R.L. (1979) Crack growth and development during creep of Barre Granite. *Int. J. Rock Mech. Min. Sci. & Geomech. Abstr.* **16**, 23-35. [Ch.6]

Kranz, R.L. (1980) The effects of confining pressure and stress difference on static fatigue of granite. *J. Geophys. Res.* **85**. B4, 1854-1866. [Ch.6]

Kranz, R.L., Harris, W.J., and Carter, N.L. (1982) Static fatigue of granite at 200°C. *Geophys. Res. Lett.* **9**, 1, 1-4. [Ch.6]

Kranz, R.L., Coughlin, J.P., and Billington, Selena. (1994). Studies of stope-scale seismicity in a hard-rock mine. Report Investigation 9525, Bureau of Mines. [Ch.6]

Kranz, R.L., and Estey, L.H. (1996) Listening to a mine relax for over a year at 10 to 1000 meter scale. *Rock Mechanics. Tools and Techniques. Proc. 2nd NARMS,* Eds. M. Aubertin, F. Hassani, and H. Mitri. Balkema, Rotterdam, 491498. [Ch.6]

Kurita, K., Swanson, P.L., Getting, I.C., and Spetzler, H. (1983) Surface deformation of Westerly granite during creep. *Geophys. Res. Letters* **10**, 1, 75-78. [Ch.6]

Kwasniewski, M.A. (1993) Mechanical behavior of anisotropic rocks. *Comprehensive Rock Engineering,* Ed-in-Chief J. A. Hudson, Vol.1, *Fundamentals,*Pergamon Press, Oxford, 285-312. [Ch.5]

Ladaniy, B., and Gill, D.E. (1983) In-situ determination of creep properties of rock salt. *Reprint 5th Int. Congr. on Rock Mechanics,* A219-A225. [Ch.4]

Ladanyi, B. (1993) Time-dependent response of rock around tunnels. *Comprehensive Rock Engineering.* Ed-in-Chief J.A. Hudson, Vol.2, *Analysis and Design Methods.* Pergamon Press, Oxford, 77-112. [Ch.4]

Lade, P.V., and Kim, M.K. (1988) Single hardening constitutive model for frictional Materials. II. Yield Criterion and Plastic Work Contours. *Computers and Geotechnique.* **6**, 13-29. [Ch.5]

Lade, P.V., and Pradel, D. (1990) Instability and plastic flow of soil. I. Experimental observations. *J. Engng Mech.* **116**, 2532-2550. [Ch.5]

Lade, P.V. (1993) Rock strength criteria: the theories and the evidence. *Comprehensive Rock Engineering.* Ed.-in-Chief J.A. Hudson, Vol.1, *Fundamentals.* Pergamon Press, Oxford, 255-284. [Ch.4].

Lade, P.V., Bopp, P.A., and Peters, J.P. (1993) Instability of dilating sand. *Mechanics of Materials,* **16,** 249-264. [Ch.4]

Lade, P.V., and Yamamuro, J.A. (1993) Stability of granular materials in postpeak softening regime. *J. Engng Mech.* **119,** 128-144. [Ch.4]

Lade, P.V. (1994) Instability and liquefaction of granular materials. *Computers and Geotechnics* **16,** 123-151. [Ch.5]

Lajtai, E.Z., and Schmidtke, R.H. (1986) Delayed failure in rock loaded in uniaxial compression. *Rock Mechanics and Rock Engineering* **19,** 11-25. [Ch.6]

Lajtai, E.Z., and Dzik, E.J. (1996) Searching for the damage threshold in intact rock. *Rock Mechanics. Tools and Techniques. Proc. 2nd NARMS,* Eds. M. Aubertin, F. Hassani, and H. Mitri. Balkema, Rotterdam, 701-708. [Ch.6]

Lama, R.D., and Vutukuri, V.S. (1978) *Handbook on Mechanical Properties of Rocks.* Vol. II and III, Trans Tech Publ., Clausthal-Zellerfeld. [Ch.1, 4]

Landolt-Börnstein, (1982) *Physikalische Eigenschaften der Gesteine.* - Gruppe 5, Band 1. Teilband b. Ed. G. Angenheister. Springer-Verlag, Heidelberg. [Ch.1]

Langer, M. (1969) Rheologie der Gesteine. *Zeitschrift deutsche Geol. Ges,* **119,** 313-425.[Ch.1]

Langer, M. (1979) Rheological behaviour of rock masses. (General report). *Proc. 4th Int. Congr. on Rock Mechanics, Montreux,* Balkema, Rotterdam, Vol.3, 29-62. [Ch.1]

Langer, M. (1986) Rheology of rock-salt and its application for radioactive waste disposal purpose. *Proc. Int. Symp. on Engineering in Complex Rock Formation, Peking,* 1-19. [Ch.1]

Langer, M. (1988) Das rheologische von Gebirgskörpern und dessen Bedeutung für den Felshohlraumbau. *Felsbau* **6,** 203-212. (Ch.4)

Langer, M. (1995) Engineering geology and waste disposal. Scientific report and recommendation (IAEG Comm. Nr.14). *IAEG Bull.* No. 51, Paris, 5-29. [Ch.1]

Langer, M. (1996) Underground disposal of wastes requiring special monitoring in salt rock masses. Recommendations of the "salt mechanics" working group of the German Soc. For Soil Mech. and Foundation Engin. (DGEG). *The Mechanical Behavior of Salt. Proc. 3rd Conf.* Eds. M. Ghoreychi, P. Berest, H.R. Hardy, Jr., and M. Langer. Trans Tech Publ,. Clausthal-Zellerfeld, 583-603. [Ch.1, 4]

Le Cleac'h, Ghazali, A., Deveughele, M., and Brulhet, J. (1996) Experimental Study of the role of humidity on the thermomechanical behaviour of various halitic rocks. *The Mechanical Behavior of Salt. Proc. 3rd Conf.,* Eds. M. Ghoreychi, P. Berest, H.R. Hardy,Jr., and M. Langer. Trans Tech Publ., Clausthal-Zellerfeld, 231-236. [Ch. 3, 4]

Leddra, M.J., Petley, D.N., and Jones, M.E. (1992) Fabric changes induced in a cemented shale through consolidation and shear. *Rock Mechanics. Proc. 33rd US Symp.,*Eds. J.R. Tillerson, and W.R. Wawersik. Balkema, Rotterdam, 917-926. [Ch..4]

Lee, M., and Haimson, B. (1993) Laboratory study of borehole breakouts in Lac du Bonnet Granite: a case of extensible failure mechanism. *Int. J. Rock Mech. Min. Sci. & Geomech. Abstr.* **30,** 1039-1045. [Ch.9]

Lee Petersen, D., and Nelson, C.R. (1993) Design and construction of a wide tunnel under shallow cover - The Lafayette Bluff Tunnel South Portal. *Rock Mechanics in the 1990s, The 34th US Symp. on Rock Mechanics,* Ed. B.C. Haimson. *Int. J. Rock Mech. Min. Sci. & Geomech. Abstr.* **30**, 1477-1483. [Ch.10]

Lemaitre, J. (1986) Local approach of fracture. *Engineering Fracture Mechanics* **25**, 523-537. [Ch.6]

Lemaitre, J., and Chaboch, J.L. (1990) *Mechanics of Solid Materials.* Cambridge University Press, Cambridge. [Ch.5]

Lembo Fazio, A.L., and Ribacchi, R. (1984) Influence of seepage on tunnel stability. *Design and Performance of Underground Excavations* ISRM/BGS, Cambridge, 173-181. [Ch.9]

Levy, A.J. (1985) A physically based constitutive equation for creep-damaging solids. *J. Appl. Mech.* **52**, 615-620. [Ch.6]

Liang, Jinhuo, and Lindblom, Ulf. (1994) Critical pressure for gas storage in unlined rock caverns. *Int. J. Rock Mech. Min. Sci.& Geomech Abstr.* **31**, 4, 377-381. [Ch.10]

Liedtke, L., and Bleich, W. (1985) Convergence calculation for backfilled tunnels in rock salt. *Proc. 5th ADINA Conf., Massachusetts,* Ed. K.J. Bathe, Pergamon Press, Oxford, 353-378. [Ch.2]

Liu, S.I. (1982) On representation of anisotropic invariants. *Int. J. Engng Sci.,* **20**, 1099-1109. [Ch.5]

Ljunggren, C., and Amadei, B. (1989) Estimation of virgin rock stresses from horizontal hydrofractures. *Int. J. Rock Mech. Min. Sci. & Geomech. Abstr.* **26**, 69-78. [Ch.7]

Lliboutry, L. (1987) *Very Slow Flows of Solids.* Martinus Nijhoff, Dordrecht. [Ch.1, 2]

Lo, T.-W., Coyner, K.B., and Toksöz, M.N. (1986) Experimental determination of elastic anisotropy of Berea sandstone, Chicopee shale, and Chelmsford granite. *Geophysics* **51**, 1, 164-171. [Ch.6]

Lockner, D. (1993) The role of acoustic emission in the study of rock fracture. *Rock Mechanics in the 1990s, The 34th US Symp. on Rock Mechanics,* Ed. B.C.Haimson. *Int. J. Rock Mech. Min. Sci.& Geomech. Abstr.* **30**, 883-899. [Ch.6]

Loe, N.M., Leddra, M.J., and Jones, M.E. (1992) Strain states during stress path testing of the chalk. *Rock Mechanics. Proc. 33rd US Symp.,* Eds. J.R. Tillerson and W.R. Wawersik. Balkema, Rotterdam, 927-936. [Ch.5]

Lubliner, J. (1990) *Plasticity Theory.* MacMillan, New York. [Ch.5]

Lux, K.H., and Heusermann, S. (1983) Creep tests on rock salt with changing load as a basis for the verification of theoretical material laws. *Proc. 6th Int. Symp on Salt.* Eds. B.C. Schreiber and H.L. Harner. The Salt Institut, Alexandria, USA. Vol.1, 417-435. [Ch.2]

Lux, K.-H. and Schmidt, T. (1996) Optimising underground gas storage operations in salt caverns, with special reference to the rock mechanics and thermodynamics involved. *The Mechanical Behavior of Salt. Proc. 3rd Conf.* Eds. M. Ghoreychi, P. Berest, H.R. Hardy, Jr., and M. Langer. Trans Tech Publ., Clausthal-Zellerfeld, 445-458. [Ch.10]

Maier, G., Nappi, A., and Papa, E. (1991) Damage models for masonry as a composite material: a numerical and experimental analysis. *Constitutive Laws for Engineering Materials. Proc. 3rd Int Conference.* Eds. C.S. Desai, E. Krempel, G. Frantziskonis, and H. Saadatmanesh. ASME Press, New York, 427-432. [Ch.6]

Malvern, L.E. (1969) *Introduction to the Mechanics of a Continuous Medium.* Prentice-Hall, Englewood Cliffs, New Jersey. [Ch.4]

Maranini, E. (1997) Comportamento Viscoplastico dei Materiali Rocciosi: Studio Sperimentale, Leggi Costitutive, Applicazioni. Tesi di Dottorato, Universita degli Studi di Ferrara, Dottorato di Ricerca in Geologia Applicata, Italia. [Ch.4]

Margolin, L.G., and Trent, B.C. (1992) A microphysical constitutive model for granular materials: II. Incorporating damage into the effective elastic moduli. *Rock Mechanics. Proc. 33rd US Symp.* Eds. J.R. Tillerson, and W.R. Wawersik, Balkema, Rotterdam, 671-679. [Ch.6]

Martin, C.D. (1990) Characterizing *in situ* stress domains at the AECL Underground Research Laboratory. *Can Geotech. J.* **27**, 631-646. [Ch.8]

Martin, C.D., ans Chandler, N.A. (1994) The progressive fracture of Lac du Bonnet Granite. *Int. J. Rock Mech. Min. Sci. & Geomech. Abstr.* **31**, 643-659. [Ch.6]

Martin, C.D., Martino, J.B., and Dzik, E.J. (1994) Comparison of borehole breakouts from laboratory and field tests. *Rock Mechanics in Petroleum Engineering, EUROCK '94.* Balkema, Rotterdam, 183-190. [Ch.9]

Martin, C.D., Read, R.S. (1996) AECL's Mine-by Experiment: A test tunnel in brittle rock. *Rock Mechanics. Tools and Technique. Proc. 2nd NARMS Symp.* Eds. M. Aubertin, F. Hassani, and H. Mitri. Balkema, Rotterdam, 13-24. [Ch.9]

Martin, R.J.,III, and Haupt, R.W. (1994) Static and dynamic elastic moduli in granite: The effect of strain amplitude. *Rock Mechanics. Models and Measurements. Challenges from Industry. Proc.1st NARMS.* Balkema, Rotterdam, 473-480. [Ch.6]

Martinetti, S., and Ribacchi, R. (1980). *In situ* stress measurements in Italy. *Rock Mechanics.* **9**, 31-47. [Ch.7]

Massier, D. and Cristescu, N. (1981) *In-situ* creep of rocks. *Rev. Roum. Sci. Techn., Ser. Mec. Appl.*, **26**, 5, 687-702. [Ch.9]

Massier, D. (1989) On the creep of rocks. Ph.D. Thesis. University of Bucharest, Romania. [Ch.5, 10].

Massier, D. (1995a) *In-situ* creep of a linear viscoelastic rock around some noncircular tunnels. Part I. Theoretical solution. *Rev. Roum.Sci. Techn., Ser. Mec. Appl.* **40**, 2-3, 413-423. [Ch.10]

Massier, D. (1995b) *In-situ* creep of a linear viscoelastic rock around some noncircular tunnels. Part II. Boundary stress and displacement concentrations. *Rev. Roum.Sci. Techn., Ser. Mec. Appl.* **40**, 4-5-6, 485-502. [Ch.10]

Massier, D. (1996) Rate-type constitutive equations with applications in the calculus of the stress state and of the displacement field around cavities in rock salt. *The Mechanical Behavior of Salt. Proc. 3rd Conf.* Eds. M. Ghoreychi, P.Berest, H.R. Hardy, Jr., and M. Langer. Trans Tech Publ., Clausthal-Zellerfeld 545-558. [Ch.10]

Mateescu, M., Teodorescu, S., Fota, D., and Cristescu, N. (1983) Rheological properties of a consolidated filler. *Mine, Petrol, Gaze.* **34**, 1, 20-29. [Ch.4]

Maury, V. (1993) An overview of tunnel, underground excavation and borehole collapse mechanisms. *Comprehensive Rock Engineering.* Ed.-in-Chief J.A. Hudson,Vol.4. *Excavation, Support and Monitoring*, Pergamon Press, Oxford, 369-412. [Ch.8, 9, 10]

McClintock, F.A., and Walsh, J.B. (1962) Friction on Griffith cracks in rocks under pressure. *Proc 4th Nat. Congr. of Applied Mechanics (ASME)*, Vol.II, 1015-1021. [Ch.5]

McCreath, D.R., and Diederichs, M.S. (1994) Assessment of near-field rock mass fracturing around a potential nuclear fuel repository in the Canadian Shield. *Int. J. Rock Mech.Min. Sci. & Geomech. Abstr.* **31**, 5, 457-470. [Ch.10]

McLamore, R. and Gray, K.E. (1967) The mechanical behavior of anisotropic sedimentary rocks. *J. Engineering for Industry, Trans.ASME*, **89**, 62-73. [Ch.5]

Mecking, W., and Estrin, Y. (1987) Microstructure-related constitutive modelling of plastic deformatin. *Proc. 8th Risø Int. Symp. of Metallurgy and Material Science: Constitutive Relations and Their Physical Basis. Risø National Laboratory, Roskilde, Denmark*, Ed. S.I. Andersen *et al.* 123-145. [Ch.2]

Mellegard, K.D., Senseny, P.E., and Hansen, F.D. (1981) Quasi-static strength and creep characteristics of 100-mm diameter specimens of salt from Avery Island, Louisiana. RE/SPEC, ONWI-250.[Ch.5]

Menzel, W., and Schreiner, W. (1977) Zum geomechanischen Verhalten von Steinsalz verschiedener Lagerstatten der DDR. Teil II. Das Verformungsverhalten. *Neue Bergbautechnik* **7**(8) 565-571. [Ch.2]

Mier, J,G.M., van (1986) Multiaxial strain-softening of concrete. Part I: Fracture. *Matériaux et Constructions,* **19**, 111, 179-200. [Ch.4]

Misbahi, A., Shao, J.-F., and Henry J.-P. (1994) Application de la méthode d'inversion à la détermination des contraintes *in situ* en milieux anisotropes. *Rev. Franç. Géotechn.* **67**, 41-48. [Ch.7]

Misra, A.K., and Murrell, S.A.F. (1965). An experimental study on the effect of temperature and stress on the creep of rocks. *Geophys. J. Roy. Astr. Soc.* **9**, 5, 509-535. [Ch.2, 4]

Mogi, K. (1972) Effect of the triaxial stress system on fracture and flow of rocks. *Phys. Earth Planet. Interiors,* **5**, 318-324. [Ch.9]

Möhring-Erdmann, G., and Rummel, F. (1987) Borehole breakout mechanisms and stress estimation. *Geol. Jb.* **E 39**, 109-123. [Ch.8]

Mokhel, A.N., Kulinich, Yu.V., Germanovich, L.N., Dyskin, A.V., and Galybin, A.N.(1996) Modeling borehole instabilities associated with natural crack systems. *Rock Mechanics. Tools and Technique. Proc. 2nd NARMS Symp.* Eds. M. Aubertin, F. Hassani, and H. Mitri. Balkema, Rotterdam, 1285-1294.   [Ch.8]

Moos, D., and Zoback, M.D. (1993) Near-surface, "Thin Skin" reverse faulting stresses in the Southeastern United States. *Rock Mechanics in the 1990s, The 34th US Symp. on Rock Mechanics,* .Ed. B.C. Haimson. *Int. J. Rock Mech. Min. Sci. & Geomech. Abstr.* **30**, 965-971. [Ch.7]

Morales, R.H., Abou-Sayed, A., and Jones, A.H. (1989a) Laboratory investigation of wellbore breakouts as an indicator of *in-situ* stresses. *Rock Mechanics as a Guide for Efficient Utilization of Natural Resources. Proc. 30th Symp. Rock Mechanics.* Balkema, Rotterdam, 741-748. [Ch.8]

Morales, R.H., Abou-Sayed, A., and Jones, A.H. (1989b) Field investigation of wellbore breakouts as an indicator of in-situ stress orientation. *Rock Mechanics as a Guide for Efficient Utilization of Natural Resources. Proc. 30th Symp. Rock Mechanics.* Balkema, Rotterdam, 877-881. [Ch.8]

Moretto, O., Sarra Pistone, R.E., and Del Rio, J.C. (1993) A case history in Argentina - Rock mechanics for the underground works in the pumping storage development of Rio Grande No.1. *Comprehensive Rock Engineering.* Ed.-in-Chief J.A. Hudson,, Vol.5. *Surface and Underground Case Histories.* Pergamon Press, Oxford 159-192. [Ch.10]

Mowar, S., Zaman, M., Stearns, D.W., and Roegiers, J.-C. (1994) Pore collapse mechanisms in Cordoba Cream limestone. *Rock Mechanics Models and Measurements. Challenges from Industry. Proc. 1st NARMS,* Eds. P.P. Nelson and S.E. Laubach. Balkema, Rotterdam 767-773. [Ch.4]

Mraz, D., Rothenburg, L., and Dusseault, M.B. (1993) Design of mines in salt and potash. *Proc. 7th Int. Congr. on Rock Mechanics.* Balkema, Rotterdam, 127-134.   [Ch.10]

Mróz, Z., and Angelillo, M. (1982). Rate-dependent degradation model for concrete and rock. *Numerical Models in Geomechanics. Proc. Int. Symp.,* Balkema, Rotterdam, 208-218 [Ch.6]

Munson, D.E., and Dawson, P.R. (1984) Salt constitutive model using mechanism maps. *The Mechanical Behavior of Salt. Proc. 1st Conf.* Eds. H.R. Hardy, Jr., and M. Langer, Trans Tech Publ., Clausthal-Zellerfeld, 717-737. [Ch.2]

Munson, D.E., Fossum, A., and Senseny, P. (1989) Approach to first principles model prediction of measured WIPP in situ room closure in salt.*Rock Mechanics as a Guide for Efficient Utilization of Natural Resources. Proc. 30th US Symp.on Rock Mechanics,* Balkema, Rotterdam, 673-680. [Ch.2, 3]

Munson, D.E., DeVries, K., Schiermeister, D.M., DeYonge, W.F., and Jones, R.L. (1992) Measured and calculated closure of open and brine filled shafts and deep vertical bore-holes in salt. *Rock Mechanics. Proc. 33rd US Symp.* Eds. J.R. Tillerson, W.R.Wawersik. Balkema, Rotterdam, 439-448. [Ch.8]

Munson, D. and DeVries, K. (1993) Development and validation of a predictive technology for creep closure of underground rooms in salt. *Proc. 7th Int. Congr. on Rock Mechanics.* Balkema, Rotterdam, 127-134. [Ch.10]

Munson, D. and Wawersik, W. (1993) Constitutive modeling of salt behavior - state of the technology. *Proc 7th Int. Congr. on Rock Mechanics.* Ed. W. Wittke. Balkema, Rotterdam, **3**, 1797-1810. [Ch.2]

Munson, D.E., DeVries, K., Fossum, A.F., and Callahan, G.D. (1996) Extension of the M-D model for treating stress drops in salt. *The Mechanical Behavior of Salt. Proc.3rd Conf.* Eds. M. Ghoreychi, P. Berest, H.R. Hardy, Jr., and M. Langer. Trans Tech Publ., Clausthal-Zellerfeld. 31-44. [Ch.2, 4]

Munson, D.E. (1997) Constitutive model of creep in rock salt applied to underground room closure. *Int. J. Rock Mech. Min. Sci. & Geomech Abstr.* **34**, 233-247.   [Ch.4]

Murakami, S. (1990) A continuum mechanics theory of anisotropic damage. *Yielding, Damage and Failure of Anisotropic Solids.* Mechanical Engineering Publ., 465-482. [Ch. 5]

Muskhelishvili, N. I. (1953) *Some Basic Problems of the Mathematical Theory of Elasticity.* Noordhoff, Groningen. [Ch.10]

Nawrocki, P.A. and Mróz, Z. (1992) Constitutive model for rocks accounting for viscoplastic deformation and damage. *Rock Mechanics. Proc. 33rd US Symp.* Eds. J.R. Tillerson, R.W. Wawersik. Balkema, Rotterdam. 691-700. [Ch.4]

Nawrocki, P.A. (1993) A semi-analytical technique to study deformations of viscoplastic seam caused by a moving longwall face. *Proc. 1st Can. Symp. Num. Mod. Appl. Min. Geomech.,* Ed. H. S. Mitri. McGill Univ., Montreal, Quebec, 129-139. [Ch. 10]

Nawrocki, P.A. (1995) One-dimensional, semi-analytical solution for time-dependent behaviour of a seam. *Int. J. Numer. Analyt. Meth. Geomech.* **19**, 59-74. [Ch. 10]

Nawrocki, P.A., Cristescu, N.D., Dusseault, M.B., and Bratli, R.K. (1997) Experimental methods for determining constitutive parameters for non-linear rock modelling. *Proc. 9th Int. Conf. on Computer Methods and Advances in Geomechanics.* Balkema, Rotterdam. (In press). [Ch. 4]

Nelson, P.P. (1996) Rock engineering for underground civil construction. *Rock Mechanics. Tools and Technique. Proc. 2nd NARMS Symposium.* Eds. M. Aubertin, F. Hassani, and H. Mitri. Balkema, Rotterdam, 3-12. [Ch.10]

Nguyen-Minh, D., and Pouya, A. (1992) Une méthode d'étude des excavations souterraines en milieu visco-plastique. Prise en compte d'un état stationaire des contraintes. *Rev. Franç. Géotech.* **59**. [Ch.10]

Niandou, H. (1994) Etude du comportement rhéologique et modélisation de l'argilite de Tournemire. Application à la stabilité d'ouvrages souterrains. Thése de Doctorat, Lab. de Mécanique de Lille URA 1441 CNRS. [Ch.4]

Niandou, H., Shao, J.F., Henry, J.P., and Fourmaintraux, D. (1997) Laboratory Investigation of the Mechanical Behaviour of Tournemire Shale. *Int. J. Rock Mech. Min. Sci. & Geomech. Abstr.* **34**, 3-16. [Ch.4]

Nicolae, M. (1996) Research work concerning the characterization of the rheological behavior of salt in Slanic-Prahova zone. *The Mechanical Behavior of Salt. Proc. 3rd Conf.* Eds. M. Ghoreychi, P. Berest, H.R. Hardy, Jr.,and M. Langer. Trans Tech Publ., Clausthal-Zellerfeld, 559-570. [Ch.5]

Nicolas, A., and Poirier, J.P. (1976) *Crystalline Plasticity and Solid State Flow in Metamorphic Rocks.* Wiley, Chichester. [Ch.3]

Niitsuma, H., and Chubachi, N. (1986) AE monitoring of well-drilling process by using a downhole AE measurement system. *Proc. 8th Int. AE Symp., Tokyo.* [Ch.6]

Niitsuma, H., Chubachi, N. and Takanohashi, M. (1987) Acoustic emission analysis of a geothermal reservoir and its application to reservoir control. *Geothermics,* **16**, 1, 47-60. [Ch.6]

Nix, W.D., and Ilschner, B. (1979) Mechanisms controlling creep of single phase metals and alloys. *Proc. 5th Int. Conf. on the Strength of Metals and Alloys.* Eds. P.Haasen, V. Gerold, and G. Kostorz. Pergamon Press, Oxford. Vol.3, 1503-1530. [Ch.2]

Nonaka, T. (1981) A time-independent analysis for the final state of an elasto-visco-plastic medium with internal cavities. *Int. J. Solids Structures* **17**, 961-967. [Ch.9]

Nordmark, A.M., and Franzen T. (1993) Subsurface space - An important dimension in Swedish construction. *Comprehensive Rock Engineering.* Ed.-in-Chief J.A. Hudson, vol.5. *Surface and Underground Case Histories*, Pergamon Press, Oxford, 29-54. [Ch.10]

Nova, R. (1980) The failure of transversely anisotropic rocks in triaxial compression. *Int. J. Rock Mech. Min. Sci. & Geomech. Abstr.* **17**, 325-332. [Ch.5]

Nova, R. (1986) An extended Cam-Clay model for soft anisotropic rocks. *Computers and Geotechnics,* **2**, 69-88. [Ch. 5]

Nova, R., and Zaninetti, A. (1990) An investigation into the tensile behaviour of a schistose rock. *Int. J. Rock Mech. Min. Sci. & Geomech. Abstr.* **27**, 231-242. [Ch.5]

Nur, A. and Simmons, G. (1969) The effect of saturation on velocity in low porosity rocks. *Earth Planet. Sci. Lett.* **7**, 183-195. [Ch.6]

Obert, L., and Duvall, W.I. (1967) *Rock Mechanics and the Design of Structures in Rock.* Wiley. New York. [Ch.1, 6, 9, 10]

Ofoegbu, G.I., and Curran, J.H. (1991) Yielding and damage of intact rock. *Can. Geotech. J.* **28**, 503-516. [Ch.6]

Oikawa, H., and Langdon, T.G. (1985) The creep characteristics of pure metals and metallic solid solutions. *Creep Behaviour of Crystalline Solids. 3. Progress in Creep and Fracture,* Eds. B. Wilshire and R.W. Evans. Pineridge Press, Swansea, 33-82. [Ch.2, 3]

Olszak, W. and Urbanowschi, W. (1956) The plastic potential and the generalized energy in theory of non-homogeneous and anisotropic elasto-plastic bodies. *Arch. Mech. Stos.*, **8**. [Ch.5]

Pahl, A. and Heusermann, S. (1991) Determination of stress in rock salt taking time-dependent behavior into consideration. *7th Int. Congr. on Rock Mechanics*. Balkema, Rotterdam, 1713-1718. [Ch.7]

Pahl, A. and Heusermann, S. (1993) *In-situ* determination of initial stress in rock salt - test methods, results, and data evaluation. *ISRM Int. Symp. EUROCK '93, Lisboa, Portugal, June 1993.* (Preprint) [Ch.7]

Pan, X.D., and Hudson, J.A. (1988a) Plane strain analysis in modelling three-dimensional tunnel excavations. *Int. J. Rock Mech. Min. Sci. & Geomech. Abstr.*, **25**, 331-337. [Ch.9]

Pan, X.D., and Hudson, J.A. (1988b) A simplified three dimensional Hoek-Brown yield criterion. *Rock Mechanics and Power Plants*. Ed. M. Romana, Balkema, Rotterdam, 95-103. [Ch.4]

Pan, X.D., and Reed, M.B. (1991) Effects of longitudinal axial stress and rock mass dilation on analysis of circular tunnels. *Proc. 7th Int. Congr. on Rock Mechanics*. Balkema, Rotterdam, 785-791. [Ch.9]

Pan, Y.-W., and Dong, J.-J. (1991) Time-dependent tunnel convergence -I. Formulation of the model. *Int. J. Rock Mech. Min. Sci. & Geomech. Abstr.* **28**, 469-475. [Ch.9].

Panet, M. (1969) Quelques problèmes de mécanique des roches posés par le tunnel du Mont Blanc. *Bull. Liaison Labo. Routiers P. et Ch.* 42, 115-145. [Ch.9]

Paraschiv-Munteanu, I. (1997) Metode Numerice in Geomecanica. Ph.D. Dissertation, University of Bucharest, Romania. [Ch.8]

Pariseau, W.G. (1972) Plasticity theory for anisotropic rocks and soils. *Proc. 10th Symp. on Rock Mechanics (AIME)*, 267-295. [Ch.5]

Passaris, E.K.S. (1979) The rheological behaviour of rocksalt as determined in an *in situ* pressurized test cavity. *Proc. 4th Int. Congr. on Rock Mechanics*, Balkema, Rotterdam, 257-264. [Ch.10]

Paterson, M.S. (1978) *Experimental Rock Deformation - The Brittle Field.* Springer-Verlag, Berlin. [Ch.1, 2]

Paterson, M.S. (1989) The interaction of water with quartz and its influence in dislocation flow - an overview. *Rheology of Solids and of the Earth,* Eds. S. Karato and M. Toriumi. Oxford University Press, New York, 107-142. [Ch.3]

Paul, B. (1968) Macroscopical criteria of plastic flow and brittle fracture. *Fracture. An Advanced Treatise.* Ed. H. Liebowitz, Academic Press, New York., vol.2, Ch.4..[Ch.4]

Peach, C.J. (1991) Influence of deformation on the fluid transport properties of salt rocks. Ph.D. Dissertation, University of Utrecht, Netherlands. [Ch.2]

Peach, C.J. (1996) Deformation, dilatancy and permeability development in halite/anhydrite composites. *The Mechanical Behavior of Salt. Proc. 3rd Conf.* Eds.M. Ghoreychi, P. Berest, H.R. Hardy, Jr., and M. Langer. Trans Tech Publ., Clausthal-Zellerfeld, 153-166. [Ch.4]

Pellegrino, A., Sulem, J., and Barla, G. (1994) Nonlinear effects in the study of borehole stability. *Rock Mechanics in Petroleum Engineering. EUROCK '94.* Balkema, Rotterdam, 231-238. [Ch.8]

Pelli, F., Kaiser, P.K., and Morgenstern, N.R. (1991) The influence of near face behaviour on monitoring of deep tunnels. *Can. Geotech. J.* **28**, 226-238. [Ch.9]

Pharr, G.M., and Ashby, M.F. (1983) On creep enhanced by a liquid phase. *Acta metall.* **31**, 129-138. [Ch.3]

Plischke, I., and Hunsche, U. (1989) In Situ-Kriechversuche unter kontrollierten Spannungsbedingungen an grosse n Steinsalzpfeilern. *Rock at Great Depth. Proc. ISRM-SPE Int. Symp.*, Eds. V. Maury and D. Fourmaintraux. Balkema, Rotterdam, Vol.1, 101-108.[Ch.1]

Plischke, I. (1996) Statistical analysis of the influence of geological-mineralogical parameters on the creep behavior of rock salt. *EUROCK'96, Proc. ISRM Int. Symp.Torino, Italy*, Ed.G. Barla. Balkema, Rotterdam, 49-52. [Ch.2]

Plumb, R.A. (1994) Variation of the least horizontal stress magnitude in sedimentary rocks. *Rock Mechanics. Models and Measurements. Challenges from Industry. Proc.1st NARMS Symp.* Eds. P.P.Nelson, and S.E.Laubach. Balkema, Rotterdam, 71-78. [Ch.7]

Poirier, J.-P. (1985). *Creep of Crystals. High-Temperature Deformation Processes in Metals, Ceramics and Minerals.* Cambridge University Press, Cambridge. [Ch.2 , 3]

Popp, T. (1994) Der Einfluss von Gesteinsmatrix, Mikrorissgefügen und intergranularen Fluiden auf die elastischen Wellengeschwindigkeiten und die elektrische Leitfähigkeit krustenrelevanter Gesteine unter PT-Bedingungen. Ph.D. Dissertation. Universität Kiel. [Ch.2]

Popp, T., and Kern, H. (1994) The influence of dry and water saturated cracks on seismic velocities of crustal rocks - A comparison of experimental data with theoretical model. *Surveys in Geophysics,* **15**, 443-465. [Ch. 2, 6]

Prij, J., Heijdra, J.J., and Horn, B.A. van den (1996) Convergence and compaction of backfilled openings in rocksalt. *The Mechanical Behavior of Salt. Proc.3rd Conf.* Eds. M. Ghoreychi, P. Berest, H.R. Hardy, Jr., and M. Langer. Trans Tech Publ., Clausthal-Zellerfeld. 337-350. [Ch.8]

Pudewills, A., and Hornberger, K. (1996) A unified viscoplastic model for rock salt. *The Mechanical Behavior of Salt. Proc. 3rd Conference,* Eds. M. Ghoreychi, P. Berest, H.R. Hardy, Jr.,and M. Langer. Trans Tech Publ., Clausthal-Zellerfeld, 45-52.   [Ch.5]

Ramamurthy, T. (1993) Strength and modulus responses of anisotropic rocks. *Comprehensive Rock Engineering,* Ed-in-Chief J.A. Hudson,, Vol.1, *Fundamentals*, Pergamon Press , Oxford, 313-329 [Ch.5]

Ranalli, G. (1995) *Rheology of the Earth.* 2nd Ed., Chapman and Hall, London. [Ch.1-3]

Rao, M.V.M.S., Sun, X., and Hardy, H.R., Jr. (1989) An evaluation of the amplitude distribution of AE activity in rock specimens stressed to failure. *Rock Mechanics as a Guide for Efficient Utilization of Natural Resources, Proc. 30th Symp. Rock Mechanics.* Balkema, Rotterdam, 261-268. [Ch.6]

Reed, M.B. (1986) Stresses and displacements around a cylindrical cavity in soft rock. *IMA J. Appl. Math.* **36**, 223-245. [Ch.8]

Reed, M.B. (1988) The influence of out-of-plane stress on a plane strain problem in rock mechanics. *Int. J. Numer. Analyt. Meth. Geomech.* **12**, 173-181. [Ch.8]

Reed, M.B., and Cassie, J. (1988) Non-associated flow rules in computational plasticity. *Numerical Methods in Geomechanics,* Ed. G. Swoboda. Balkema, Rotterdam, 481-488. [Ch.5]

Ribacchi, R. (1988) Rock mass deformability: In situ tests, their interpretation and typical results in Italy. *Proc. 2nd Int. Symp. on Field Measurements in Geomechanics,* Ed. S.Sakurai, Balkema, Rotterdam, 171-192. [Ch.7]

Rice, J.R. (1975). On the stability of dilatant hardening for saturated rock masses. *J. Geophys. Res.* **80**, 11, 1531-1537. [Ch.4]

Rindorf, H.J. (1984) Location of microseismic activity. *Acoustic Emission/Microseismic Activity in Geologic Structures and Materials. Proc. 3rd Conf.*, Eds. H.R.Hardy, Jr., and F.W. Leighton. Trans Tech Publ., Clausthal-Zellerfeld, 695-706. [Ch.6]

Roatesi, S. (1997) Numerical Methods in Mathematical Modelling of Mining Supports. Ph. D.Dissertation, University of Bucharest, Romania. [Ch.8]

Roegiers, J.-C. (1993) The use of rock mechanics in petroleum engineering: General overview. *Comprehensive Rock Engineering.* Ed.-in-Chief J.A. Hudson. Vol.5, *Surface and Underground Case Histories.* Pergamon Press, Oxford, 605-616. [Ch.8]

Roegiers, J-C. (1994) Petroleum geomechanics. *Computer Methods and Advances in Geomechanics.* Eds. H.J. Siriwardane, and M.M. Zaman. Balkema, Rotterdam, 269-277. [Ch.4]

Rokahr, R.B., and Staudtmeister, K. (1983). Creep rupture criteria for rock salt. *Proc. 6th Int. Symp. on Salt,* Eds.B.C. Schreiber and H.L. Harner. The Salt Institute, USA Vol.1, 455-462. [Ch.6]

Rokahr, R.B., and Staudtmeister, K. (1996). Laboratory tests with hollow cylinders for the dimensioning of gas caverns in rock salt. *Rock Mechanics. Tools and Techniques. Proc. 2nd NARMS.,* Eds. M. Aubertin, F. Hassani, and H. Mitri. Balkema, Rotterdam, 61-68. [Ch.8]

Rudnicki, J.W., and Chen, C.-H. (1988) Stabilization of rapid frictional slip on a weakening fault by dilatant hardening. *J. Geophys. Res.* **93**, B5, 4745-4757. [Ch.4]

Rummel, F. and Baumgärtner, J. (1985) Hydraulic fracturing in-situ stress and permeability measurements in the research Borehole Kronzen, Hohen Venn (West Germany), *N. Jb. Geol. Palaont. Abh.* **171**, 183-193. [Ch.7]

Rummel, F. (1986) Stresses and tectonics of the upper continental crust - a review. *Proc. Int. Symp. on Rock Stress and Rock Stress Measurements,* 177-186. [Ch.7]

Rummel, F., Möhring-Erdmann, G., and Baumgärtner, J. (1986) Stress constraints and hydrofracturing stress data for the continental crust. *PAGEOPH.* **124**, 4/5, 875-895. [Ch.7]

Rutter, E.H. (1976) The kinetics of rock deformation by pressure-solution. *Phil. Trans. Roy. Soc. Lond.* **A283**, 203-219. [Ch.3]

Rutter, E.H. (1983) Pressure solution in nature, theory and experiment. *J. Geol. Soc. Lond.* **140**, 725-740. [Ch.3]

Saari, K. (Ed.) (1987) *Proc. Int. Symp. on Large Rock Caverns* (3 vols), Pergamon Press. Oxford. [Ch.10].

Salzer, K.and Schreiner, W. (1991) Der Rechencode MKEN zur Ermittlung der Zeitab-hängigkeit des Spannungs-Verformungs-Zustandes um Hohlräume im Salzgebirge.*Kali und Steinsalz* 10(12), 416-423. [Ch.2]

Sano, O. (1981) A note on the source of acoustic emissions associated with subcritical crack growth. *Int. J. Rock Mech. Min. Sci. & Geomech. Abstr.* **18**, 259-263. [Ch.6]

Sano, O., Itô, I. and Terada, M. (1981) Influence of strain rate on dilatancy and strength of Oshima granite under uniaxial compression. *J. Geophys. Res.* **86**, B10, 9299-9311. Ch.4, 6]

Sano, O., Terada, M., and Ehara, S. (1982) A study on the time-dependent microfracturing and strength of Oshima granite. *Tectonophysics* **84,** 343-362. [Ch.6]

Savin, G.N. (1961) *Stress Concentration around Holes.* McGraw-Hill, New York. [Ch.10]

Schmidtke, R.H., and Lajtai, E.Z. (1985) The long-term strength of Lac du Bonnet granite. *Int. J. Rock Mech. Min. Sci. & Geomech. Abstr.,* **22**, 461-465. [Ch.6]

Schock, R.N., Heard, H.C., and Stephens, D.R. (1973) Stress-strain behavior of a grano-diorite and two graywackes on compression to 20 kilobars. *J. Geophys.Res.* **78**, 26, 5922-5941. [Ch. 6]

Schock, R. N., and Heard, H. C. (1974) Static mechanical properties and shock loading response of granite. *J. Geophys. Res.,* **79**, 11, 1662-1666. [Ch. 5, 6]

Schock, R.N. (1976) A constitutive relation describing dilatant behavior of Climax Stock granodiorite. *Int. J. Rock Mech. Min. Sci.& Geomech. Abstr.* **13**, 221-223. [Ch.4]

Schock, R. N. (1977) The response of rocks to large stresses. *Impact and Explosion Cratering.* Eds. D. J. Roddy, R. O. Pepin, and R. B. Merrill. Pergamon Press, New York, 657-668. [Ch.4]

Scholz, C. H. (1968) Mechanics of creep in brittle rock. *J. Geophys. Res.* **73**, 3295-3302 [Ch.6]

Schreiber, E., Anderson, O.L., and Soga, N.(1973). *Elastic Constants and Their Measurement.* McGraw-Hill, New York. [Ch.6]

Schreyer, H. L., and Babcock, S. M. (1985) A third-invariant plasticity theory for low-strength concrete. *J. Engng Mech.,* **111**, 545-558. [Ch.2]

Schulze, O. (1984) Thermomechanical properties of irradiated rock salt. *Nucl. Instrum. Meth. Phys. Research,* B1, Elsevier, Amsterdam, 542-548. [Ch.2]

Schulze, O. (1986) Der einfluss radioaktiver Strahlung auf das mechanische Verhalten von Steinsalz. *Zeitsch. der Deutschen Geol. Gesellschaft,* **137**, 47-69. [Ch.2]

Schulze, O. (1993) Effect of humidity on creep of rock salt. *Geotechnik-Sonderheft 1993,* Essen, 169-172. [Ch.3]

Schutjens, P.M.T.M., de Ruig, H., van Munster, J.G., Sayers, C.M., and Whitworth, J.L. (1994) Compressibility measurement and acoustic characterisation of quartz-rich consolidated reservoir rock (Brent Field, North Sea). *Rock Mechanics in Petroleum Engineering. EUROCK '94.* Balkema, Rotterdam, 557-571. [Ch.6]

Scott, T.E., Jr., Ma, Q., and Roegiers, J.-C. (1993) Acoustic velocity changes during shear enhanced compaction of sandstone. *Rock Mechanics in the 1990s, The 34th US Symp. on Rock Mechanics.* Ed. B.C.Haimson. *Int. J. Rock Mech. Min. Sci. & Geomech. Abstr.* **30**, 763-769. [Ch.6]

Sedlacek, R.(1995) Glide dislocation shapes and long-range internal stresses in dislocation wall structures. *Phys. Stat. Sol. (a)* **149**, 85-93. [Ch.2]

Seeger, A. (1956) The mechanism of glide and work hardening in face-centered cubic and hexagonal close-packed materials. *Dislocations and Mechanical Properties of Crystals.* Eds. J.C. Fisher, W.G. Johnston, R. Thomson, and T. Vreeland. John Wiley, New York, 243-329. [Ch.2]

Senseny, P.E. (1985) Determination of a constitutive law for salt at elevated temperature and pressure. *Measurement of Rock Properties at Elevated Pressures and Temperatures. ASTM STP 869,* Eds. H. J. Pincus and E.R. Hoskins. ASTM, Philadelphia, 55-71. [Ch.2]

Senseny, P.E. (1986) Triaxial compression creep tests on salt from the Waste Isolation Pilot Plant. SANDIA Rep. SAND85-7261. [Ch.5]

Senseny, P.E., Brodsky, N.S., and De Vries, K.L. (1993) Parameter evaluation for a unified constitutive model. *Trans ASME. J. Engng Mater. Technol.* **115**, 157-162. [Ch. 5]

Shao, J.F., and Henry, J.P. (1990) Development of an elastoplastic model for porous rock. *Int. J. Plasticity* **7**, 1-13. [Ch.5]

Shao, J.F., and Khazraei, R. (1994). Wellbore stability analysis in brittle rocks with continuous damaged model. *Rock Mechanics in Petroleum Engineering. EUROK '94.* Balkema, Rotterdam, 215-222. [Ch. 8]

Shao, J.F., Kondo, D., and Ikogou, S. (1994) Stress-induced microcracking in rock and its influence on wellbore stability analysis. *Int. J. Rock Mech. Min. Sci. & Geomech.Abstr.* **31**, 149-155. [Ch.6, 8]

Shao, J.F., and Khazraei, R. (1996). A continuum damage mechanics approach for time independent and dependent behaviour of brittle rock. *Mechanics Research Communications* **23**, 257-265. [Ch.6]

Sheorey, P.R. (1994) A theory for *in situ* stresses in isotropic and transversely isotropic rock. *Int. J. Rock Mech. Min. Sci. & Geomech. Abstr.* **31**, 23-34. [Ch.7]

Siggings, A. F. (1993) Dynamic elastic tests for rock engineering. *Comprehensive Rock Engineering.* Ed.-in-Chief J.A. Hudson., Vol.3, *Rock Testing and SiteCharacterization,* Pergamon Press, Oxford, 601-619. [Ch.4]

Sondergeld, C.H., and Estey, L.H. (1981) Acoustic mission study of micro-fracturing during the cyclic loading of Westerly granite. *J. Geophys. Res.* **86**, B4, 2915-2924 [Ch.6]

Skrotzki, W. (1984) An estimate of the brittle to ductile transition in salt. *The Mechanical Behavior of Salt. Proc.1st Conf.* Eds. H.R. Hardy, Jr.,and M. Langer. Trans Tech Publ, Clausthal-Zellerfeld, 381-388. [Ch.2]

Sorace, S. (1996) Creep in building stones under tensile conditions. *Trans. ASME,J.Engng. Mater. Technol.* **118**, 456-462. [Ch.6]

Spetzler, H., Mizutani, H., and Rummel, F. (1982) A model for time-dependent rock failure. *High-Pressure Researches in Geoscience,* Ed. W. Schreyer. E. Schweizer-bart'sche Verlagsbuchhandlung, Stuttgart, 85-93. [Ch.6]

Spiers, C.J., Peach, C.J., Brzesowsky, R.H., Schutjens, P.M, Liezenberg, J.L., and Zwart, H.J. (1989). Long-term rheological and transport properties of dry and wet salt rocks. Final Report, Nuclear Science and Technology, Commission of the European Communities, EUR 11848 EN. [Ch.2]

Spiers, C.J., Schutjens, P.M.T.M., Brzesowsky, R.H., Peach, C.J., Liezenberg , J.L., and Zwart, H.J. (1990) Experimental determination of constitutive pasrameters governing creep of rocksalt by pressure solution. *Deformation Mechanisms. Rheology and Tectonics.* Eds. R.J. Knipe, and E.H. Rutter, Geological Society Special Publication No.54, 215-227. [Ch.3]

Spiers, C.J., and Brzesowky, R.H. (1993) Densification behaviour of wet granular salt:theory versus experiment. *Proc. 7th Symp. on Salt.* Elsevier, Amsterdam, vol.1, 83-92. [Ch.2]

Spies, Th., Meister, D., and Eisenblätter, J. (1997). Acoustic emission measurements as a contribution for the evaluation of stability in rock salt. *Proc. Int. Symp. on Rockbursts and Seismicity in Mines, Krakow,* Ed.: S.J. Gibowicz. Balkema, Rotterdam. [Ch.1]

Staupendahl, G.., Schmidt, M.W., Meister,M.W., and Wallner, M. (1979). Geotechnical investigations in the prototype cavity in the ASSE salt mine. *Proc. 4th Intern. Congr. on Rock Mechanics.* Balkema, Rotterdam, Vol.III, 645-653. [Ch.8]

Steiger, R.P., and Leung, P.K. (1993) Advances in shale mechanics - the key to wellbore stability predictions. *Comprehensive Rock Engineering.* Ed.-in-Chief J.A. Hudson. Vol.5, *Surface and Underground Case Histories..* Pergamon Press, Oxford, 629-639. [Ch.8]

Stephansson, O. (1993) Rock stress in the Fennoscandian Shield. *Comprehensive Rock Engineering*. Ed.-in-Chief J.A. Hudson., Vol.3, *Rock Testing and Site Characterization*, Pergamon Press, Oxford 445-459. [Ch.7]

Sterling, R.L. (1993) The expanding role of rock engineering in developing national and local infrastructures. *Comprehensive Rock Engineering*. Ed-in-Chief J.A. Hudson., Vol. 5, *Surface and Underground Project Case Histories*. Pergamon Press, Oxford, 1- 27. [Ch.10]

Stocker, R.L., and Ashby, M.F. (1973) On the rheology of the upper mantle. *Reviews of Geophysics and Space Physics*. **11**, 391-426. [Ch.3]

Stormont, J.C., and Daemen, J.J.K. (1992) Laboratory study of gas permeability changes in rock salt during deformation. *Int. J. Rock Mech. Min. Sci. & Geomech. Abstr.* **29**, 325-342. [Ch.2]

Stormont, J.C., Daemen, J.J.K., and Desai, C.S. (1992) Prediction of dilation and permeability changes in rock salt. *Int. J. Numer. Analyt. Metho. Geomech.* **16**, 545-569. [Ch.5]

Stormont, J.C., and Fuenkajorn, K. (1994) Dilation-induced permeability changes in rock salt. *Computer Methods and Advances in Geomechanics*, Eds. H.J. Siriwardane and M.M. Zaman, Balkema, Rotterdam, 1269-1273. [Ch.4].

Stührenberg, D., and Zhang, C. (1996) Kompaktionsverhalten von Salzgrus. *Kali und Steinsalz.* **12**(3) 103-111. [Ch.2]

Stumvoll, M., and Swoboda, G. (1993) Deformation behavior of ductile solids containing anisotropic damage. *J. Enging Mech.* **119**, 1331-1352. [Ch.6]

Sugawara, K., and Jang, H.K. (1996) Evaluation of initial rock stress by the orthotropic spherical shell model. *Rock Mechanics. Tools and Techniques. Proc. 2nd NARMS*, Eds. M. Aubertin, F. Hassani and H. Mitri. Balkema, Rotterdam, 905-912. [Ch.7]

Sulem, J., Panet, M., and Guenot, A. (1987) An analytical solution for time-dependent displacements in a circular tunnel. *Int. J. Rock Mech. Min. Sci. & Geomech. Abstr.* **24**, 155-64. [Ch.9]

Swoboda, G., Merty, W., and Beer, G. (1987) Rheological analysis of tunnel excavations by means of coupled finite element (FEM) - Boundary element (BEM) analysis. *Int. J. Numer. Analyt. Meth. Geomech.* **11**, 115-129. [Ch.9]

Takeuchi, S. (1989) Motion of dissociations: characteristics of high-temperature deformation. *Rheology of Solids and of the Earth*. Eds. S.I.Karato and M. Toriumi. Oxford Science Publications, 3-14. [Ch.2]

Tanimoto, K., and Nakamura, J. (1984) Use of AE technique in field investigation of soil. *Acoustic Emission/Microseismic Activity in Geologic Structures and Materials. Proc. 3rd Conf.* Eds H.R. Hardy, Jr., and F.W.Leighton. Trans Tech Publ., Clausthal-Zellerfeld, 601-612. [Ch.6]

Tapponier, P., and Brace, W.F. (1976) Development of stress-induced microcracks in Westerly granite. *Int. J. Rock Mech. Min. Sci. & Geomech. Abstr.* **13**, 103-112. [Ch.6]

Taylor, L.M., Chen, E.P., and Kuszmaul, J.S. (1986). Microcrack-induced damage accumulation in brittle rock under dynamic loading. *Comput. Meth. Appl. Mech. Engng.* **55**, 301-320. [Ch.6]

Terada, M., Yanagidani, T., and Ehara, S. (1984) A.E. rate controlled compression test of rocks. *Acoustic Emission/Microseismic Activity in Geologic Structures and Materials. Proc. 3rd Conf.* Eds. H.R. Hardy, Jr., and F.W.Leighton, Trans Tech Publ., Clausthal-Zellerfeld. 159-171. [Ch.6]

Terada, M., and Yanagidani, T. (1986) Application of Ultrasonic Computer Tomography to Rock Mechanics. Ultrasonic Spectroscopy and its Applications to Materials Science, Report of Stecial Project Research, Kyoto. 205-210. [Ch.6]

Thallak, S.G., and Gray, K.E. (1993) Discrete particle modelling for analysis of borehole stability. Applications in petroleum geomechanics. *Numerical Modelling Applications in Mining and Geomechanics.* Ed. H.S. Mitri. McGill University, 277-286. [Ch.8]

Tharp, T.M. (1996) A fracture mechanics analysis of stand-up time for mine roof beams. *Rock Mechanics. Tool and Techniques.Proc. 2nd NARMS*, Eds. M. Aubertin, F. Hassani, and H. Mitri. Balkema, Rotterdam, 1177-1184. [Ch.6]

Theocaris, P.S. (1989a) The elliptic paraboloid failure surface for transversely isotropic materials off-axis loaded. *Rheologica Acta.* **28**, 154-165. [Ch.5]

Theocaris, P.S. (1989b) The paraboloid failure surface for the general orthotropic material. *Acta Mechanica.* **79**, 53-79. [Ch.5]

Theocaris, P.S. (1991) The elliptic paraboloid failure criterion for cellular solids and brittle foams. *Acta Mechanica.* **89**, 93-121. [Ch.5]

Thorel, L., and Ghoreychi, M. (1996). Rock salt damage - Experimental results and interpretation. *The Mechanical Behavior of Salt. Proc.3rd Conf.* Eds. M. Ghoreychi, P. Berest, H.R. Hardy, Jr., and M. Langer. Trans Tech Publ., Clausthal-Zellerfeld.175-190. [Ch.2, 6]

Truesdell, C., and Noll, W. (1965) *The Non-linear Field Theories of Mechanics.* Handbuch der Physik, III/3, Springer-Verlag, Berlin. [Ch.4]

Tsai, S.W., and Wu, E. (1971) A general theory of strength of anisotropic materials. *J. Composite Materials.* **5**, 58-80. [Ch.5]

van den Hoek, Smit, D.-J., Kooijman, A.P., Bree, Ph.de, Kenter, C.J., and Khodaverdian, M. (1994) Size dependency of hollow-cylinder stability. *Rock Mechanics in Petroleum Engineering. EUROCK '94.* Balkema, Rotterdam, 191-197. [Ch.9]

Van Sambeek, L.L., Fossum, A., Callahan, G., and Ratigan, J. (1993) Salt mechanics: empirical and theoretical developments. *Proc. 7th Int. Symp. on Salt, Kyoto, April 1992.* Elsevier, Amsterdam. [Ch..2, 4]

Vardoulakis, I., Sulem, J., and Guenot A. (1988) Borehole instabilities as bifurcation phenomena *Int. J. Rock Mech. Min. Sci. & Geomech. Abstr.,* **25**, 159-170. [Ch.8]

Varo, L., and Passaris, E.K.S. (1977) The role of water in the creep properties of halite. *Proc. Conf. Rock Engineering,* Brit. Geotech. Soc./Univ. Newcastle upon Tyne. 85-100. [Ch.3]

Vincké, O. (1994) An estimation of bulk moduli of sandstones as a function of confining pressure using their petrographic and petrophysic description. *Rock Mechanics in Petroleum Engineering. EUROCK '94.* Balkema, Rotterdam, 139-146. [Ch.4]

Vogler, S., and Blum, W. (1990) Micromechanical modelling of creep in terms of the composite model. *Proc. 4th Int. Conf. on Creep and Fracture of Engineering Materials and Structures,* Eds. B. Wilshire and R.W. Evans. The Institute of Materials, London, 65-79. [Ch.2, 3]

Vogler, S. (1992) Kinetik der plastischen Verformung von natürlichem Steinsalz und ihre quantitative Beschreibung mit dem Verbundmodell. Dissertation. Universität Erlangen-Nürnberg. [Ch.1-3]

Volarovich, M.P., Bayuk, E.I., Levykin, A.I., and Tomashevskaya, I.S. (1974) *Physicomechanical Properties of Rocks and Minerals at High Pressures and Temperatures.* Nauka, Moscow. [Ch.4, 6]

Vouille, G., Bergues, J., Durup, J.-G. and You, T. (1996) Study of the stability of caverns in rock salt created by solution mining proposal for a new design criterion. *The Mechanical Behavior of Salt. Proc. 3rd Conf.* Eds. M. Ghoreychi, P. Berest, H.R. Hardy, Jr., and M. Langer. Trans Tech Publ., Clausthal-Zellerfeld, 427-444. [Ch.10]

Voyiadjis, G.Z., and Abu-Lebdeh, T.M. (1993) Damage model for concrete using bounding surface concept. *J. Engng Mech.* **119**, 1865-1885, [Ch.6]

Vutukuri, V.S., Lama, R.D. and Saluja, S.S. (1974). *Handbook on Mechanical Properties of Rocks.* vol.I, Trans Tech Publ., Clausthal-Zellerfeld. [Ch.1, 4]

Wallner, M. (1983). Stability calculations concerning a room and pillar design in rock salt. *Proc. 5th Int. Congr. on Rock Mechanics,* Balkema, Rotterdam, D9-D15. [Ch.2 , 3]

Wallner, M., Caninenberg, C., and Gonther, H. (1979) Ermittlung zeit- und temperaturabhangiger mechanischer Kennwerte von Salzgesteinen. *Proc. 4th Int. Congress on Rock Mech.* Montreux, Balkema, Rotterdam, Vol.1, 313-318. [Ch.2]

Walsh, J.B., and Brace, W.F. (1964) A fracture criterion for brittle anisotropic rock. *J. Geophys.Res.,* **69**, 16, 3449-3456. [Ch.5]

Walsh, J.B. (1965). The effect of cracks on the compressibility of rock. *J. Geophys. Res.,* **70**, 2, 381-389. [Ch.4]

Walsh, J.B. (1981) Effect of pore pressure and confining pressure on fracture permeability. *Int. J. Rock Mech Min Sci & Geomech. Abstr.* **18**, 429-435. [Ch.2]

Wan, R.G. (1996) Modelling the viscoplastic behaviour of rocks. *Rock Mechanics. Tools and Techniques. Proc. 2nd NARMS.*, Eds. M.Aubertin, F. Hassani and H. Mitri. Balkema, Rotterdam, 125-130. [Ch.4]

Wang, Y., Don Scott, J., and Dusseault, M.B. (1992) Thermal stresses near a circular opening in an unconsolidated reservoir. *Rock Mechanics, Proc. 33rd US Symp.,* Eds. J.R. Tillerson, and W.R. Wawersik , Balkema, Rotterdam, 387-396. [Ch.9]

Wang, Y., and Dusseault, M.B. (1994) Stress around a circular opening in an elastoplastic porous medium subjected to repeated loading. *Int. J. Rock Mech. Min. Sci. & Geomech. Abstr.,* **31**, 597-616. [Ch.8]

Wang, Y., Papamichos, E., and Dusseault, M.B. (1996) Thermal stresses and borehole stability. *Rock Mechanics. Tools and Techniques. Proc. 2nd NARMS,* Eds. M.Aubertin, F. Hassani and H. Mitri. Balkema, Rotterdam, 1121-1126. [Ch.8]

Wanten, P.H., Spiers, C.J., and Peach, C.J. (1996) Deformation of NaCl single crystal at 0.27 $T_m < T < 0.44$ T_m.*The Mechanical Behavior of Salt. Proc. 3rd Conf.* Eds. M. Ghoreychi, P. Berest, H.R. Hardy, Jr., and M. Langer. Trans Tech Publ, Clausthal-Zellerfeld,117-128. [Ch.2, 3, 4]

Wawersik, W., and Hannum, D. W. (1980) Mechanical behavior of New Mexico rock salt in triaxial compression up to 200° C. *J. Geophys. Res., B85,* 891-900. [Ch.1]

Wawersik, W. (1985) Determination of steady state creep rates and actuvation parameters for rock salt. *Measurement of Rock Properties at Elevated Pressures and Temperatures. ASTM STP 869.* Eds. H.J. Pincus and E.R. Hoskin., ASTM, Philadelphia, 72-92. [Ch.2]

Wawersik, W.R., and Zeuch, D.H. (1986) Modeling and mechanistic interpretation of creep of rock salt below 200°C. *Tectonophysics* **121**, 125-152. [Cap.5]

Wawersik, W. (1988) Alternatives to a power-law creep model for rock salt at temperatures below 160°C. *The Mechanical Behavior of Salt. Proc. 2nd Conf.* Eds. H.R.Hardy, and M. Langer. Trans Tech Publ., Clausthal-Zellerfeld, 103-128. [Ch.2,3]

Weertman, J., and Weertman, J.R. (1983a) Mechanical properties, strongly temperature-dependent. *Physical Metallurgy*, Eds.R.W. Cahn, and P. Haasen. North-Holland., Amsterdam, Vol.2, Ch. 20, 1309-1340. [Ch.3]

Weertman, J., and Weertman, J.R. (1983b) Mechanical properties, mildly temperature dependent. *Physical Metallurgy*, Eds. R.W. Chan, and P.Haasen. North-Holland, Amsterdam, Vol.2, Ch.19, 1259-1307. [Ch.2, 3]

Weertman, J., and Weertman, J.R. (1987) Constitutive equation and diffusion-dislocation controlled creep. *Constitutive Relations and Their Physical Basis. Proc. 8th Risø Int. Symp. on Metallurgy and Materials Science* Eds. S.I. Anderson *et al.* Risø National Laboratories, Roskilde, Denmark, 191-203. [Ch.3]

Weidinger, P., Blum, W., Hampel, A., and Hunsche, U. (1996a) Description of the creep of rock salt with the composite model -I. salts. *The Mechanical Behavior of Salt. Proc.4th Conf.* Eds. P. Habib, H.R. Hardy, B. Ladanyi, and M. Langer. Trans Tech Publ., Clausthal-Zellerfeld, (in Press). [Ch.2, 3]

Weidinger, P., Blum, W., Hunsche, U., and Hampel, A. (1996b) The influence of friction on plastic deformation in compression tests. *Phys. Stat. Sol.* **156**, 305-315. [Ch.1,3]

Whittaker, B.N., and Reddish, D.J. (1991) Study of tunnel shape and invert design in relation to stability in stratified rock conditions prone to heaving behaviour. *Tunnelling '91.* Institution of Mining and Metallurgy, Elsevier Applied Science, London, 79-90. [Ch.9]

Whittaker, R.N, Singh, R.N., and Sun, G. (1992). *Rock fracture Mechanics. Principles, Design and Applications.* Elsevier, Amsterdam, [Ch.2]

Wong, H. (1995) Thermoplastic and thermo-viscoplastic behaviours of underground cavities. *Proc. Int. Congr. on Rock Mechanics, Tokyo,* Balkema, Rotterdam, 479-483. [Ch. 8]

Wong, H., and Simionescu, O. (1995) Thermoplastic behaviour of tunnels for a cohesive-frictional material. *THERMAL STRESSES '95", J. Thermal Stresses,* 549-552. [Ch.9]

Wong, H., and Simionescu, O. (1996) An analytical solution of thermoplastic thick-walled tube subject to internal heating and variable pressure, raking into account corner flow and nonzero initial stress. *Int. J. Engng Sci.* **34**, 1259-1269. [Ch.8]

Wong, H., and Simionescu, O. (1997) Closed form solution on the thermoplastic behaviour of a deep tunnel in a thermal-softening material. *Mechanics of Cohesive-frictional Materials* **2** (submitted). [Ch.10]

Wong, T.-F. (1982) Shear fracture energy in Westerly granite from post-failure behavior. *J. Geophys. Res.* **87**, B2, 990-1000. [Ch.6]

Xiong, L., and Hunsche, U. [1988] Creep behavior of salt under true triaxial stress.*The Mechanical Behavior of Salt. Proc. 2nd Conf. 1984.* Eds. H.R.Hardy, Jr., and M. Langer. Trans Tech Publ., Clausthal-Zellerfeld, 235-243. [Ch.1]

Yamada, I., Masuda, K., and Mizutani, H. (1989). Electromagnetic and acoustic emission associated with rock fracture. *Phys. Earth Planet. Interiors,* **57**, 157-168. [Ch.1]

Yamatomi, J., Yamashita, S., Ogata, Y., and Kawabe, K. (1987) Analysis of time-dependent deformation of rock mass around a circular tunnel subjected to a hydrostatic field stress. Scientific and Technical Reports of the Mining College, Akita University, No.8, 9-16. [Ch.9]

Yanagidani, T., Ehara, S., Nishizawa, O., Kusunose, K., and Terada, M. (1985). *J. Geophys. Res.,* **90**, No.B8, 6840-6858. [Ch.6]

Yassir, N.A., and Bell, J.S. (1994) Abnormally high fluid pressures and associated porosities and stress regimes in sedimentary basins. *Rock Mechanics in Petroleum Engineering.* Balkema, Rotterdam, 879-886. [Ch. 7]

Young, R.P., and Martin, C.D. (1993) Potential role of acoustic emission/microseismicity investigations in the site characterization and performance monitoring of nuclear waste repositories. *Int. J. Rock Mech. Min. Sci. & Geomech. Abstr.* **30**, 797-803. [Ch.9[

Zaman, M., Faruque, M.O., and Hossain, M.I. (1992) Modeling creep behavior of rock salt. *Proc. Composite Material Technology Conf. Petroleum Div. ASME* **45**, 51-56. [Ch.5]

Zhang, C., Schmidt, M.M., and Staupendahl, G. (1996) Experimental and modelling results for compaction of crushed salt. *The Mechanical Behavior of Salt. Proc.3rd Conf.* Eds. M. Ghoreychi, P. Berest, H.R. Hardy, Jr., and M. Langer. Trans Tech Publ.,Clausthal-Zellerfeld, 391-404. [Ch.2, 4]

Zeuch, D.H. (1990) Isostatic hot-pressing mechanism maps for pure and natural sodium chloride: applications to nuclear waste isolation in bedded and domal salt formations. *Int. J. Rock Mech. Min. Sci. & Geomech. Abstr.* **27**, 505-524. [Ch.2]

Zheng, Q.S. (1994) Theory of representations for tensor functions - A unified invariant approach to constitutive equations. *App. Mech. Rev.* **47**, 11, 545-587. [Ch.5]

Zheng, Z., and Khodaverdian, M. (1996) Utilizing failure characteristics to create stable underground opening. *Rock Mechanics. Tools and Techniques. Proc. 2nd NARMS*, Eds. M.Aubertin, F. Hassani, and H. Mitri. Balkema, Rotterdam, 787-793. [Ch.9]

Zimmer, U., and Yaramanci, U. (1993) In-Situ-Bestimmung der seismischen Dämpfung des Steinsalzes. *Kali und Steinsalz* **11**, 168-175. [Ch.2]

Zimmerman, R.W., King, M.S., and Monteiro, P.J.M. (1986) The elastic moduli of mortar as a porous-granular material. *Cement and Concrete Research* **16**, 239-245 [Ch.6]

Zoback, M.D., Moos, D., Mastin, L., and Anerson, R.N. (1985) Well bore breakouts and in situ stress. *J. Geophys. Res.,* **90**, B7, 5523-5530. [Ch.8]

Zoback, M.D., Moos, D., and Stephenson, D.E. (1989) State of stress and the relation to tectonics in the central Savannah River area of South Carolina. *Rock Mechanics as a Guide for Efficient Utilization of Natural Resources. Proc. 30th US Symp. Rock Mechanics.* Balkema, Rotterdam, 553-560. [Ch.7]

Zoback, M.D., Apel, R., Baumgärtner, J., Brudy, M., Emmermann, R., Engeser, B., Fuchs, K., Kessels, W., Rischmüller, H., Rummel, F., and Vernik, L. (1993) Upper-crustal strength inferred from stress measurements to 6 km depth in the KTB borehole. *Nature.* **365**, 633-635. [Ch.7]

INDEX